Biographical Information.

Myron Evans is currently the only scientist on the British Civil List and is the third chemical physicist to be appointed in British history, the other two being John Dalton and Michael Faraday. He is currently the Director of the Alpha Foundation's Institute for Advanced Studies (A.I.A.S.) and the author or editor of some seven hundred scientific papers and forty monographs. These are to be collected by the Library of Congress and a list of these publications has been collected by the Nils Bohr Library of the American Institute of Physics. He was educated at the then University College of Wales Aberystwyth and is sometime Junior Research Fellow of Wolfson College Oxford. Numerous honours include the Harrison Memorial Prize and Meldola Medal of the Royal Society of Chemistry. He has made scientific inferences and discoveries, including the explanation of the far infra red spectra of materials; the first computer simulation of the far infra red; the discovery of the gamma or boson peak of the far infra red; the development of non-equilibrium and non-linear molecular dynamics simulation methods at Aberystwyth, IBM and elsewhere; the development of fundamental methods in statistical mechanics; the application of computer simulation to non-linear optics; the inference of the fundamental Evans spin field B(3) of electromagnetism at Cornell in 1991; and the development of Einstein Cartan Evans unified field theory as described for the first time in this book.

GENERALLY COVARIANT UNIFIED FIELD THEORY

THE GEOMETRIZATION OF PHYSICS

VOLUME IV

Myron W. Evans

Published 2007 by abramis

www.abramis.com

ISBN 978-1-84549-248-9

Printed and bound in the United Kingdom & USA

abramis is an imprint of arima publishing

arima publishing
ASK House, Northgate Avenue
Bury St Edmunds, Suffolk IP32 6BB, UK
t: (+44) 01284 700321

www.arimapublishing.com

I dedicate this book to my dear wife Larisa.

Preface

In this fourth volume of "Generally Covariant Unified Field Theory" (Abramis Academic, 2005, 2006) sixteen papers are published on the applications of Einstein Cartan Evans (ECE) field theory. This is a field theory based on Cartan geometry and rigorously completes the Einstein Hilbert (EH) field theory of gravitation based on Riemann geometry without torsion. In ECE the electromagnetic sector of the unified field is recognized to be the Cartan torsion within a C negative scalar factor proportional to the primordial voltage of the universe. The two Cartan structure equations and two Bianchi identities of differential geometry define the unified field in ECE theory. The gravitational sector is the Cartan curvature form, the electromagnetic sector is the Cartan torsion form. The weak and strong sectors are based on the $SU(2)$ and $SU(3)$ representation spaces of the Cartan torsion form.

Fermions and bosons are described by the ECE Lemma, which is the subsidiary proposition that leads to the ECE wave equation. The ECE Lemma is an identity based on the tetrad postulate of differential geometry, a foundational postulate which states that the complete vector field is independent of its components and basis elements.

In the first volumes of this monograph it has been shown that the wave equations of physics can be derived from the ECE wave equation. Thus, quantum mechanics and general relativity have been unified in the necessary causal and objective manner. General relativity is a statement of objectivity in science. Without objectivity there would be no science at all.

The point of view offered by ECE theory has been studied intensively for the past three years by essentially the entire physics community worldwide: in universities, institutes, government laboratories and departments, military facilities, organisations and by individual scholars. It has also been studied in related areas such as chemistry and electrical engineering, and is therefore accepted physics. This conclusion is an objective one, based on detailed analysis of the feedback sites of *www.aias.us*. These feedback sites are *www.aias.us/weblogs/log.html*, *www.aias.us/weblogs/log.files.html* and *www.aias.us/new_stats/*. They give detailed information on all aspects of interest in the *www.aias.us* website and the related *www.atomicprecision.com* website. Currently these two web-sites attract about two million hits a year worldwide, well over one hundred thousand individual visits a year of uniformly high quality from the leading physics departments worldwide. The center of gravity of theoretical physics is now firmly European by a margin of four to one over the United States. The study of advanced theoretical physics is concentrated in these two regions of the world, with the rest of the world being represented mainly by Canada, Australia and the Far East. However, visits from a total of one hundred and sixteen countries have been recorded, illustrating a thirst for ECE theory world-wide. This is one of the many things that can been seen from a systematic study of these feedback activity sites managed by David Burleigh, Chief Executive Officer of Annexa Inc. of Rochester, New York State. This type of feedback information is available in real time and is far more accurate and representative than science citation indices. It can be seen from the feedback sites that the ECE theory has attracted a permanent following among essentially the whole physics world. This is not discernible from citation indices, which are therefore incomplete and inaccurate. There is plenty of reason therefore to go on developing ECE theory systematically, and for indicating its areas of technological application.

In chapter one of volume four (55th paper of ECE theory) generally covariant translational and rotational dynamics are developed systematically. Their mutual interaction is defined by Cartan geometry and the equations of generally covariant dynamics are given in the same form as those of generally covariant electrodynamics. In chapter two geodesics and the Aharonov Bohm effects are studied with the concept of covariant exterior derivative of the potential form along a path in ECE space-time. The class of Aharonov Bohm effects are shown to be due to parallel transport of the covariant exterior derivative. In chapter three ECE field theory is shown to be a rigorous quantum field theory through considerations of canonical and second quantization. In chapter four the effect of Cartan torsion is considered on the Schwarzschild metric and on the well known phenomenon of light bending by gravitation. In standard EH cosmologies the Cartan torsion is not considered. In chapter five it is shown that the spin connection in the Coulomb law produces a resonant amplification of voltage from ECE space-time. This is potentially important as a source of energy. This theme is developed for atoms and molecules in chapter six to show that free electrons may be produced by ionization at resonance. In chapter seven the relevant spin connection in the Coulomb Law is shown to be of the form $1/r$, where r is the radial coordinate. In chapter eight the ECE Lemma is illustrated with fermionic and electromagnetic fields and is shown to be rigorously self consistent. In chapter nine spin connection resonance (SCR) is developed analytically and numerically, showing the presence of resonance peaks of theoretically infinite amplitude. In chapter ten SCR is shown to be present in EH theory itself (i.e. in gravitational general relativity). In chapter eleven SCR is developed for magneto-statics, showing that intense magnetic fields can be obtained theoretically at resonance. In chapter twelve the vector boson character of the electric field is developed using angular momentum theory and showing that ECE theory gives greater insight to the fundamental concepts than the standard model. Chapter thirteen shows that the polarization is changed of light deflected by gravitation, as observed experimentally. This is a property of ECE theory that is not given by Einstein Hilbert (EH) field theory. In chapter fourteen it is shown that spin connection resonance occurs in gravitational theory and is probably observed in a recent European Space Agency experiment which gives the gravitational equivalent of the Faraday law of induction. Chapter fifteen discusses the effect of gravitation on radiatively induced fermion resonance (RFR), the resonant equivalent of the inverse Faraday effect. Finally chapter sixteen introduces the concepts of chirality and spin vectors in ECE theory.

There are many experimentally tested advantages of ECE theory over the standard model, and thirty of these are summarized in the News section of *www.aias.us*. Several of the basic concepts of the standard model are now known to be flawed or non-existent - these include Big Bang, black holes, the Higgs boson, the Higgs mechanism, spontaneous symmetry breaking, asymptotic freedom and any development that is based on the Higgs boson. This has not been found at CERN at the predicted energy. Crothers (*www.aias.us*) has shown that the mathematics of the Big Bang model are flawed fundamentally and it has been admitted by the string theorists of the standard model that string theory cannot be tested experimentally. It is therefore quite an elegant mathematical model, but irrelevant to physics or natural philosophy.

Below, the differences between the Standard Model and ECE are listed:

ECE	**Standard Model**
All fields are objective, causal and generally covariant in both classical and quantum manifestations.	Only gravity is generally covariant, causal and objective and only in its classical limit.
Torsion is considered	Neglects torsion
Quantum mechanics is objective and causal, just like gravity	Quantum mechanics and gravity cannot be unified.
Electrodynamics and gravity are two geometrical aspects of space-time, the former being torsion- and the latter curvature.	Electrodynamics and gravity are philosophically different.
No concepts, which have been disproven experimentally, are used. For example, Heisenberg Uncertainty Principle, strings, superstrings, Higgs mechanism, renormalization, asymptotic freedom, spontaneous symmetry breaking, dark matter, singularities such as the Big Bang and black hole theory and abstract internal space of Yang Mills gauge theory.	Several concepts which are experimentally unproven and/or do not exist in relativity theory.
Quantum electrodynamics is developed from simultaneous ECE wave equations and is generally covariant objective and causal.	Hugely over-elaborate and contains adjustable parameters.
All the wave equations are generally covariant, are derived from the tetrad postulate of geometry, and all are objective and causal.	Wave equations of physics such as the Dirac and Proca equations are postulates of special relativity unrelated to geometry, objectivity and causality
Photon mass and the Proca equation are derived geometrically	Photon mass is asserted to be zero, causing problems of several kinds
Inter-relation of fields is geometrical and can be developed relatively easily.	The inter-relation of fields cannot be developed.
Derives all of physics geometrically from two basic postulates	Cannot describe all of physics geometrically
No assumptions are made regarding virtual photons or virtual electron-positron pairs. Is rigorously consistent with causality and general relativity.	QED assumes existence of virtual photons and virtual electron-positron pairs. These cannot be observed and violate special relativity and causality.
Derives the Lamb shift for all orbitals of all atoms and molecules in terms of the Evans-Eckardt radius (which is a new characteristic parameter of atoms and molecules). In addition, no claims of precision are made.	Claims of precision have been shown to be false in paper 85.
No removal of infinities	Removal of infinities in QED results in an artificial change in the mass and charge of the electron
Calculation of Lamb shift for Hydrogen is exact.	Bethe calculation of the Lamb shift of Hydrogen relied on perturbation theory (i.e. approximation)
No renormalization is used so the g factor is much more straightforwardly calculated. There are no infinities in the ECE method and it is general enough to be applied to all atoms and molecules.	The g factor of the electron is ascribed to a convergent vertex on QED because renormalization is used.

The following table of comparative impact shows that AIAS is the leading theoretical physics institute in the world at present, and that ECE theory is mainstream physics by actual usage.

Institution	Hits	Years
AIAS	5 million plus	2003 to date
Tulane chemistry	80,934	1998 to date
Bonaventure physics	9,601	2001 to date
Madison physics	24,027	2003 to date
Syracuse physics	31,525	Several
Brigham Young physics students	11,000	2003 to date
Colgate physics and astronomy	69,712	Several
Groningen computational physics	77,648	1998 to present
British Columbia Institute of Tech.	82,934	2000 to present
Bristol Univ.,School of Chem(5*)	2,621,606	1996 to present
UNC Asheville, chemistry	62,124	2000 to present
Penn State Winograd Group (physics)	23,915	2003 to present
Univ Copenhagen, chemistry	5,105	2004 to present
UN Kansas City chemistry	33,252	Several
Polomar Univ Cal.,chemistry	25,076	1998 to present
La Salle Univ., Philadelphia	7,586	2000 to present

The above table shows that the impact being made by the two AIAS sites (*www.aias.us* and *www.atomicprecision.com*) is a hundred to a thousand times greater than the average good department in academia over the same time span of a few years. The impact being made by AIAS is regularly checked in great detail for quality (see feedback files on *www.aias.us*) and there is no doubt that AIAS is the highest impacting chemical physics institution in the world, both in terms of quality and quantity. This impact can be measured in many ways, here it is restricted to files downloaded ("hits") in order to make a comparison as in the table. I will go on researching more university feedback sites where I can find them, but the pattern is already stunningly clear.

The British Head of State (Queen Elizabeth II), Parliament, Prime Minister and Royal Society are thanked for appointing the author to the British Civil List in Spring 2005 in recognition of distinguished contributions to Britain and the Commonwealth in science. This is a high honor commensurate with Order of Merit or Companion of Honour and is also an appointment voted in by Parliament. So the author is most grateful to the British Government for this recognition of his life-work. Predecessors in astronomy, physics and chemistry on the Civil list or Master of the Mint include Sir Isaac Newton, Sir William Herschel, John Dalton, Michael Faraday, James Joule, Mary Fairfax-Somerville, Oliver Heaviside, W. F. Denning and Sir George Airy.

The AIAS Fellows and intellectual environment are thanked for many interesting discussions.

Franklin Amador and Gianni Giacchetta are thanked for the important task of typesetting these papers and also volumes one to three. This was done entirely voluntarily in their spare time and also very accurately to a high standard.

Craigcefnparc, Wales
July 2007

Myron W. Evans
The British and Commonwealth Civil List scientist

Contents

Chapter 1

Generally covariant dynamics

(Paper 55)
by
Myron W. Evans,
Alpha Institute for Advanced Study (AIAS).
(emyrone@aol.com, www.aias.us, www.atomicprecision.com)

Abstract

Generally covariant translational and rotational dynamics are developed on the basis of Einstein Cartan Evans (ECE) field theory. Translational or central dynamics are defined as the limit of vanishing Cartan torsion where the Einstein Hilbert (EH) theory of gravitation is recovered. Rotational dynamics are defined in the limit where the translational Riemann curvature form vanishes and where the rotational Riemann form is dual to the Cartan torsion form. The mutual influence of translation and rotation is defined by the two Cartan structure equations and the two Bianchi identities of differential geometry. The equations of generally covariant rotational and translational dynamics are developed in the same form as the equations of generally covariant electrodynamics.

Keywords: Einstein Cartan Evans (ECE) field theory, generally covariant dynamics, generally covariant electrodynamics.

1.1 Introduction

Classical dynamics has been developed continuously for more than four hundred years. Major advances occurred in the sixteenth and seventeenth centuries [1], notably by Galileo, Brahe, Kepler and Newton, who synthesized the laws of classical translational dynamics. Later, rotational dynamics were developed by Euler and Coriolis, who inferred accelerations not present in Newtonian dynamics. Notable contributions came from Lagrange, Laplace and Hamilton using variational calculus. Following upon the results of the Michelson Morley experiment, length contraction was suggested by Fitzgerald, and developed into the theory of special relativity with notable contributions from Lorentz, Poincaré, Einstein and several others. Einstein inferred the theory of translational special relativity in 1905 and developed it into the theory of translational general relativity. The Einstein Hilbert (EH) field equation of translational dynamics was inferred independently by Einstein and Hilbert, and published in 1916. Later, Einstein and Cartan corresponded on the need for incorporating Cartan torsion into general relativity. In this paper the two Cartan structure equations and the two Bianchi identities of differential or Cartan geometry are inferred to be the equations of generally covariant dynamics, in which both translational and rotational motions are considered. These dynamics are valid in any frame of reference moving arbitrarily with respect to any other frame. The equations of these dynamics are therefore generally covariant as required by the theory of relativity and by objective natural philosophy.

In Section 1.2 the two Cartan structure equations and the two Bianchi identities are developed in a form which is identical to the equations of generally covariant electrodynamics [2]– [15] within a scalar factor $A^{(0)}$, essentially a primordial voltage. EH translational dynamics are defined as the limit where the Cartan torsion form vanishes and rotational dynamics are defined as the limit where the translational Riemann form vanishes. For rotational dynamics, the rotational Riemann form is dual to the Cartan torsion form, and the spin connection is dual to the tetrad. In Section 1.3 the Newtonian equations and principle of equivalence are inferred from the second Bianchi identity with zero Cartan torsion and the Euler equation is inferred from the first Cartan structure equation. In Section 1.4 the equations of generally covariant translational and rotational dynamics are developed in vector notation in the same form as the equations of generally covariant electrodynamics. Generally covariant dynamics provides several new inferences and suggests several phenomena not present in the EH limit. These may be tested with respect to cosmological anomalies where EH theory is not sufficient. There appear resonance solutions in both generally covariant dynamics and electrodynamics, and the ECE theory also gives the equations needed to describe the interaction of gravitation and electrodynamics.

1.2 The equations of generally covariant dynamics

The equations of classical dynamics are given by ECE theory in any frame of reference moving arbitrarily with respect to any other frame. In a condensed notation with all indices suppressed for clarity [2]– [15] the equations of motion are given by Cartan geometry:

$$T = D \wedge q \tag{1.1}$$

$$R = D \wedge \omega \tag{1.2}$$

$$D \wedge T = R \wedge q \tag{1.3}$$

$$D \wedge q = 0, \tag{1.4}$$

$$D\wedge = d \wedge + \omega \wedge . \tag{1.5}$$

Here T is the torsion form, q is the tetrad form, ω is the spin connection, R is the Riemann form and D denotes the covariant exterior derivative of Cartan [2]–[15]. This notation is fully explained and developed elsewhere [2]– [15] in form, tensor and vector notation. The condensed indexless notation of Eq.(1.1) to (1.4) gives the basic structure most clearly. These equations of geometry are transformed into equations of classical dynamics using the Einstein Ansatz:

$$R = -kT \tag{1.6}$$

where in Eq.(1.6), R denotes scalar curvature, k is the Einstein constant and T is the index contracted canonical energy momentum tensor [16]). In general T contains contributions from all four fundamental fields (gravitational, electromagnetic, weak and strong). In the EH theory only the gravitational contribution is considered.

In the notation of Eqs.(1.1) to (1.5) the EH field theory of 1916 is:

$$T = 0 \tag{1.7}$$

$$R \wedge q = 0 \tag{1.8}$$

$$D \wedge R = 0 \tag{1.9}$$

In this limit therefore the torsion form vanishes. Eq.(1.8) is the Ricci cyclic equation of EH theory. In tensor notation Eq.(1.8) is the familiar cyclic combination of Riemann tensors:

$$R_{\sigma\mu\nu\rho} + R_{\sigma\rho\mu\nu} + R_{\sigma\nu\rho\mu} = 0 \tag{1.10}$$

Eq.(1.9) is the second Bianchi identity. In tensor notation it becomes:

$$D^{\mu} G_{\mu\nu} = 0 \tag{1.11}$$

where $G_{\mu\nu}$ is the Einstein tensor:

$$G_{\mu\nu} = R_{\mu\nu} - \frac{1}{2} R g_{\mu\nu} \tag{1.12}$$

In Eq.(1.12) $R_{\mu\nu}$ is the Ricci tensor and $g_{\mu\nu}$ is the symmetric metric tensor. The EH field equation is obtained from the second Bianchi identity (1.11) and the Noether Theorem:

$$D^{\mu} T_{\mu\nu} = 0 \tag{1.13}$$

From Eqs.(1.11) and (1.13) we obtain the EH field equation:

$$D \wedge R = 0 \tag{1.14}$$

Here $T_{\mu\nu}$ is the canonical energy momentum tensor, whose index contracted form in EH theory is:

$$T = g^{\mu\nu} T_{\mu\nu} \tag{1.15}$$

It can be seen that the EH theory is limited by the omission of the Cartan torsion form and is therefore confined to the pure translational part of the general equations of dynamics, Eqs.(1.1) to (1.5). Pure rotational dynamics are defined as [2]– [15]:

$$T = D \wedge q \tag{1.16}$$

$$d \wedge T = 0 \tag{1.17}$$

and it is seen that the translational part of the Riemann or curvature form is zero. Pure rotation is defined by:

$$R \wedge q = \omega \wedge T \tag{1.18}$$

which means that the Cartan torsion is dual to the rotational part of the Riemann form. In this limit the Cartan torsion is the vector valued two-form dual to the rotational Riemann form, a tensor valued two-form. This duality is defined in the tangent (Minkowski) spacetime [2]– [15] of differential geometry and is analogous to the type of duality between for example an anti-symmetric tensor in three dimensions and an axial vector in three dimensions. The rotational Riemann form is the dual of the Cartan torsion form. The translational Riemann form on the other hand is not dual to the Cartan torsion form. The translational Riemann form may therefore be zero when the Cartan form is non-zero and vice-versa. In EH theory only the translational Riemann form is considered, the rotational Riemann form and the Cartan torsion form are not considered in EH theory. In the latter the connection is the symmetric Christoffel connection and EH theory does not consider the interaction between rotation and translation. The equations needed to describe this interaction are Eqs.(1.1) to (1.5).

In the notation [15] of standard differential geometry Eqs.(1.1) to (1.4) become:

$$T^a = d \wedge q^a + \omega^a{}_b \wedge q^b \tag{1.19}$$

$$R^a{}_b = d \wedge \omega^a{}_b + \omega^a{}_c \wedge \omega^c{}_b \tag{1.20}$$

$$d \wedge T^a + \omega^a{}_b \wedge T^b = R^a{}_b \wedge q^b \tag{1.21}$$

$$d \wedge R^a{}_b + \omega^a{}_c \wedge R^c{}_b - R^a{}_c \wedge \omega^c{}_b = 0 \tag{1.22}$$

in which the indices are those of the tangent (Minkowski) spacetime at point P to the base manifold. The indices of the base manifold are Greek indices which are always the same on both sides of any equation of Cartan geometry. So in the standard notation [15] of Cartan geometry the Greek indices are not written out. In refs. [2]– [14] however the equations of Cartan geometry are written out in full in form, tensor and vector notation.

The equations of Cartan geometry may be written as follows in the same overall format as the ECE equations of electrodynamics [2]– [14]. The first and second Cartan structure equations can be written as differential field equations in which the left hand side involves only the Cartan exterior derivative, and the right hand side is a combination of terms defining currents or source terms. Thus Eqs.(1.21) and (1.22) can be written as:

$$d \wedge T^a = j^a = R^a{}_b \wedge q^b - \omega^a{}_b \wedge T^b \tag{1.23}$$

$$d \wedge R^a{}_b = j^a{}_b = R^a{}_c \wedge \omega^c{}_b - \omega^a{}_c \wedge R^c{}_b \tag{1.24}$$

Here j^a is the homogeneous current of ECE electrodynamics within the factor $A^{(0)}$. Therefore dynamics and electrodynamics both originate in Cartan geometry, and are unified. The Hodge duals [2]– [14] of Eqs.(1.23) and (1.24) are:

$$d \wedge \widetilde{T}^a = J^a \tag{1.25}$$

$$d \wedge \widetilde{R}^a{}_b = J^a{}_b \tag{1.26}$$

and J^a is the inhomogeneous current of ECE electrodynamics within the factor $A^{(0)}$. Similarly the two structure equations (1.19) and (1.20) may be written as follows:

$$d \wedge q^a = j_1^a = T^a - \omega^a{}_b \wedge q^b \tag{1.27}$$

$$d \wedge \omega^a{}_b = j_{1b}^a = R^a{}_b - \omega^a{}_c \wedge \omega^c{}_b \tag{1.28}$$

whose Hodge duals are:

$$d \wedge \widetilde{q}^a = J_1^a \tag{1.29}$$

$$d \wedge \widetilde{\omega}^a{}_b = J_{1b}^a \tag{1.30}$$

Eqs.(1.23),(1.24), (1.25) and (1.26) may be written in tensor and vector notation, producing much novel information on dynamics and cosmology. Later in this paper it will be shown that Newtonian dynamics is a limit of the vector equation:

$$\boldsymbol{\nabla} \cdot \mathbf{R}^a{}_b\,(\text{orbital}) = j^{a(0)}{}_b \tag{1.31}$$

an equation which is one out of two vector equations given by the single form equation (1.26).

1.3 Newton and Euler equations

In Eq.(1.31), $R^a{}_b$ is the orbital part of the translational Riemann form, a quantity which is defined fully in Section 1.4. In this section the Newton inverse square law is derived straightforwardly from Eq.(1.31). First recognize that the Newtonian force $\mathbf{F}$ is derived from the force form $F^a{}_b$ defined by:

$$\mathbf{F}^a{}_b = -m_1 m_2 G R^a{}_b \mathbf{F} \tag{1.32}$$

Here R is the scalar curvature of Cartan geometry [2]– [14] and c is the speed of light in vacuo, the universal constant of relativity theory. The force form is:

$$F^a{}_b = m_1 g^a{}_b \tag{1.33}$$

where m_1 is a scalar quantity with the units of kilograms. Thus m_1 is recognized as mass and $g^a{}_b$ is a tensor valued two-form with the units of acceleration. Eq.(1.31) is mathematically equivalent to:

$$\mathbf{F}^a{}_b = -G\frac{m_1 m_2}{r^2}\mathbf{k}^a{}_b = m_1 \mathbf{g}^a{}_b \tag{1.34}$$

where G is the Newton gravitational constant defined by:

$$k = \frac{8\pi G}{c^2} \tag{1.35}$$

and where r is the distance between two masses m_1 and m_2. Restoring the indices of the base manifold to equation (1.33) gives:

$$F^a{}_{b\mu\nu} = m_1 g^a{}_{b\mu\nu} \tag{1.36}$$

an equation which shows that the force ${F^a}_{b\mu\nu}$ and the acceleration ${g^a}_{b\mu\nu}$ are both proportional to the orbital part of the translational Riemann form. In the Newtonian limit the base manifold approaches a Minkowski spacetime, so:

$$\mathbf{F}^a{}_b = \mathbf{F}, \quad \mathbf{g}^a{}_b = \mathbf{g} \tag{1.37}$$

and Eq.(1.34 becomes the Newton inverse square law:

$$\mathbf{F} = -G\frac{m_1 m_2}{r^2}\mathbf{k} \tag{1.38}$$

and Eq.(1.33) becomes the Newton force law:

$$\mathbf{F} = m_1 \mathbf{g} \tag{1.39}$$

Both laws are derived from the orbital part of the translational Riemann form of Cartan geometry, and this is the principle of equivalence of inertia and acceleration. The principle of equivalence states that must exist a scalar quantity m_1. It may be seen from Eq.(1.32) that mass does not enter into the relation between force and curvature, and this is the result of the well known Galileo experiment where two different masses dropped from the same height in the Earth's gravitational field hit the ground at the same time from the apocryphal Tower of Pisa. (In fact [1] Galileo proved this result using inclined planes.)

This simple derivation of Newtonian dynamics is confirmed as follows with reference to the well known Schwarzschild metric (SM), which by Birkhoff's Theorem is the unique spherically symmetric vacuum solution of the EH field equation (1.14). The SM is thus a solution of pure translational dynamics and takes no account of torsion in cosmology.

The SM in in spherical polar coordinates [2]– [15] is:

$$ds^2 = \left(1 - \frac{2GM}{rc^2}\right)c^2 dt^2 - \left(1 - \frac{2GM}{rc^2}\right)^{-1} dr^2 - r^2 d\theta^2 - r^2 \sin^2\theta d\phi^2 \tag{1.40}$$

where M is a parameter with units of kilograms identified with mass. If this parameter is identically zero Eq.(1.40) becomes the Minkowski metric in spherical polar coordinates:

$$ds^2 = c^2 dt^2 - dr^2 - r^2 d\theta^2 - r^2 \sin^2\theta d\phi^2 \tag{1.41}$$

The six non-zero elements of the Riemann tensor in the SM are as follows [2]– [15]:

$$R^0{}_{101} = e^{2(\beta-\alpha)}\left(\partial_0^2\beta + (\partial_0\beta)^2 - \partial_0\alpha\partial_0\beta\right) + \partial_1\alpha\partial_1\beta - \partial_1^2\alpha - (\partial_1\alpha)^2 \tag{1.42}$$

$$R^0{}_{202} = -re^{-2\beta}\frac{\partial_1\alpha}{r^2} \tag{1.43}$$

$$R^0{}_{303} = \sin^2\theta R^0{}_{202} \tag{1.44}$$

$$R^1{}_{212} = re^{-2\beta}\frac{\partial_1\beta}{r^2} \tag{1.45}$$

$$R^1{}_{323} = R^2{}_{323} = \left(1 - e^{-2\beta}\right)\frac{\sin^2\theta}{r^2} \tag{1.46}$$

where the parameters are defined by:

$$e^{2\alpha} = e^{-2\beta} = 1 - \frac{2GM}{rc^2} \tag{1.47}$$

The three orbital components of the SM are ${R^0}_{101}$, ${R^0}_{202}$, and ${R^0}_{303}$ and the three spin components are ${R^1}_{212}$, ${R^1}_{313}$, and ${R^2}_{323}$. It is seen that:

$${R^0}_{101} = e^{-4\alpha}\left(-(\partial_1\alpha)^2 - \partial_1^2\alpha - (\partial_1\alpha)^2\right) \tag{1.48}$$

and that:

$${R^0}_{202} + {R^0}_{303} = -\frac{GM}{c^2r^3}\left(1+\sin^2\theta\right) \tag{1.49}$$

The Newtonian limit is recovered from:

$$\nabla \cdot {\mathbf{g}^a}_b = -c^2\left({R^0}_{202} + {R^0}_{303}\right) = \frac{2GM}{r^3} \tag{1.50}$$

This result allows us to infer:

$${\mathbf{g}^a}_b = \frac{1}{m}{\mathbf{F}^a}_b = -\frac{GM}{r^2}{\mathbf{k}^a}_b \tag{1.51}$$

and to recover Eq.(1.32) in the form:

$${\mathbf{F}^a}_b = -\frac{GmM}{r^2}{\mathbf{k}^a}_b \tag{1.52}$$

Thus the scalar curvature R of Eq.(1.32) may be identified as:

$${\mathbf{R}^a}_b = -\frac{{\mathbf{k}^a}_b}{r^2} \tag{1.53}$$

More generally the SM produces well known departures from Newtonian dynamics, giving for example the perihelion advance of planets and the result of the Eddington experiment [2]– [15], now verified to one part in one hundred thousand accuracy. The above derivation of the Newton inverse square law from the orbital components of the Riemann tensor of the SM assumes that the ${R^0}_{101}$ element goes to zero faster than the sum of the ${R^0}_{202}$ and ${R^0}_{303}$ elements as M goes to zero and uses the geometry $\theta = 0$ in spherical polar coordinates.

The Riemann form in the SM is therefore the antisymmetric tensor:

$${R^a}_{b\mu\nu} = \begin{bmatrix} 0 & {R^0}_{101} & {R^0}_{202} & {R^0}_{303} \\ {R^0}_{110} & 0 & {R^1}_{212} & {R^1}_{313} \\ {R^0}_{220} & {R^1}_{221} & 0 & {R^2}_{323} \\ {R^0}_{330} & {R^1}_{331} & {R^2}_{332} & 0 \end{bmatrix} \tag{1.54}$$

and consists of orbital and spin components. The six non vanishing elements of the SM are precisely the three orbital elements and the three spin elements. The Riemann tensor is obtained from the Riemann form as follows [2]– [15]:

$${R^\rho}_{\sigma\mu\nu} = {q^b}_\sigma {q^\rho}_a {R^a}_{b\mu\nu} \tag{1.55}$$

The orbital Riemann vector is defined from Eq.(1.54) as:

$$\mathbf{R}^a{}_b\,(\text{orbital}) = R^0{}_{101}\mathbf{i}^a{}_b + R^0{}_{202}\mathbf{j}^a{}_b + R^0{}_{303}\mathbf{k}^a{}_b \tag{1.56}$$

and the spin Riemann vector as:

$$\mathbf{R}^a{}_b\,(\text{Spin}) = R^2{}_{323}\mathbf{i}^a{}_b + R^1{}_{331}\mathbf{j}^a{}_b + R^1{}_{212}\mathbf{k}^a{}_b \tag{1.57}$$

This derivation can be developed and summarized as follows. Starting with the second Bianchi identity in form notation:

$$D \wedge R^a{}_b = 0 \tag{1.58}$$

i.e.

$$d \wedge R^a{}_b = j^a{}_b = R^a{}_c \wedge \omega^c{}_b - \omega^a{}_c \wedge R^c{}_b \tag{1.59}$$

the homogeneous field equation of translational dynamics is obtained in the tensor notation:

$$\partial_\mu \widetilde{R}^a{}_b{}^{\mu\nu} = \widetilde{j}^a{}_b{}^\nu \tag{1.60}$$

The Hodge dual gives the inhomogeneous field equation of translational dynamics:

$$\partial_\mu R^a{}_b{}^{\mu\nu} = J^a{}_b{}^\nu \tag{1.61}$$

The operators ∂_μ are the Minkowski differential operators:

$$\partial_\mu = \left(\frac{1}{c}\frac{\partial}{\partial t}, \ \frac{\partial}{\partial X}, \ \frac{\partial}{\partial Y}, \ \frac{\partial}{\partial Z}\right) \tag{1.62}$$

and their space part refers therefore to a Cartesian frame of reference with Cartesian unit vectors $\mathbf{i}, \mathbf{j}, \mathbf{k}$. Thus in vector notation in the SM:

$$\mathbf{R}^a{}_b\,(\text{orbital}) = R^a{}_{b01}\mathbf{i} + R^a{}_{b02}\mathbf{j} + R^a{}_{b03}\mathbf{k} \tag{1.63}$$

and

$$\mathbf{R}^a{}_b\,(\text{Spin}) = R^2{}_{323}\mathbf{i} + R^1{}_{331}\mathbf{j} + R^1{}_{212}\mathbf{k} \tag{1.64}$$

It is also possible to define:

$$\mathbf{R}^a{}_b\,(\text{orbital}) = R^a{}_b{}^{01}\mathbf{i} + R^a{}_b{}^{02}\mathbf{j} + R^a{}_b{}^{03}\mathbf{k} \tag{1.65}$$

and

$$\mathbf{R}^a{}_b\,(\text{Spin}) = R^2{}_3{}^{23}\mathbf{i} + R^1{}_3{}^{31}\mathbf{j} + R^1{}_2{}^{12}\mathbf{k} \tag{1.66}$$

In the SM:

$$\mathbf{R}^0{}_1 = R^0{}_1{}^{01}\mathbf{i}, \quad \mathbf{R}^0{}_2 = R^0{}_2{}^{02}\mathbf{j}, \quad \mathbf{R}^0{}_3 = R^0{}_3{}^{03}\mathbf{k} \tag{1.67}$$

Now define:

$$\mathbf{R} = R^0{}_1{}^{01}\mathbf{i} + R^0{}_2{}^{02}\mathbf{j} + R^0{}_3{}^{03}\mathbf{k} \tag{1.68}$$

to obtain:

$$\mathbf{R}\,(\text{orbital}) = \mathbf{R} \tag{1.69}$$

and similarly:

$$\mathbf{R}\,(\text{Spin}) = R^2{}_3{}^{23}\mathbf{i} + R^1{}_3{}^{31}\mathbf{j} + R^1{}_2{}^{12}\mathbf{k} \tag{1.70}$$

The translational equations of motion are therefore:

$$\partial_\mu \widetilde{R}^{\mu\nu} = \widetilde{j}^\nu, \quad \partial_\mu R^{\mu\nu} = J^\mu \tag{1.71}$$

where the translational field tensor can be defined as:

$$R^{\mu\nu} = \begin{bmatrix} 0 & R^0{}_1{}^{01} & R^0{}_2{}^{02} & R^0{}_3{}^{03} \\ R^0{}_1{}^{10} & 0 & R^1{}_2{}^{12} & R^1{}_3{}^{13} \\ R^0{}_2{}^{20} & R^1{}_2{}^{21} & 0 & R^2{}_3{}^{23} \\ R^0{}_3{}^{30} & R^1{}_3{}^{31} & R^2{}_3{}^{32} & 0 \end{bmatrix} \tag{1.72}$$

In vector notation Eq.(1.70) are:

$$\nabla \cdot \mathbf{R}\,(\text{Spin}) = \widetilde{j}^0 \tag{1.73}$$

$$\nabla \times \mathbf{R}\,(\text{orbital}) + \frac{1}{c}\frac{\partial \mathbf{R}}{\partial t}\,(\text{Spin}) = \widetilde{\mathbf{j}} \tag{1.74}$$

$$\nabla \cdot \mathbf{R}\,(\text{orbital}) = J^0 \tag{1.75}$$

$$\nabla \times \mathbf{R}\,(\text{Spin}) - \frac{1}{c}\frac{\partial \mathbf{R}}{\partial t}\,(\text{orbital}) = \mathbf{J} \tag{1.76}$$

Eq.(1.73) is analogous to the generally covariant Gauss law of magnetism [2]–[14], Eq.(1.74) to the generally covariant Faraday law of induction, Eq.(1.75) to the generally covariant Coulomb Law and Eq.(1.76) to the generally covariant Ampere Maxwell law.

The equations of pure rotational dynamics are Eqs.(1.16) and (1.17). To end this section Eq.(1.16) is developed by defining the angular momentum tetrad [2]–[14] as:

$$J^a{}_\mu = J^{(0)} q^a{}_\mu \tag{1.77}$$

The first Cartan structure equation (1.16) then defines the generally covariant torque equation:

$$N^a = c\left(d \wedge J^a + \omega^a{}_b \wedge J^b\right) \tag{1.78}$$

In the classical non-relativistic limit (1.17) the Euler equation of motion is:

$$\mathbf{N} = \left(\frac{d\mathbf{J}}{dt}\right)_{\text{lab}} = \left(\frac{d\mathbf{J}}{dt}\right)_{\text{moving}} + \boldsymbol{\omega} \times \mathbf{J} \tag{1.79}$$

where:

$$\mathbf{N} = \left(\frac{d\mathbf{J}}{dt}\right)_{\text{lab}} \tag{1.80}$$

is valid only in an inertial Cartesian frame. Eq.(1.78) on the other hand is generally covariant, it is valid in any frame of reference moving arbitrarily with respect to any other frame of reference. It defines the generally covariant torque N^a as the Cartan torsion T^a within a scalar factor $cJ^{(0)}$. Therefore the Cartan torsion always defines torque in general relativity. In Eulerian dynamics the torque is still constructed from classical Newtonian dynamics, (classical non-relativistic torque ($\mathbf{N}$) is the arm ($\mathbf{r}$) cross multiplied by Newtonian force ($\mathbf{F}$)).

It has been demonstrated already that source of Newtonian dynamics is the Riemann form, not the torsion form. Therefore ECE theory gives new insight to the nature of both rotational and translational dynamics in the required generally covariant formulation. Einstein Hilbert (EH) field theory deals only with translational dynamics and is restricted to a particular Riemann geometry where the Christoffel connection is symmetric [2]– [15].

Eq.(1.78) is closely analogous to the equation defining the electromagnetic field:

$$F^a = d \wedge A^a + \omega^a{}_b \wedge A^b \tag{1.81}$$

wherein the $\omega^a{}_b \wedge A^b$ term originates in a spinning frame of reference, i.e. spinning space-time itself. In general relativity (ECE theory) there is no Cartan torsion if space-time is not spinning, so the generally covariant torque vanishes in a static or purely curving but not spinning space-time:

$$N^a = 0 \tag{1.82}$$

This limit of zero torque is the generally covariant description of translational dynamics, which reduces to Newtonian dynamics as we have argued in this section. The classical non-relativistic torque [17] has therefore been inferred historically from:

$$\mathbf{N} = \dot{\mathbf{J}} = \mathbf{r} \times \mathbf{F} \tag{1.83}$$

The classical, non-relativistic angular momentum $\mathbf{J}$ is defined as in Eq.(1.80). The total time derivative of $\mathbf{J}$ is the classical non-relativistic torque $\mathbf{N}$. Therefore classical non-relativistic rotational dynamics, as inferred historically by Euler and Coriolis and others, is not a self-consistent theory of general relativity, because it attempts to describe torque without Cartan torsion.

The self-consistent and generally covariant description of rotational dynamics in objective physics must be Eq.(1.78). It follows that there are dynamical effects in ECE theory that do not exist in EH theory, and so do not exist in classical non-relativistic rotational dynamics. When rotational and translational motions are mutually influential, the generally covariant torque interacts with a gravitational field through the COMPLETE first Bianchi identity (1.21) to give:

$$\begin{aligned} d \wedge N^a + \omega^a{}_b \wedge N^b &= cJ^{(0)} R^a{}_b \wedge q^b \\ &= cR^a{}_b \wedge J^b \end{aligned} \tag{1.84}$$

i.e.

$$D \wedge N^a = cR^a{}_b \wedge J^b \tag{1.85}$$

In the absence of a gravitational field:

$$R^a{}_b = 0 \tag{1.86}$$

so we recover Eq.(1.85) for generally covariant rotational dynamics:

$$D \wedge N^a = 0 \tag{1.87}$$

On the other hand, the source of the classical non-relativistic [17] torque (1.83) is not the first Bianchi identity (1.21), but the second Bianchi identity:

$$D \wedge R^a{}_b = 0 \tag{1.88}$$

which reduces to Newtonian dynamics as argued already in this section. The reason is that the classical non-relativistic torque is derived through Eq.(1.32) from the Newtonian force, and the latter derives from the Riemann curvature Eq.(1.20), a curvature which obeys the second Bianchi identity. The Riemann or curvature form enters into the first Bianchi identity if and only if there is interaction between rotation and translation. If there is no such interaction the motion is either pure rotational Eq.(1.87) or pure translational as in the Einstein Hilbert theory of 1916.

1.4 The generally covariant field equations of rotational dynamics and electrodynamics

In analogy with the Riemann form Eq.(1.54), the torsion form may also be developed as follows into an anti-symmetric tensor in four dimensions with orbital and spin components:

$$T^{\mu\nu} = \begin{bmatrix} 0 & -\frac{T_L^1}{c} & -\frac{T_L^2}{c} & -\frac{T_L^3}{c} \\ \frac{T_L^1}{c} & 0 & -T_S^3 & T_S^2 \\ \frac{T_L^2}{c} & T_S^3 & 0 & -T_S^1 \\ \frac{T_L^3}{c} & -T_S^2 & T_S^1 & 0 \end{bmatrix} \tag{1.89}$$

The factor c has been introduced into the definition of the orbital components for ease of comparison with generally covariant ECE electromagnetic theory [2]–[14]. Therefore there are orbital and intrinsic (or spin) components of torque in general relativity. The orbital component is proportional to the electric field and the spin component is proportional to the magnetic field in generally covariant electrodynamics. In vector notation:

$$\mathbf{T}_L^a = -\frac{\partial \mathbf{q}^a}{\partial t} - c\nabla q^{0a} - c\omega^{0a}{}_b\mathbf{q}^b + c\boldsymbol{\omega}^a{}_b q^{0b} \tag{1.90}$$

$$\mathbf{T}_S^a = \nabla \times \mathbf{q}^a - \boldsymbol{\omega}^a{}_b \times \mathbf{q}^b \tag{1.91}$$

The generally covariant equation of rotational motion is obtained using:

$$\mathbf{J}^a = J^{(0)}\mathbf{q}^a \tag{1.92}$$

$$\mathbf{N}^a = cJ^{(0)}\mathbf{T}^a \tag{1.93}$$

thus defining the generally covariant angular momentum $\mathbf{J}^a$ and the generally covariant torque $\mathbf{N}^a$. Therefore the generally covariant orbital torque is:

$$\mathbf{N}_L^a = -\frac{\partial \mathbf{J}^a}{\partial t} - c\nabla J^{0a} - c\omega^{0a}{}_b\mathbf{J}^b + c\boldsymbol{\omega}^a{}_b J^{0b} \tag{1.94}$$

and the generally covariant intrinsic or spin torque is:

$$\mathbf{N}_S^a = c\left(\nabla \times \mathbf{J}^a - \boldsymbol{\omega}^a{}_b \times \mathbf{J}^b\right) \tag{1.95}$$

Eqs.(1.94) and (1.95) are the generally covariant equations of rotational dynamics. They are valid in any reference frame moving arbitrarily with respect to any other reference frame, as required by the principle of relativity and thus of objective physics. The Eulerian limit of classical non-relativistic dynamics is obtained as follows:

$$\mathbf{N}_L^a \longrightarrow -\frac{\partial \mathbf{J}^a}{\partial t}, \quad \mathbf{N}_S^a \longrightarrow -c\boldsymbol{\omega}^a{}_b \times \mathbf{J}^b \tag{1.96}$$

Assuming that:

$$\nabla \times \mathbf{J}^a \longrightarrow \mathbf{0} \tag{1.97}$$

(i.e. that the angular momentum is ir-rotational) the total torque is:

$$\mathbf{N}^a = \mathbf{N}_L^a + \mathbf{N}_S^a = -\frac{\partial \mathbf{J}^a}{\partial t} - c\boldsymbol{\omega}^a{}_b \times \mathbf{J}^b \tag{1.98}$$

and this has the same form as the Euler equation (1.79). However, as argued in Section 1.3, Eq.(1.98) comes from the Cartan torsion form, and the Euler equation comes from the Cartan curvature form. Therefore ECE theory has a great deal more inherent information than the classical and non-relativistic Euler theory. The task is to reveal such information experimentally, using high accuracy experiments in the laboratory or in astronomy. These would amount to rigorous experimental tests of Einsteinian philosophy itself, because ECE theory completes the EH theory of 1916. They would therefore be important experiments.

For pure rotational motion unaffected by the Cartan curvature but defined by the Cartan torsion, it has been shown [2]– [14] that the spin connection is dual to the tetrad and that the form $R^a{}_b$ (torsion) is dual to the Cartan torsion:

$$\omega^a{}_b = -\frac{\kappa}{2}\epsilon^a{}_{bc}q^c \tag{1.99}$$

$$R^a{}_b\,(\text{torsion}) = -\frac{\kappa}{2}\epsilon^a{}_{bc}T^c \tag{1.100}$$

Here κ has the units of inverse meters and can be identified as a scalar wave-number. Eqs.(1.99) and (1.100) are written in the tangent space-time and are valid for all indices of the base manifold. They mean that for pure rotational motion, the spin connection is the two index quantity dual to the one index tetrad. The tangent space-time of Cartan geometry is a Minkowski space-time [2]– [15] so the duality of Eqs.(1.99) and (1.100) occurs through the totally anti-symmetric unit tensor $\epsilon^a{}_{bc}$ in the Minkowski space-time. If we restrict consideration to three space indices and dimensions, this type of duality is analogous to the well known three-dimensional duality between an anti-symmetric tensor and an axial vector [2]– [15]. Note that $R^a{}_b$(torsion) in Eq.(1.100) is not the Riemann curvature form. The latter is not dual to the torsion form because the Cartan curvature (or Riemann) form can be non-zero when the Cartan torsion form is zero and vice-versa. Thus one cannot be the dual of the other. The quantity $R^a{}_b$(torsion) is the two-index representation of the Cartan torsion and the existence of $R^a{}_b$(torsion) had not been inferred or clearly defined prior to the development [2]– [14] of ECE theory. For a complete understanding of generally covariant dynamics, and also of generally covariant electrodynamics, it is necessary to realize that $R^a{}_b$(torsion) exists in Cartan geometry and that it is different from the curvature form $R^a{}_b$. This is also important in devising new tests of relativity, as argued already in this section, and for coding purposes in computation.

The generally covariant angular velocity for pure rotational motion may now be defined as:

$$\Omega^a{}_b = c\omega^a{}_b = -\frac{\omega}{2}\epsilon^a{}_{bc}q^c \tag{1.101}$$

where

$$\Omega^c = \omega q^c, \quad \omega = \kappa c \tag{1.102}$$

So the two index $\Omega^a{}_b$ is the dual of the one index ω^c as follows:

$$\Omega^a{}_b = -\frac{1}{2}\epsilon^a{}_{bc}\omega^c \tag{1.103}$$

It is now possible to define a generally covariant two index quantity with the units of force:

$$F^a{}_b = cJ^{(0)}R^a{}_b\,(\text{torsion}) \tag{1.104}$$

This definition is analogous to Eq.(1.93) for the torque. Therefore $F^a{}_b$ is the two-index dual of the generally covariant torque:

$$F^a{}_b = -\frac{\kappa}{2}\epsilon^a{}_{bc}N^c_L \tag{1.105}$$

Inverting Eq.(1.105) gives an expression for torque:

$$N^c_L = -2r\epsilon^{bc}{}_aF^a{}_b \tag{1.106}$$

in terms of the distance r defined by:

$$r = \frac{1}{\kappa}. \tag{1.107}$$

Eq.(1.106) is the correctly and generally covariant formulation of the classical and non-relativistic definition of orbital torque as being the arm cross multiplied by the Newtonian force:

$$\mathbf{N}_L = \mathbf{r} \times \mathbf{F} \tag{1.108}$$

Eq.(1.106) is again valid in any reference frame moving arbitrarily with respect to any other frame, while Eq.(1.108) is non-relativistic and therefore incomplete. Eq.(1.106) originates in the spinning of space-time while Eq.(1.108) originates in Newtonian concepts of force. In Newtonian dynamics the spin connection does not exist as a concept. The intrinsic or spin torque:

$$\mathbf{N}^a{}_S = -c\boldsymbol{\omega}^a{}_b \times \mathbf{J}^b \tag{1.109}$$

originates again in the spinning of space-time, while its Eulerian predecessor, $\boldsymbol{\omega} \times \mathbf{J}$ derives from considerations of one Cartesian frame moving with respect to another Cartesian frame in absolute three dimensional space, using the separate concept of absolute time in the manner of Newton. In ECE theory these are merged into a four dimensional space-time whose curvature and spin are defined by Cartan geometry (Eqs.(1.1) to (1.5)). In ECE theory there is no absolute frame of reference. This is the major philosophical advance made in dynamics between the seventeenth and twenty first centuries and many of the experimental consequences of this advance remain to be evaluated. Added to this advance is the unification of dynamics with electrodynamics made possible by ECE theory [2]– [15]. It is now known how the gravitational field interacts with the electromagnetic field on both classical and quantum levels. This interaction is governed entirely by geometry in both classical and wave mechanics. For physics, Cartan geometry appears to be sufficient in our current state of knowledge, but in mathematics there are more abstract geometries which may contain more information than Cartan geometry in the same way that Cartan geometry contains more information [2]– [15] than Riemann geometry. These more abstract geometries may turn out to contain physical information, so should be borne in mind.

The orbital torque (1.94) may be simplified using certain approximations, given here as examples only. The most general orbital torque is always Eq.(1.95). If angular momentum is considered [18] to be a space-like property with no time-like component the scalars J^{0a} and J^{0b} vanish because the 0 index is time-like and the a and b indices have been restricted by definition to space-like. Using this assumption, Eq.(1.94) simplifies to:

$$\mathbf{N}_L^a = -\frac{\partial \mathbf{J}^a}{\partial t} - c\omega^{0a}{}_b \mathbf{J}^b \tag{1.110}$$

The components of Eq.(1.110) are:

$$\mathbf{N}_L^1 = -\frac{\partial \mathbf{J}^1}{\partial t} - c\omega^{01}{}_b \mathbf{J}^b \tag{1.111}$$

$$\mathbf{N}_L^2 = -\frac{\partial \mathbf{J}^2}{\partial t} - c\omega^{02}{}_b \mathbf{J}^b \tag{1.112}$$

$$\mathbf{N}_L^3 = -\frac{\partial \mathbf{J}^3}{\partial t} - c\omega^{03}{}_b \mathbf{J}^b \tag{1.113}$$

For rotational motion not affected by central gravitation:

$$\omega^a{}_b = -\frac{\kappa}{2}\epsilon^a{}_{bc} q^c \tag{1.114}$$

$$\epsilon^a{}_{bc} = \eta^{ad}\epsilon_{dbc} \tag{1.115}$$

where η^{ab} is the Minkowski metric [2]– [15], so

$$\omega^1{}_{\mu 2} = -\frac{\kappa}{2} q^3{}_\mu \tag{1.116}$$

$$\omega^3{}_{\mu 1} = -\frac{\kappa}{2} q^2{}_\mu \tag{1.117}$$

$$\omega^3{}_{\mu 2} = -\frac{\kappa}{2} q^1{}_\mu \tag{1.118}$$

It follows that:

$$\omega^1{}_{02} = -\frac{\kappa}{2} q^3{}_0 = 0 \tag{1.119}$$

$$\omega^3{}_{01} = -\frac{\kappa}{2} q^2{}_0 = 0 \tag{1.120}$$

$$\omega^3{}_{02} = -\frac{\kappa}{2} q^1{}_0 = 0 \tag{1.121}$$

and:

$$\mathbf{N}_L^a = -\frac{\partial \mathbf{J}^a}{\partial t} \tag{1.122}$$

This equation has the same form as the well known non-relativistic and inertial part of torque:

$$\mathbf{N} = \frac{\partial \mathbf{J}}{\partial t} \tag{1.123}$$

The sign change in Eq.(1.123) is a convention, and the indices a in Eq.(1.122) refer to the Cartesian space-like part of Minkowski space-time. Thus Eqs.(1.122) and (1.123) contain the same physical information but are philosophically different as argued already. The classical result (1.123) is a well defined limit of the generally covariant result (1.94) and the latter contains more observable information. Some of this information may already have been observed and described as "anomalies" which cannot be described by Einstein Hilbert theory or its Newtonian limit. These anomalies are therefore due to Cartan torsion, an important result of ECE theory [2]- [14].

If central gravitation affects rotational motion then Eqs.(1.114) to (1.121) are no longer true in general and there is an additional torque term as in Eq.(1.110). This term leads to additional physical and observable effects on a rotating object in a gravitational field. If for some reason J^{0a} and J^{0b} are non-zero, then all four terms on the right hand side of Eq.(1.94) contribute in general to the generally covariant orbital torque, and there are several effects which may for example affect the orbit of a planet or satellite. In addition there are the effects of the generally covariant intrinsic torque or spin torque, Eq.(1.95). Both orbital and intrinsic torque are present in general in subjects such as astronomy and cosmology as well as in precise laboratory experiments. Prior to this paper such effects have not been realized to exist, and they are not considered in EH theory, because in EH theory there is no torsion at all. These conclusions require an extensive re-evaluation of cosmology, notably planetary cosmology and satellite technology, and Big Bang theory, dark matter theory and so forth. In none of these theories is Cartan torsion adequately considered, or considered at all. Standard model textbooks in general relativity usually give at best only a cursory mention of Cartan torsion and restrict development to Cartan curvature in its Riemann limit (EH theory) in an un-unified field theory of gravitation only. Thus ECE theory open up a large new area of physics hitherto unexplored. ECE theory is a generally covariant unified field theory [2]- [14], the first of its kind, and completes the search of Einstein, Cartan and others for this end.

The generally covariant spin torque is the required objective and generally covariant description of the well known Coriolis and centripetal accelerations of the classical, non relativistic limit of general relativity. In order to introduce these classical accelerations (usually known as the "non-inertial" accelerations) it is convenient firstly to give a summary of the original derivation of Coriolis, (1835), a summary which is based on ref. [17].

Consider two sets of Caretsian coordinate axes. One is fixed ("inertial") and the other is in arbitrary motion with respect to the first. These are designated "fixed" and "rotating" as in the following figure [17]: For any point P:

$$\mathbf{r}' = \mathbf{R} + \mathbf{r} \tag{1.124}$$

as in Fig. 1.1. It is always possible [17] to represent an arbitrary infinitesimal displacement by a pure rotation and an axis called the instantaneous axis of

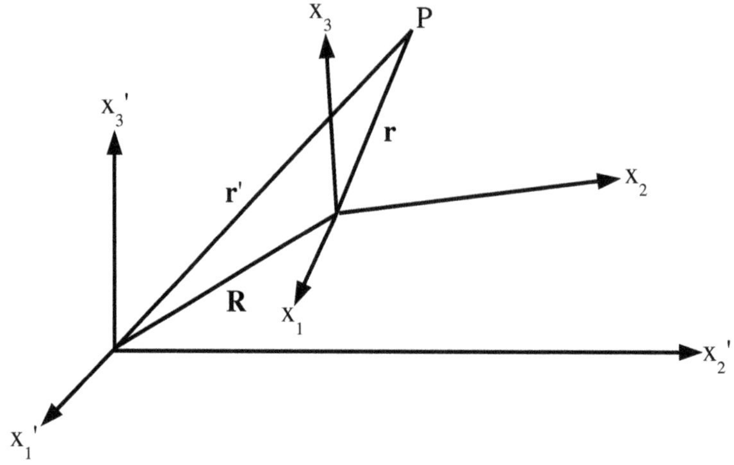

Figure 1.1: Fixed and Rotating Coordinate Axis

rotation. For example, if a disk rolls down an inclined plane, the motion is a rotation about the point of contact of the disk with the plane. Therefore if the x_i system undergoes an infinitesimal rotation $\partial\theta$, corresponding to an arbitrary infinitesimal displacement,

$$(d\mathbf{r})_{\text{fixed}} = d\boldsymbol{\theta} \times \mathbf{r} \tag{1.125}$$

Note that Eq.(1.125) is the result of geometry and so must be a limit of Cartan geometry, specifically a limit of Cartan torsion. The quantity $d\mathbf{r}$ is measured in the $x_i^{'}$, or fixed, coordinate system. The point P is considered to be at rest with respect to the x_i system, but P is moving with respect to the $x_i^{'}$ system. Now divide Eq.(1.124) by the time interval during which the infinitesimal displacement takes place, using the Newtonian fluxions (differential calculus) to give:

$$\left(\frac{d\mathbf{r}}{dt}\right)_{\text{fixed}} = \frac{d\boldsymbol{\theta}}{dt} \times \mathbf{r} \tag{1.126}$$

Identify the angular velocity as:

$$\omega = \frac{d\boldsymbol{\theta}}{dt} \tag{1.127}$$

so:

$$\left(\frac{d\mathbf{r}}{dt}\right)_{\text{fixed}} = \boldsymbol{\omega} \times \mathbf{r} \tag{1.128}$$

Now make the theory more general by considering P to have a velocity with respect to the x_i system:

$$\left(\mathbf{v} = \frac{d\mathbf{r}}{dt}\right)_{\text{moving}} \tag{1.129}$$

The total velocity is therefore:

$$\left(\frac{d\mathbf{r}}{dt}\right)_{\text{fixed}} = \left(\frac{d\mathbf{r}}{dt}\right)_{\text{moving}} + \boldsymbol{\omega} \times \mathbf{r} \tag{1.130}$$

Eq.(1.130) is valid for any vector $\mathbf{Q}$ [17]:

$$\left(\frac{d\mathbf{Q}}{dt}\right)_{\text{fixed}} = \left(\frac{d\mathbf{Q}}{dt}\right)_{\text{moving}} + \boldsymbol{\omega} \times \mathbf{Q} \tag{1.131}$$

If for example Q denotes linear velocity then:

$$\left(\frac{d\mathbf{v}}{dt}\right)_{\text{fixed}} = \left(\frac{d\mathbf{v}}{dt}\right)_{\text{moving}} + \boldsymbol{\omega} \times \mathbf{v} \tag{1.132}$$

where from Eq.(1.129):

$$\mathbf{v}_{\text{fixed}} = \mathbf{v}_{\text{moving}} + \boldsymbol{\omega} \times \mathbf{r} \tag{1.133}$$

The quantity $\boldsymbol{\omega} \times \mathbf{v}$ is proportional to the non-relativistic Coriolis acceleration. Now differentiate Eq.(1.133) to obtain:

$$\left(\frac{d\mathbf{v}}{dt}\right)_{\text{fixed}} = \left(\frac{d\mathbf{v}}{dt}\right)_{\text{moving}} + \frac{d\boldsymbol{\omega}}{dt} \times \mathbf{r} + \boldsymbol{\omega} \times \frac{d\mathbf{r}}{dt} \tag{1.134}$$

From Eq.(1.130):

$$\boldsymbol{\omega} \times \left(\frac{d\mathbf{r}}{dt}\right)_{\text{fixed}} = \boldsymbol{\omega} \times \left(\frac{d\mathbf{r}}{dt}\right)_{\text{moving}} + \boldsymbol{\omega} \times (\boldsymbol{\omega} \times \mathbf{r}) \tag{1.135}$$

and the quantity $\boldsymbol{\omega} \times (\boldsymbol{\omega} \times \mathbf{r})$ is proportional to the non-relativistic centripetal acceleration. The third type of "non-inertial" and non-relativistic acceleration which emerges form this analysis is $\dot{\boldsymbol{\omega}} \times \mathbf{r}$. The non-inertial accelerations link a translating object in the fixed frame with the same translating object in the rotating frame. The three non-inertial accelerations are therefore observable in both frames: the observer or fixed frame and the rotating frame. Therefore a link with relativity theory is clearly indicated, but in a general context where both translations and rotations are considered, and where accelerations are considered. In other words general relativity with torsion is needed, and ECE theory.

The original derivation by Coriolis is given in ref. [17] as follows. Coriolis considered an observer in a rotating Cartesian coordinate system using the concepts of absolute space and absolute time. Ref. [17] reproduces this 1835 derivation by differentiating Eq.(1.124) to give:

$$\begin{aligned}\left(\frac{d\mathbf{r}'}{dt}\right)_{\text{fixed}} &= \left(\frac{d\mathbf{R}}{dt}\right)_{\text{fixed}} + \left(\frac{d\mathbf{r}}{dt}\right)_{\text{fixed}} \\ &= \left(\frac{d\mathbf{R}}{dt}\right)_{\text{fixed}} + \left(\frac{d\mathbf{r}}{dt}\right)_{\text{moving}} + \boldsymbol{\omega} \times \mathbf{r}\end{aligned} \tag{1.136}$$

Now define:

$$\mathbf{v}_f = \dot{\mathbf{r}}_f = \left(\frac{d\mathbf{r}}{dt}\right)_{\text{fixed}} \tag{1.137}$$

$$\mathbf{V} = \dot{\mathbf{R}}_f = \left(\frac{d\mathbf{R}}{dt}\right)_{\text{fixed}} \tag{1.138}$$

$$\mathbf{v}_r = \dot{\mathbf{r}}_r = \left(\frac{d\mathbf{r}}{dt}\right)_{\text{rotating}} \tag{1.139}$$

so:

$$\mathbf{v}_f = \mathbf{V} + \mathbf{v}_r + \boldsymbol{\omega} \times \mathbf{r} \tag{1.140}$$

The Newtonian acceleration is the term [17]:

$$\mathbf{a}_f = m\left(\frac{d\mathbf{v}_f}{dt}\right)_{\text{fixed}} \tag{1.141}$$

$$\mathbf{F} = m\mathbf{a}_f \tag{1.142}$$

while differentiation of Eq.(1.140) gives extra terms:

$$\left(\frac{d\mathbf{v}_f}{dt}\right)_{\text{fixed}} = \left(\frac{d\mathbf{V}}{dt}\right)_{\text{fixed}} + \left(\frac{d\mathbf{v}_r}{dt}\right)_{\text{fixed}} + \dot{\boldsymbol{\omega}} \times \mathbf{r} + \boldsymbol{\omega} \times \left(\frac{d\mathbf{r}}{dt}\right)_{\text{fixed}} \tag{1.143}$$

Denote:

$$\ddot{\mathbf{R}}_f = \left(\frac{d\mathbf{V}}{dt}\right)_{\text{fixed}} \tag{1.144}$$

The second term in Eq.(1.143) is evaluated by substituting $\mathbf{v}_r$ for $\mathbf{Q}$ in Eq.(1.131):

$$\left(\frac{d\mathbf{v}_r}{dt}\right)_{\text{fixed}} = \left(\frac{d\mathbf{v}_f}{dt}\right)_{\text{rotating}} + \boldsymbol{\omega} \times \mathbf{v}_r = \mathbf{a}_r + \boldsymbol{\omega} \times \mathbf{v}_r \tag{1.145}$$

The last term in Eq.(1.143) is obtained from:

$$\begin{aligned}&\boldsymbol{\omega} \times \left(\frac{d\mathbf{r}}{dt}\right)_{\text{fixed}} \\ =&\boldsymbol{\omega} \times \left(\frac{d\mathbf{v}_f}{dt}\right)_{\text{rotating}} + \boldsymbol{\omega} \times (\boldsymbol{\omega} \times \mathbf{r}) \\ =&\boldsymbol{\omega} \times \mathbf{v}_r + \boldsymbol{\omega} \times (\boldsymbol{\omega} \times \mathbf{r})\end{aligned} \tag{1.146}$$

So:

$$\mathbf{F} = m\mathbf{a}_f = m\ddot{\mathbf{R}}_f + m\mathbf{a}_r + m\dot{\boldsymbol{\omega}} \times \mathbf{r} + m\boldsymbol{\omega} \times (\boldsymbol{\omega} \times \mathbf{r}) + 2m\boldsymbol{\omega} \times \mathbf{v}_r \tag{1.147}$$

To an observer in a rotating coordinate system:

$$\mathbf{F}_r = m\mathbf{a}_r = \mathbf{F} - m\ddot{\mathbf{R}}_f - m\dot{\boldsymbol{\omega}} \times \mathbf{r} - m\boldsymbol{\omega} \times (\boldsymbol{\omega} \times \mathbf{r}) - 2m\boldsymbol{\omega} \times \mathbf{v}_r \tag{1.148}$$

The standard centripetal force is thus $-m\boldsymbol{\omega} \times (\boldsymbol{\omega} \times \mathbf{r})$, and is directed outward from the center of rotation. The Coriolis force is $-2m\boldsymbol{\omega} \times \mathbf{v}_r$. Thus:

$$\mathbf{F}_r = m\mathbf{a}_f + \text{non-inertial terms.} \tag{1.149}$$

The classical derivation by Coriolis is incomplete because it is not relativistic, i.e. it is not generally covariant. The latter theory of the Coriolis and centripetal forces must be based on the Cartan structure equation:

$$T^a = d \wedge q^a + \omega^a{}_b \wedge q^b \tag{1.150}$$

and the first Bianchi identity:

$$d \wedge T^a = R^a{}_b \wedge q^b - \omega^a{}_b \wedge T^b \tag{1.151}$$

In an inertial or central treatment such as that of Einstein and Hilbert there is no Cartan torsion and:

$$T^a = 0, \quad R^a{}_b \wedge q^b = 0 \tag{1.152}$$

However in the required ECE theory there are non-zero terms:

$$R^a{}_b \wedge q^b = d \wedge T^a + \omega^a{}_b \wedge q^b \neq 0 \tag{1.153}$$

$$T^a \neq 0, \tag{1.154}$$

which come from the rotational motion or spinning of space-time itself. It is important to realize that this spinning motion is not that of one frame with respect to another, it is a generally covariant motion with no preferred frame of reference. The space-time ITSELF is both spinning and curving. Analogously, in EH theory, the space-time itself is curving but not spinning.

Now define the generally covariant version of the vector **Q**. This is the tetrad within a dimensional scalar $Q^{(0)}$ Invariant under frame transformation. Thus:

$$Q^a{}_\mu = Q^{(0)} q^a{}_\mu \tag{1.155}$$

For example there is a position tetrad and velocity tetrad:

$$r^a{}_\mu = r^{(0)} q^a{}_\mu, \quad V^a_{\ \mu} = V^{(0)} q^a{}_\mu \tag{1.156}$$

Thus:

$$\begin{aligned} R^a{}_b \wedge r^b =& d \wedge \left(d \wedge r^a + \omega^a{}_b \wedge r^b\right) \\ &+ \omega^a{}_b \wedge \left(d \wedge r^b + \omega^b{}_c \wedge r^c\right) \end{aligned} \tag{1.157}$$

$$\begin{aligned} R^a{}_b \wedge V^b =& d \wedge \left(d \wedge V^a + \omega^a{}_b \wedge V^b\right) \\ &+ \omega^a{}_b \wedge \left(d \wedge V^b + \omega^b{}_c \wedge V^c\right) \end{aligned} \tag{1.158}$$

From Eq.(1.157) we find the generally covariant centripetal term:

$$\left(R^a{}_b \wedge r^b\right)_{\text{centripetal}} = \omega^a{}_b \wedge \left(\omega^b{}_c \wedge r^c\right) \tag{1.159}$$

and from Eq.(1.158) the generally covariant Coriolis term:

$$\left(R^a{}_b \wedge V^b\right)_{\text{coriolis}} = d \wedge \left(\omega^a{}_b \wedge V^b\right) + \omega^a{}_b \wedge \left(\omega^b{}_c \wedge V^c\right) \tag{1.160}$$

When there is no space-time spin:

$$\omega^a{}_b = 0 \tag{1.161}$$

$$\omega^a{}_b \wedge \left(\omega^b{}_c \wedge r^c\right) = 0 \tag{1.162}$$

$$\omega^a{}_b \wedge V^b = 0 \tag{1.163}$$

$$d \wedge q^a = 0 \tag{1.164}$$

$$d \wedge V^a = d \wedge r^a = 0 \tag{1.165}$$

and so:

$$T^a = 0 \tag{1.166}$$

In this limit the Newtonian acceleration is recovered from a limit of the second Bianchi identity as shown in Section 1.3. In the generally covariant development of the non-inertial accelerations there are several terms and phenonema not present in the non-relativistic development of Coriolis. The computational task is to solve for these in order to compare with data from novel or known experiments.

In summary therefore the equations of generally covariant rotational dynamics are:

$$d \wedge T^a = j^a \tag{1.167}$$

$$d \wedge \widetilde{T}^a = J^a \tag{1.168}$$

where

$$j^a = R^a{}_b \wedge q^b - \omega^a{}_b \wedge T^b \tag{1.169}$$

$$J^a = \widetilde{R}^a{}_b \wedge q^b - \omega^a{}_b \wedge \widetilde{T}^b \tag{1.170}$$

In tensor notation Eqs.(1.167) and (1.168) become:

$$\partial_\mu \widetilde{T}^{a\mu\nu} = \widetilde{j}^{a\nu} \tag{1.171}$$

$$\partial_\mu T^{a\mu\nu} = \widetilde{J}^{a\nu} \tag{1.172}$$

where

$$\widetilde{j}^{a\nu} = \widetilde{R}^a{}_b{}^{\mu\nu} - \omega^a{}_{\mu b} \widetilde{T}^{b\mu\nu} \tag{1.173}$$

$$\widetilde{J}^{a\nu} = R^a{}_b{}^{\mu\nu} - \omega^a{}_{\mu b} T^{b\mu\nu} \tag{1.174}$$

In vector notation the currents are defined by

$$\widetilde{j}^{a\mu} = \left(\frac{1}{c}\widetilde{j}^{a0}, \widetilde{\mathbf{j}}^a\right) \tag{1.175}$$

$$\widetilde{J}^{a\mu} = \left(\frac{1}{c}\widetilde{J}^{a0}, \widetilde{\mathbf{J}}^a\right) \tag{1.176}$$

and Eqs.1.171 and 1.172 become four vector equations:

$$\nabla \cdot \mathbf{T}^a_s = \frac{\widetilde{j}^{a0}}{c} \tag{1.177}$$

$$\nabla \times \mathbf{T}^a_L + \frac{1}{c}\frac{\partial \mathbf{T}^a_s}{\partial t} = \widetilde{\mathbf{j}}^a \tag{1.178}$$

$$\nabla \cdot \mathbf{T}^a_L = \widetilde{J}^{a0} \tag{1.179}$$

$$\nabla \times \mathbf{T}^a_s - \frac{1}{c^2}\frac{\partial \mathbf{T}^a_L}{\partial t} = \widetilde{\mathbf{J}}^a \tag{1.180}$$

Analogously, the equations of generally covariant translational dynamics are:

$$d \wedge R^a{}_b = j^a{}_b \tag{1.181}$$

$$d \wedge \widetilde{R}^a{}_b = J^a{}_b \tag{1.182}$$

where

$$j^a{}_b = R^a{}_c \wedge \omega^c{}_b - \omega^a{}_c \wedge R^c{}_b \tag{1.183}$$

$$J^a{}_b = \widetilde{R}^a{}_c \wedge \omega^c{}_b - \omega^a{}_c \wedge \widetilde{R}^c{}_b \tag{1.184}$$

The equations of generally covariant electrodynamics are obtained from the ECE Ansatzen:

$$A^a = A^{(0)} q^a \tag{1.185}$$

$$F^a = A^{(0)} T^a \tag{1.186}$$

In form notation they are:

$$d \wedge F^a = \mu_0 j^a \tag{1.187}$$

$$d \wedge \widetilde{F}^a = \mu_0 J^a \tag{1.188}$$

where:

$$j^a = \frac{A^{(0)}}{\mu_0} \left(R^a{}_b \wedge q^b - \omega^a{}_b \wedge T^b \right) \tag{1.189}$$

$$J^a = \frac{A^{(0)}}{\mu_0} \left(\widetilde{R}^a{}_b \wedge q^b - \omega^a{}_b \wedge \widetilde{T}^a \right) \tag{1.190}$$

In tensor notation:

$$\partial_\mu \widetilde{F}^{a\mu\nu} = \mu_0 \widetilde{j}^{a\nu} \tag{1.191}$$

$$\partial_\mu F^{a\mu\nu} = \mu_0 \widetilde{J}^{a\nu} \tag{1.192}$$

where:

$$\widetilde{j}^{a\nu} = \frac{A^{(0)}}{\mu_0} \left(\widetilde{R}^a{}_\mu{}^{\mu\nu} - \omega^a{}_{\mu b} \widetilde{T}^{b\mu\nu} \right) \tag{1.193}$$

$$\widetilde{J}^{a\nu} = \frac{A^{(0)}}{\mu_0} \left(R^a{}_\mu{}^{\mu\nu} - \omega^a{}_{\mu b} T^{b\mu\nu} \right) \tag{1.194}$$

and in vector notation:

$$\nabla \cdot \mathbf{B}^a = \mu_0 \widetilde{j}^{a0} \tag{1.195}$$

$$\nabla \times \mathbf{E}^a + \frac{\partial \mathbf{B}^a}{\partial t} = \mu_0 \widetilde{\mathbf{j}}^a \tag{1.196}$$

$$\nabla \cdot \mathbf{E}^a = \mu_0 c \widetilde{J}^{a0} \tag{1.197}$$

$$\nabla \times \mathbf{B}^a - \frac{1}{c^2} \frac{\partial \mathbf{E}^a}{\partial t} = \frac{\mu_0}{c} \widetilde{\mathbf{J}}^a \tag{1.198}$$

It is seen that the generally covariant equations of electrodynamics and dynamics have the same overall structure and are inter-influential in general. This means that gravitation may influence electrodynamics as well as rotational dynamics. Many types of novel cross influences of this kind are indicated by ECE theory, on both classical and quantum levels [2]– [18].

Acknowledgements The British Parliament and Head of State, Queen Elizabeth II, are thanked for a Civil List pension. The voluntary staff of AIAS are thanked for many interesting discussions.

Bibliography

[1] A. Koestler, The Sleepwalkers (Hutchison, London, 1968).

[2] M. W. Evans, Generally Covariant Unified Field Theory (Abramis, 2005), volume 1.

[3] M. W. Evans, Generally Covariant Unified Field Theory (Abramis, 2006), volume 2.

[4] M. W. Evans, Generally Covariant Unified Field Theory (Abramis, 2006), volume 3, preprint on www.aias.us and www.atomicprecision.com.

[5] L. Felker, The ECE Equations of Unified Field Theory (preprint on www.aias.us and www.atomicprecision.com).

[6] M. W. Evans, Found. Phys. Lett., 16, 367, 507 (2003).

[7] M. W. Evans, Found. Phys. Lett., 17, 25, 149, 267, 301, 393, 433, 535, 663 (2004).

[8] M. W. Evans, Found. Phys. Lett., 18, 139, 259, 519 (2005).

[9] M. W. Evans and L. B. Crowell, Classical and Quantum Electrodynamics and the $\boldsymbol{B}^{(3)}$ Field (World Scientific, Singapore, 2001).

[10] M. W. Evans (ed.), Modern Non-Linear Optics, a special topical issue of I. Prigogine and S. A. Rice, Advances in Chemical Physics (Wiley Interscience, 2001, 2nd ed.), vols 119(1) to 119(3).

[11] M. W. Evans and J.-P. Vigier, The Enigmatic Photon (Kluwer, Dordrecht, 1994 to 2002, hardback and softback), in five volumes.

[12] M. W. Evans and A. A. Hasanein, The Photomagneton in Quantum Field Theory (World Scientific, Singapore, 1994).

[13] M. W. Evans and S. Kielich (eds.), first edition of ref. (10) (Wiley Interscience, New York, 1992, 1993, 1997, hardback and softback), vols. 85(1) to 85(3).

[14] M. W. Evans, Physica B, 182, 227, 237 (1992), the original B(3) papers.

[15] S. P. Carroll, Space-time and Geometry: an Introduction to General Relativity, (Addison-Wesley, New York, 2004).

[16] A. Einstein, The Meaning of Relativity (Princeton Univ. Press, 1921-1953).

[17] J. B. Marion and S. T. Thornton, Classical Dynamics of Particles and Systems (HBJ, New York, 1988, 3rd. Ed.).

[18] L. H. Ryder, Quantum Field Theory (Cambridge, 1996, 2nd ed.).

Chapter 2

Geodesics and the Aharonov Bohm Effects in ECE Theory

(Paper 56)
by
Myron W. Evans,
Alpha Institute for Advanced Study (AIAS).
(emyrone@aol.com, www.aias.us, www.atomicprecision.com)

Abstract

Geodesics are considered in the Einstein Cartan Evans (ECE) unified field theory. The concept of parallel transport along any curve is extended to the covariant exterior derivative of Cartan geometry. The Lorentz force is thus recognized as the covariant exterior derivative of the potential form along any path in ECE space-time. The class of Aharonov Bohm effects are due to parallel transport of the exterior covariant derivative. The equations of the electromagnetic and gravitational Aharonov Bohm effects are developed in ECE theory.

Keywords: Einstein Cartan Evans (ECE) unified field theory, geodesics, Lorentz force equation, Aharonov Bohm effects.

2.1 Introduction

In Einsteinian philosophy every equation of physics must be rigorously objective to all observers. This is the most fundamental principle of relativity theory, the latter must be generally covariant, covariant under the general coordinate transformation. This principle must apply both on the classical and quantum levels. Recently [1]– [15], a generally covariant unified field theory has been developed which meets these fundamental philosophical requirements, and is known as Einstein Cartan Evans (ECE) field theory. It has been applied to many aspects of physics [1] [2] and has been tested extensively against experimental data. It also meets the fundamental philosophical requirement known as Ockham's Razor, that a theory of physics must be as simple as possible. In this respect the ECE theory needs only the four dimensions of the original theory of relativity, and is preferred to string theory. The latter uses many unphysical dimensions as is well known. ECE theory is preferred to gauge theory because the latter is not generally covariant in three of its sectors: the electromagnetic, weak and strong sectors. Gauge theory in these sectors superimposes an abstract set of numbers on a Minkowski spacetime, while ECE theory is rigorously covariant in all sectors [1]- [15], being based directly on standard Cartan geometry. Gauge theory runs into trouble in trying to explain the well known Aharonov Bohm (AB) effects, while ECE theory explains them straightforwardly. The standard model runs into trouble in quantum mechanics, because of the lack of objectivity in its quantum mechanical sector, while ECE theory rigorously retains objectivity on the classical and quantum levels. Therefore ECE theory is preferred to the standard model, and explains data in a generally covariant manner under all circumstances.

In Section 2.2 the theory of geodesics in ECE theory is developed by extending the concept of parallel transport to the covariant exterior derivative of Cartan geometry. It is shown that the potential form [1]- [15] is parallel transported along any curve in ECE theory. It follows that the Cartan structure equations are also parallel transported along any curve, and that the Lorentz force equation is the covariant exterior derivative of the potential form along any path in ECE space-time. All AB effects are due to parallel transport of the exterior covariant derivative. These concepts extend the concept of the geodesic in relativity theory. The geodesic is the path that parallel transports it own tangent vector [15] and in gravitational relativity (Einstein Hilbert (EH) field theory) is the path followed by unaccelerated particles. In Section 2.3 the equations of the electromagnetic AB effect are given, and in Section 2.4, the equations are given of the gravitational AB effect.

2.2 Parallel transport and geodesics in ECE Theory, applications to the Lorentz Force equation and Aharonov Bohm effects

In standard general relativity (EH theory [15]) the geodesic is defined as the path that parallel transports its own tangent vector. If $x^{\mu}(\lambda)$ is any curve, the tangent vector is defined [15] as:

$$t^{\mu}=\frac{dx^{\mu}}{d\lambda} \tag{2.1}$$

The covariant derivative along the path is defined [15] as:

$$\frac{D}{d\lambda}:=\frac{dx^{\mu}}{d\lambda}D_{\mu} \tag{2.2}$$

where D_{μ} is the covariant derivative. The parallel transport of any tensor is then defined [15] as:

$$\left(\frac{DT}{d\lambda}\right)^{\mu_1\mu_2\cdot\mu_k}_{\nu_1\nu_2\cdot\mu_l}=\frac{dx^{\sigma}}{d\lambda}D_{\sigma}T^{\mu_1\mu_2\cdot\mu_k}_{\nu_1\nu_2\cdot\nu_l}=0 \tag{2.3}$$

In EH theory the connection is always metric compatible [15]:

$$D_\mu g_{\rho\nu} = 0 \tag{2.4}$$

where $g_{\rho\nu}$ is the symmetric metric of EH field theory. Such a connection is always parallel transported [15]:

$$\frac{Dg_{\mu\nu}}{d\lambda} = \frac{dx^\sigma}{d\lambda} D_\sigma g_{\mu\nu} = 0 \tag{2.5}$$

The equation of the geodesic in EH theory is [15]:

$$\frac{D}{d\lambda}\left(\frac{dx^\mu}{d\lambda}\right) = 0 \tag{2.6}$$

and parallel transports the tangent vector. Eq.(2.6) can be rewritten as:

$$\frac{d^2x^\mu}{d\lambda^2} + \Gamma^\mu{}_{\rho\sigma}\frac{dx^\rho}{d\lambda}\frac{dx^\sigma}{d\lambda} = 0 \tag{2.7}$$

which is the best known form of the geodesic equation of EH field theory. The geodesic in EH field theory is the path followed by unaccelerated particles. In flat space-time the connection vanishes, and Eq.(2.7) becomes:

$$\frac{d^2x^\mu}{d\lambda^2} = 0 \tag{2.8}$$

which is the equation of a straight line. Newton's first law is therefore regained, unaccelerated particles move in a straight line. The parameter λ can be related [15] to the proper time τ by:

$$\lambda = a\tau + b \tag{2.9}$$

An affine parameter is related to the proper time in this way. If:

$$\lambda = \tau \tag{2.10}$$

the geodesic equation becomes:

$$\frac{d^2x^\mu}{d\tau^2} + \Gamma^\mu{}_{\rho\sigma}\frac{dx^\rho}{d\tau}\frac{dx^\sigma}{d\tau} = 0 \tag{2.11}$$

and in the Newtonian limit this equation means that:

$$\mathbf{a} = \mathbf{0} \tag{2.12}$$

where **a** denotes acceleration. Therefore in the Newtonian limit the geodesic equation is the Newton force law for:

$$\mathbf{f} = m\mathbf{a} = \mathbf{0} \tag{2.13}$$

The Newton force law is therefore the limit of a more general geometry than Euclidean geometry. For EH theory this is well known to be Riemann geometry. For ECE theory Riemann goometry is generalized to the well known Cartan geometry [1]– [15].

When there is a force present the right hand side of Eq.(2.11) is no longer zero. The standard model Lorentz force for example [15] is:

$$f^\mu = eU^\lambda F_\lambda{}^\mu = eF^\mu{}_\nu \frac{dx^\nu}{d\tau} \tag{2.14}$$

where $F^\mu{}_\nu$ is the standard model electromagnetic field tensor and U^λ is the four velocity. In the presence of a Lorentz force, the electron no longer moves along a geodesic. The standard model Lorentz force, however, is Lorentz covariant, and is therefore the result of a theory of special relativity, not of general relativity as required. In this section the required generally covariant Lorentz force equation is developed with ECE theory. In so doing the effect of gravitation on the Lorentz force may be estimated. This is not possible in the standard model.

Parallel transport is defined for the arbitrary tensor T (indices suppressed for clarity) as:

$$\frac{dT}{d\lambda} = \frac{dx^\mu}{d\lambda}\frac{\partial T}{\partial x^\mu} = 0 \tag{2.15}$$

This result comes from the rule for differentiation [16] of a function of a function. If:

$$T = T\left(x^\mu\left(\lambda\right)\right) \tag{2.16}$$

then:

$$\frac{dT}{d\lambda} = \frac{dT}{dx^\mu}\frac{dx^\mu}{d\lambda} \tag{2.17}$$

The rule (2.17) may be extended to the covariant derivative of a tensor of any rank:

$$\frac{DT}{d\lambda} = \frac{dx^\mu}{d\lambda} D_\mu T \tag{2.18}$$

where T is defined by Eq.(2.16). The rule (2.18) is extended in this Section to the covariant exterior derivative of Cartan [1]– [15], used in the first and second Cartan structure equations:

$$T^a = D \wedge q^a = d \wedge q^a + \omega^a{}_b \wedge q^b \tag{2.19}$$

and

$$R^a{}_b = D \wedge \omega^a{}_b + \omega^a{}_c \wedge \omega^c{}_b \tag{2.20}$$

Here T^a is the torsion form, $R^a{}_b$ is the Riemann or curvature form, q^a is the tetrad form, $\omega^a{}_b$ is the spin connection, $D\wedge$ denotes the covariant exterior derivative and $d\wedge$ denotes the exterior derivative. It will be proven that:

$$\frac{dx^\mu}{d\lambda}\left(D \wedge A\right)_{\mu\nu} = \frac{dx^\mu}{d\lambda} F^a{}_{\mu\nu} \tag{2.21}$$

is the origin of the generally covariant Lorentz force and also the origin of the class of AB effects. Here $F^a{}_{\mu\nu}$ is the generally covariant electromagnetic field tensor of ECE theory, defined in standard notation [1]– [15] as:

$$F^a = D \wedge A^a \tag{2.22}$$

Consider:

$$A^a{}_\mu = A^a{}_\mu\left(x^\nu\left(\lambda\right)\right) \tag{2.23}$$

and

$$A^a{}_\nu = A^a{}_\nu \left(x^\mu \left(\lambda\right)\right) \tag{2.24}$$

with:

$$F^a{}_{\mu\nu} = \partial_\mu A^a{}_\nu - \partial_\nu A^a{}_\mu + \omega^a{}_{\mu b} A^b{}_\nu - \omega^a{}_{\nu b} A^b{}_\mu \tag{2.25}$$

Then:

$$\frac{DA^a{}_\mu}{d\lambda} = \frac{dx^\mu}{d\lambda} D_\nu A^a{}_\mu \tag{2.26}$$

$$\frac{DA^a{}_\nu}{d\lambda} = \frac{dx^\nu}{d\lambda} D_\mu A^a{}_\nu \tag{2.27}$$

Using the tetrad postulate [1]– [15]:

$$D_\nu A^a{}_\mu = D_\mu A^a{}_\nu = 0, \quad A^a{}_\mu = A^{(0)} q^a{}_\mu \tag{2.28}$$

it follows that:

$$\frac{DA^a{}_\mu}{d\lambda} = \frac{DA^a{}_\nu}{d\lambda} = 0 \tag{2.29}$$

a result which means that

$$A^a{}_\mu = A^{(0)} q^a{}_\mu \tag{2.30}$$

is parallel transported along any curve $x^\mu\left(\lambda\right)$. It follows that:

$$\frac{dx^\mu}{d\lambda}\left(D_\nu A^a{}_\mu - D_\mu A^a{}_\nu\right) = 0 \tag{2.31}$$

The exterior covariant derivative is defined as:

$$\left(D \wedge A\right)^a_{\mu\nu} = F^a{}_{\mu\nu} \tag{2.32}$$

with:

$$D_\mu A^a{}_\nu - D_\nu A^a{}_\mu = \partial_\mu A^a{}_\nu - \partial_\nu A^a{}_\mu + \omega^a{}_{\mu b} A^b{}_\nu - \omega^a{}_{\nu b} A^b{}_\mu - F^a{}_{\mu\nu} = 0 \tag{2.33}$$

It follows from Eqs.(2.33) and (2.35) that the Cartan structure equation is parallel transported along any curve $x^\mu\left(\lambda\right)$:

$$\frac{dx^\mu}{d\lambda}\left(F^a - D \wedge A^a\right)_{\mu\nu} = 0 \tag{2.34}$$

and so:

$$\frac{dx^\mu}{d\lambda}\left(D \wedge A^a\right)_{\mu\nu} = \frac{dx^\mu}{d\lambda} F^a{}_{\mu\nu} \tag{2.35}$$

Q.E.D. The generally covariant Lorentz force is the right hand side of Eq. (2.35) multiplied by e, the charge on the electron.

In the absence of the Lorentz force and thus in the absence of acceleration:

$$\frac{dx^\mu}{d\lambda} F^a{}_{\mu\nu} = \frac{dx^\mu}{d\lambda}\left(D \wedge A^a\right)_{\mu\nu} = 0 \tag{2.36}$$

and the exterior covariant derivative of Cartan [1]– [15] is parallel transported along any curve.

Similarly:

$$\frac{dx^\mu}{d\lambda} R^a{}_{b\mu\nu} = \frac{dx^\mu}{d\lambda}\left(D \wedge \omega^a{}_b\right)_{\mu\nu} \tag{2.37}$$

and in the absence of force, the exterior covariant derivative of the spin connection is parallel transported along any curve:

$$\frac{dx^\mu}{d\lambda} R^a{}_{b\mu\nu} = \frac{dx^\mu}{d\lambda} \left(D \wedge \omega^a{}_b\right)_{\mu\nu} = 0 \tag{2.38}$$

The class of AB effects may now be defined in terms of these equations. The electromagnetic AB effects are defined by Eq.(2.36). When there is no field there may still be a potential present in ECE theory, a potential defined by:

$$A^{(0)} \left(d \wedge q^a + \omega^a{}_b \wedge q^b\right) = 0 \tag{2.39}$$

where $cA^{(0)}$ is a primordial voltage. Eq.(2.39) is true for all $A^{(0)}$ and all paths $x^\mu(\lambda)$. The Cartan geometry of the electromagnetic AB effects is therefore always:

$$D \wedge q^a = 0 \tag{2.40}$$

Similarly the Cartan geometry of the gravitational AB effects is always:

$$D \wedge \omega^a{}_b = d \wedge \omega^a{}_b + \omega^a{}_c \wedge \omega^c{}_b = 0 \tag{2.41}$$

All AB effects are therefore due to parallel transport of the exterior covariant derivative of Cartan geometry. They are effects of a generally covariant unified field theory [1] [2], of spinning and curving space-time. In the standard model there are no AB effects because in the standard model [1] [2]:

$$F_{\mu\nu} = \left(d \wedge A\right)_{\mu\nu} \tag{2.42}$$

and if F is zero so is A and vice-versa.

2.3 The equations of the electromagnetic AB effects

The class of electromagnetic AB effects is defined by:

$$A^{(0)} D \wedge q = 0 \tag{2.43}$$

where $cA^{(0)}$ is a primordial voltage which is always non-zero. Eq.(2.43) means that when the electromagnetic field is zero the electromagnetic potential is non-zero. In the well known Chambers experiment [1] [2] the magnetic field of an iron whisker does not interact with the electron beams, but there is present a potential which causes a fringe shift in a Young interferometer. Eq.(2.43) is written in the standard notation [1]– [15] of differential geometry as:

$$d \wedge A^a + \omega^a{}_b \wedge A^b = 0 \tag{2.44}$$

and implies:

$$F^a = 0 \tag{2.45}$$

The fringe shift of the Chambers experiment is described [1] [2] in the minimal prescription by the following shift in the momentum tetrad:

$$p^a \rightarrow p^a + eA^a \tag{2.46}$$

In the standard model:

$$F = d \wedge A \tag{2.47}$$

so if F is zero so is A and vice-versa. In ECE theory A^a can be non-zero if F^a is zero, as first observed by Chambers and in many experiments since then. The mathematical task is to solve Eq.(2.44) for the potential components responsible for the class of electromagnetic AB effects. If we consider electromagnetism free from gravitational effects [1]– [14], then the spin connection is dual to the tetrad as follows:

$$\omega^a{}_b = -\frac{\kappa}{2}\epsilon^a{}_{bc}q^c \tag{2.48}$$

Here:

$$\epsilon^a{}_{bc} = \eta^{ad}\epsilon_{dbc} \tag{2.49}$$

where

$$\begin{aligned} \eta^{ad} &= \mathrm{diag}\,(1, -1, -1, -1) \\ &= \begin{bmatrix} 1 & 0 & 0 & 0 \\ 0 & -1 & 0 & 0 \\ 0 & 0 & -1 & 0 \\ 0 & 0 & 0 & -1 \end{bmatrix} \end{aligned} \tag{2.50}$$

is the Minkowski metric of the tangent space-time at point P to the base manifold of ECE theory [1]– [14]. In Eq.(2.49) ϵ_{abc} is the three index, totally antisymmetric unit tensor in four dimensions. This is defined in general as follows:

$$\left.\begin{aligned} \epsilon_{123} = \epsilon_{231} = \epsilon_{312} = 1, \\ \epsilon_{132} = \epsilon_{213} = \epsilon_{321} = -1, \\ \epsilon_{012} = \epsilon_{120} = \epsilon_{201} = 1, \\ \epsilon_{021} = \epsilon_{102} = \epsilon_{210} = -1, \\ \epsilon_{012} = \epsilon_{301} = \epsilon_{130} = \epsilon_{023} = \epsilon_{302} = \epsilon_{230} = 1, \\ \epsilon_{031} = \epsilon_{310} = \epsilon_{103} = \epsilon_{032} = \epsilon_{320} = \epsilon_{203} = -1 \end{aligned}\right\} \tag{2.51}$$

Therefore:

$$
\left.
\begin{aligned}
\epsilon^{1}{}_{23} &= g^{1d}\epsilon_{d23} = g^{11}\epsilon_{123} = -\epsilon_{123}, \\
\epsilon^{3}{}_{12} &= g^{3d}\epsilon_{d12} = g^{33}\epsilon_{312} = -\epsilon_{312}, \\
\epsilon^{2}{}_{31} &= g^{2d}\epsilon_{d31} = g^{22}\epsilon_{231} = -\epsilon_{231}, \\
\epsilon^{0}{}_{12} &= g^{00}\epsilon_{012} = \epsilon_{012}, \\
\epsilon^{2}{}_{01} &= g^{22}\epsilon_{201} = -\epsilon_{120}, \\
\epsilon^{0}{}_{13} &= g^{00}\epsilon_{013} = \epsilon_{013}, \\
\epsilon^{0}{}_{23} &= g^{00}\epsilon_{023} = \epsilon_{023}, \\
\epsilon^{3}{}_{01} &= g^{33}\epsilon_{301} = \epsilon_{301}, \\
\epsilon^{2}{}_{30} &= g^{22}\epsilon_{230} = -\epsilon_{230}.
\end{aligned}
\right\} \tag{2.52}
$$

Eq.(2.44) means that:

$$
\left.
\begin{aligned}
d \wedge A^{0} + \omega^{0}{}_{b} \wedge A^{b} &= 0, \\
d \wedge A^{1} + \omega^{1}{}_{b} \wedge A^{b} &= 0, \\
d \wedge A^{2} + \omega^{2}{}_{b} \wedge A^{b} &= 0, \\
d \wedge A^{3} + \omega^{3}{}_{b} \wedge A^{b} &= 0,
\end{aligned}
\right\} \tag{2.53}
$$

with summation implied over repeated contravariant-covariant indices as follows:

$$
\left.
\begin{aligned}
d \wedge A^{0} + \omega^{0}{}_{1} \wedge A^{1} + \omega^{0}{}_{2} \wedge A^{2} + \omega^{0}{}_{3} \wedge A^{3} &= 0, \\
d \wedge A^{1} + \omega^{1}{}_{0} \wedge A^{0} + \omega^{1}{}_{2} \wedge A^{2} + \omega^{1}{}_{3} \wedge A^{3} &= 0, \\
d \wedge A^{2} + \omega^{2}{}_{0} \wedge A^{0} + \omega^{2}{}_{1} \wedge A^{1} + \omega^{2}{}_{3} \wedge A^{3} &= 0, \\
d \wedge A^{3} + \omega^{3}{}_{0} \wedge A^{0} + \omega^{3}{}_{1} \wedge A^{1} + \omega^{3}{}_{2} \wedge A^{2} &= 0
\end{aligned}
\right\} \tag{2.54}
$$

The various spin connections are therefore:

$$\left.\begin{aligned}
\omega^0{}_1 &= -\frac{\kappa}{2}\left(\epsilon^0{}_{12}q^2 + \epsilon^0{}_{13}q^3\right),\\
\omega^0{}_2 &= -\frac{\kappa}{2}\left(\epsilon^0{}_{23}q^3 + \epsilon^0{}_{21}q^1\right),\\
\omega^0{}_3 &= -\frac{\kappa}{2}\left(\epsilon^0{}_{31}q^1 + \epsilon^0{}_{32}q^2\right),\\
\omega^1{}_2 &= -\frac{\kappa}{2}\left(\epsilon^1{}_{20}q^0 + \epsilon^1{}_{23}q^3\right),\\
\omega^1{}_3 &= -\frac{\kappa}{2}\left(\epsilon^1{}_{32}q^2 + \epsilon^1{}_{30}q^0\right),\\
\omega^2{}_3 &= -\frac{\kappa}{2}\left(\epsilon^2{}_{31}q^1 + \epsilon^2{}_{30}q^0\right),
\end{aligned}\right\} \tag{2.55}$$

which can be rewritten as:

$$\left.\begin{aligned}
\omega^0{}_1 &= -\frac{\kappa}{2}\left(\epsilon_{012}q^2 + \epsilon_{013}q^3\right),\\
\omega^0{}_2 &= -\frac{\kappa}{2}\left(\epsilon_{023}q^3 + \epsilon_{021}q^1\right),\\
\omega^0{}_3 &= -\frac{\kappa}{2}\left(\epsilon_{031}q^1 + \epsilon_{032}q^2\right),\\
\omega^1{}_2 &= \frac{\kappa}{2}\left(\epsilon_{120}q^0 + \epsilon_{123}q^3\right),\\
\omega^1{}_3 &= \frac{\kappa}{2}\left(\epsilon_{132}q^2 + \epsilon_{130}q^0\right),\\
\omega^2{}_3 &= \frac{\kappa}{2}\left(\epsilon_{231}q^1 + \epsilon_{230}q^0\right),
\end{aligned}\right\} \tag{2.56}$$

Therefore we obtain:

$$\left.\begin{aligned}
\omega^0{}_1 &= -\frac{\kappa}{2}\left(q^2 + q^3\right),\\
\omega^0{}_2 &= -\frac{\kappa}{2}\left(q^3 - q^1\right),\\
\omega^0{}_3 &= -\frac{\kappa}{2}\left(-q^1 - q^2\right),\\
\omega^1{}_2 &= \frac{\kappa}{2}\left(q^0 + q^3\right),\\
\omega^1{}_3 &= \frac{\kappa}{2}\left(-q^2 + q^0\right),\\
\omega^2{}_3 &= \frac{\kappa}{2}\left(q^1 + q^0\right)
\end{aligned}\right\} \tag{2.57}$$

Therefore there are six spin connections, each of which can be expressed as the sum of two tetrad elements. The cyclic potential equations are then:

$$d \wedge A^1 + \frac{\kappa}{2}\left(q^2 \wedge A^0 + q^3 \wedge A^0 + q^0 \wedge A^2 + q^3 \wedge A^2 + q^0 \wedge A^3 - q^2 \wedge A^3\right) = 0,$$

$$\text{i.e.} \qquad d \wedge A^1 = gA^2 \wedge A^3 \tag{2.58}$$

$$d \wedge A^2 + \frac{\kappa}{2}\left(q^3 \wedge A^0 - q^1 \wedge A^0 - q^0 \wedge A^1 - q^3 \wedge A^1 + q^1 \wedge A^3 + q^0 \wedge A^3\right) = 0,$$

$$\text{i.e.} \qquad d \wedge A^2 = gA^3 \wedge A^1 \tag{2.59}$$

$$d \wedge A^3 + \frac{\kappa}{2}\left(-q^1 \wedge A^0 - q^2 \wedge A^0 - q^0 \wedge A^1 + q^2 \wedge A^1 - q^3 \wedge A^1 - q^0 \wedge A^2\right) = 0,$$

$$\text{i.e.} \qquad d \wedge A^3 = gA^1 \wedge A^2 \tag{2.60}$$

$$d \wedge A^0 - \frac{\kappa}{2}\left(q^2 \wedge A^1 + q^3 \wedge A^1 + q^3 \wedge A^2 - q^1 \wedge A^2 + q^1 \wedge A^3 + q^2 \wedge A^3\right) = 0,$$

$$\text{i.e.} \qquad d \wedge A^0 = gA^2 \wedge A^3 = -d \wedge A^3 \tag{2.61}$$

where we have used the anti-symmetry of the wedge product, for example:

$$q^3 \wedge A^2 = -q^2 \wedge A^3 \tag{2.62}$$

and so on. In summary therefore the equations of the class of all electromagnetic AB effects are:

$$\left.\begin{aligned} d \wedge A^1 &= gA^2 \wedge A^3, \\ d \wedge A^2 &= gA^3 \wedge A^1, \\ d \wedge A^3 &= gA^1 \wedge A^2, \\ d \wedge A^0 &= -d \wedge A^3 \end{aligned}\right\} \tag{2.63}$$

in generally covariant unified field theory. These are based on the duality of Eq.(2.48). When gravitation affects electromagnetism Eq.(2.48) no longer applies, so gravitation will affect the electromagnetic AB effects, and all electromagnetic effects in general.

In regions where there are fields present:

$$F = D \wedge A \neq 0, \tag{2.64}$$

and the field components are:

$$\left.\begin{aligned} F^1 &= d \wedge A^1 - gA^2 \wedge A^3, \\ F^2 &= d \wedge A^2 - gA^3 \wedge A^1, \\ F^3 &= d \wedge A^3 - gA^1 \wedge A^2 \end{aligned}\right\} \tag{2.65}$$

These components have $O(3)$ symmetry. An example of the equations (2.65) is $O(3)$ electrodynamics [1]– [15]. The electromagnetic AB effects are caused by the potentials A^0, A^1, A^2 and A^3 in regions defined by Eqs. (2.43) to (2.63). In the Chambers experiment for example, the magnetic field inside the iron whisker is defined by Eqs.(2.65) and the potential outside the iron whisker by Eqs.(2.43) to (2.63). Eqs.(2.43) to (2.63) show that in general there is an electric AB effect and an electromagnetic AB effect [1] [2]. In regions outside a radar beam for example, there will be a discernible AB effect to second order in the potential akin to the inverse Faraday effect [1] [2] in regions where the field is non-zero. This type of electromagnetic AB effect is due to the ECE spin field [1]- [14] observed in the inverse Faraday effect. The ECE spin field is defined by:

$$F^3 = -gA^1 \wedge A^2 \tag{2.66}$$

and is to SECOND order in the potential. It is a new type of magnetic field whose origins can now be traced via ECE theory to general relativity. The ECE spin field does not exist in special relativity (Maxwell Heaviside field theory), so it shows the general covariance of electromagnetism [1]- [14]. This means that electromagnetism is a spinning frame of reference which must be described with a non-zero spin connection. The existence of the spin connection is shown experimentally by the well known inverse Faraday effect (the magnetization of any type of material by a circularly or elliptically polarized electromagnetic field at any frequency).

In the standard model [16] an attempt is made to explain the magnetic AB effect using gauge theory in special relativity. In the notation of differential geometry [1]- [15] gauge theory means that:

$$A \rightarrow A + dX \tag{2.67}$$

but by the Poincaré Lemma:

$$d \wedge dX := 0 \tag{2.68}$$

In the standard model [16] the Stokes Theorem [1]- [14] [17] is used to claim that:

$$\oint dX = \int_S d \wedge dX \neq 0 \tag{2.69}$$

However, this claim is incorrect [1] [2] because it violates the Stokes Theorem. It is well known that the latter applies [17] to non simply connected spaces, so:

$$\oint dX = \int_S d \wedge dX := 0 \tag{2.70}$$

using the Poincaré Lemma (2.68) the Stokes Theorem implies that:

$$\oint dX := 0 \tag{2.71}$$

thus contradicting the claim (2.69) of the standard model.

This error in the standard model can be illustrated for example by the error in Eq.(3.105) of a standard model textbook such as ref. [16]. The task is to evaluate:

$$\Delta\delta = \oint \boldsymbol{\nabla} X \cdot d\mathbf{r} = \int_S \boldsymbol{\nabla} \times \boldsymbol{\nabla} X := 0 \tag{2.72}$$

As we have seen, Eq.(2.72) must be the correct result by the Stokes Theorem. The function X in the example we are considering here (Eq.3.105 of ref. [16]) is

$$X = \frac{BR^2\theta}{2} \tag{2.73}$$

where B is magnetic flux density, R is a radius and θ is a cylindrical polar coordinate. The cylindrical polar coordinates are related [17] to the Cartesian coordinates by:

$$\left.\begin{array}{ll} \theta = \tan^{-1} y/x, & r = \left(x^2+y^2\right)^{1/2}, \\ x = r\cos\theta, & y = r\sin\theta \end{array}\right\} \tag{2.74}$$

as is well known. The left hand side of Eq.(2.72) is evaluated between $\theta = 0$ and $\theta = 2\pi$. Therefore:

$$\left.\begin{array}{ll} x = r\cos 2\pi = r, & y = r\sin 2\pi = 0, \\ x = r\cos 0 = r, & y = r\sin 0 = 0 \end{array}\right\} \tag{2.75}$$

and the integral is:

$$\Delta\delta = \oint_r^r \frac{\partial}{\partial x}\left(\tan^{-1}\frac{y}{x}\right)dx + \oint_0^0 \frac{\partial}{\partial y}\left(\tan^{-1}\frac{y}{x}\right)dy = 0 \tag{2.76}$$

This is the correct result as given by the Stokes Theorem (2.72). However, in ref. [16] it is incorrectly asserted that:

$$\oint \boldsymbol{\nabla} X \cdot d\mathbf{r} =? \qquad [X]_{\theta=0}^{\theta=2\pi} \neq 0 \tag{2.77}$$

The correct way to evaluate Eq.(2.77) is:

$$\oint_{\theta=0}^{\theta=2\pi} \boldsymbol{\nabla} X \cdot d\mathbf{r} = \oint_x^x \boldsymbol{\nabla} X \cdot d\mathbf{r} = 0 \tag{2.78}$$

because

$$x = r\cos\theta, \qquad r = \left(x^2+y^2\right)^{1/2} \tag{2.79}$$

and:

$$\left.\begin{array}{ll} \text{if} \ \ \theta = 2\pi, & r = x, \\ \text{if} \ \ \theta = 0, & r = x \end{array}\right\} \tag{2.80}$$

The whole of the argument on the AB effect in ref. [16] is therefore incorrect following the occurrence of this error, both mathematically and physically. The magnetic AB effect is not due to multiply connected spaces in special relativity. It is due to general relativity as we have argued. Thus ECE theory is preferred to the standard model, philosophically, mathematically and experimentally.

The electromagnetic AB effects in ECE theory are summarized as follows: In the Chambers experiment for example the observable is a phase shift:

$$\Delta\phi = \frac{e}{\hbar}\Phi \tag{2.81}$$

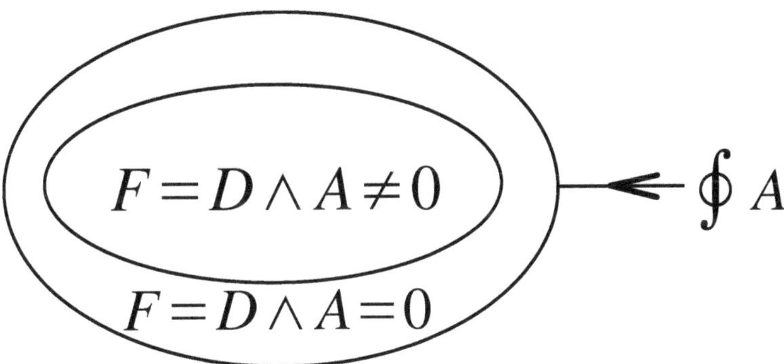

Figure 2.1: AB Effects in ECE Theory

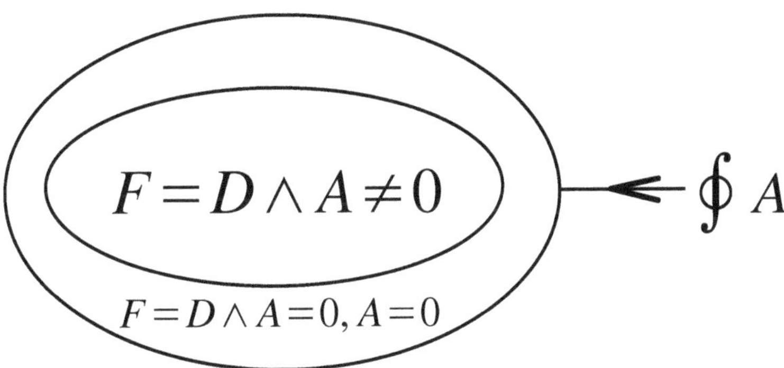

Figure 2.2: Standard Theory

where the magnetic flux is defined by:

$$\Phi = \int_S F = \int_S D \wedge A = \oint A \tag{2.82}$$

The line integral in Eq.(2.82) is around the outer contour of Fig. 2.1, the contour defined by the electron beams of the Chambers experiment [1], [2]. The magnetic flux density is enclosed in the inner area of Fig. 2.1, representing the area of the iron whisker. The latter area is much smaller than the area enclosed by the electron beams. In the area outside the whisker but inside the electron beams there is no magnetic flux density, but there is a potential defined by Eq.(2.63). The AB effect of the Chambers experiment is caused by this potential as we have argued.

In the standard model on the other hand there is no explanation for the Chambers experiment, as summarized in Fig. 2.2 below: As we have seen, the attempted gauge theoretical explanation of the standard model is mathematically incorrect.

2.4 The gravitational AB effect

The gravitational AB effect is defined in ECE theory in regions where:

$$R = D \wedge \omega = 0 \tag{2.83}$$

$$\omega \neq 0 \tag{2.84}$$

Eq.(2.83) may be developed as:

$$d \wedge \omega^a{}_b + \omega^a{}_c \wedge \omega^c{}_b = 0 \tag{2.85}$$

If there is no interaction between gravitation and electromagnetism the Cartan torsion is zero and so the gravitational AB effect in this case is further constrained by:

$$T^a = d \wedge q^a + \omega^a{}_b \wedge q^b = 0 \tag{2.86}$$

These are the conditions for the EH field theory of centrally directed gravitation. The limit of ECE theory where the Cartan torsion is zero. As is well known for EH theory, the Christoffel connection is symmetric:

$$\Gamma^\kappa{}_{\mu\nu} = \Gamma^\kappa{}_{\nu\mu} \tag{2.87}$$

and the metric is symmetric:

$$g_{\mu\nu} = g_{\nu\mu} = q^a{}_\mu q^b{}_\nu \eta_{ab} \tag{2.88}$$

The symmetry of the spin connection in EH theory is determined by the tetrad postulate [1]- [15]:

$$D_\mu q^a{}_\lambda = \partial_\mu q^a{}_\lambda + \omega^a{}_{\mu b} q^b{}_\lambda - \Gamma^\nu{}_{\mu\lambda} q^a{}_\nu = 0 \tag{2.89}$$

Interchange μ and λ:

$$D_\lambda q^a{}_\mu = \partial_\lambda q^a{}_\mu + \omega^a{}_{\lambda b} q^b{}_\mu - \Gamma^\nu{}_{\lambda\mu} q^a{}_\nu = 0 \tag{2.90}$$

and subtract Eq.(2.90) from Eq.(2.89) to find that:

$$\partial_\mu q^a{}_\lambda + \omega^a{}_{\mu b} q^b{}_\lambda - (\partial_\lambda q^a{}_\mu + \omega^a{}_{\lambda b} q^b{}_\mu) = 0 \tag{2.91}$$

This result is consistent with the fact that there is no Cartan torsion in EH theory:

$$T^a{}_{\mu\lambda} = \partial_\mu q^a{}_\lambda + \omega^a{}_{\mu b} q^b{}_\lambda = 0 \tag{2.92}$$

$$T^a{}_{\lambda\mu} = \partial_\lambda q^a{}_\mu + \omega^a{}_{\lambda b} q^b{}_\mu = 0 \tag{2.93}$$

Eqs.(2.92) to (2.93) must be symmetric under interchange of μ and λ, so

$$\omega^a{}_{\mu b} q^b{}_\lambda = \omega^a{}_{\lambda b} q^b{}_\mu \tag{2.94}$$

i.e.

$$\omega^a{}_{\mu\lambda} = \omega^a{}_{\lambda\mu} \tag{2.95}$$

Now interchange the indices a and b to find that:

$$\omega^b{}_{\mu\lambda} = \omega^b{}_{\lambda\mu} \tag{2.96}$$

Eqs.(2.95) and (2.96) are the same, so the spin connection in EH theory must be symmetric:

$$\omega^a{}_{\mu b} = \omega^b{}_{\mu a} \tag{2.97}$$

Therefore the gravitational AB equations for the diagonal elements of the spin connection are:

$$\left.\begin{aligned} d \wedge \omega^0{}_0 + \omega^0{}_c \wedge \omega^c{}_0 &= 0, \\ d \wedge \omega^1{}_1 + \omega^1{}_c \wedge \omega^c{}_1 &= 0, \\ d \wedge \omega^2{}_2 + \omega^2{}_c \wedge \omega^c{}_2 &= 0, \\ d \wedge \omega^3{}_3 + \omega^3{}_c \wedge \omega^c{}_3 &= 0 \end{aligned}\right\} \tag{2.98}$$

and there is a similar set of equations for the off diagonal elements. The gravitational AB effect has been observed experimentally [1] [2].

Acknowledgements The British Parliament and Head of State, Queen Elizabeth II, are thanked for a Civil List pension. The voluntary staff of AIAS are thanked for many interesting discussions and Franklin Amador for accurate and voluntary typesetting.

Bibliography

[1] M. W. Evans, Generally Covariant Unified Field Theory (Abramis, 2005), vol. 1, ibid., vols. 2 and 3 (Abramis, 2006), vol. 4.

[2] L. Felker, The ECE Equations of Unified Field Theory (preprint on www.aias.us and www.atomicprecision.com). H. Eckardt and L. Felker, popular article on these sites.

[3] M. W. Evans, Found. Phys. Lett., 16, 367, 507 (2003).

[4] M. W. Evans, Found. Phys. Lett., 17, 25, 149, 267, 301, 393, 433, 535, 663 (2004).

[5] M. W. Evans, Found. Phys. Lett., 18, 139, 259, 519 (2005).

[6] M. W. Evans and L. B. Crowell, Classical and Quantum Electrodynamics and the $\boldsymbol{B}^{(3)}$ Field, (World Scientific, Singapore, 2001).

[7] M. W. Evans (ed.), Modern Non-Linear Optics, a special topical issue in three parts of I. Prigogine and S. A. Rice (series eds.), Advances in Chemical Physics (Wiley Interscience, New York, 2001, 2nd ed., hardback and e book), vols. 119(1) - 119(3).

[8] M. W. Evans and J.-P. Vigier, The Enigmatic Photon, (Kluwer, Dordrecht, 1994 to 2002, hardback and softback), in five volumes.

[9] M. W. Evans and A. A. Hasanein, The Photomagneton in Quantum Field Theory (World Scientific, 1994).

[10] M. W. Evans and S. Kielich (eds.), first edition of ref. (7) (Wiley Interscience, New York, 1992, reprinted 1993, softback 1997), vols. 85(1) - 85(3).

[11] M. W. Evans, Physica B, 182, 227, 237 (first papers on the ECE spin field).

[12] M. W. Evans, Generally Covariant Dynamics, (preprint on www.aias.us and www.atomicprecision.com).

[13] M. W. Evans et al., preprints of seventy-two papers on ECE theory (www.aias.us and www.atomicprecision.com).

[14] Feedback sites for www.aias.us and www.atomicprecision.com, indicating about two million hits a year.

[15] S. P. Carroll, "Space-time and Geometry: an Introduction to General Relativity" (Addison Wesley, New York, 2004).

[16] . L. H Ryder, Quantum Field Theory, (Cambridge, 1996, 2nd ed.).

[17] E. G. Milewski (ed.), The Vector Analysis Problem Solver (REA, New York City, 1987), for example problem 17-27.

Chapter 3

Canonical and second quantization in generally covariant quantum field theory

(Paper 57)
by
Myron W. Evans,
Alpha Institute for Advanced Studyy (AIAS).
(emyrone@aol.com, www.aias.us, www.atomicprecision.com)

Abstract

Einstein Cartan Evans (ECE) field theory is shown to be a rigorous quantum field theory in which the tetrad is both the eigenfunction or wave-function and a quantized field that is generally covariant. Unification of fields is achieved with standard Cartan geometry on both the classical and quantum levels. The fundamental commutators needed for canonical quantization of the field/eigenfunction are derived self consistently from the same Cartan geometry. Second quantization proceeds straightforwardly thereafter by expanding the tetrad in terms of creation and annihilation operators. The latter are used to define the number operator in the usual way, and a generally covariant multi particle field theory obtained. The theory is illustrated with a discussion of the electromagnetic Aharonov Bohm effect.

Keywords: Einstein Cartan Evans (ECE) unified field theory, quantum field theory, canonical quantization and second quantization, Aharonov Bohm effects.

3.1 Introduction

Einstein Cartan Evans (ECE) unified field theory has been well developed analytically on the classical and single particle quantum levels [1]- [7]. In this paper it is shown that ECE field theory is a rigorous quantum field theory in which the tetrad is both the wave-function and the field. In Section 3.2, the fundamental commutators needed for canonical quantization [8], [9] are introduced from Cartan geometry and developed and in Section 3.3 second quantization [8], [9] is developed straightforwardly by expanding the tetrad in a Fourier series to give creation and annihilation operators that define the number operator. Therefore ECE theory produces a rigorous quantum field theory and can be given a multi particle interpretation as required [8], [9]. The theory in this paper is illustrated in Section 3.4 with the electromagnetic Aharonov Bohm effect.

Canonical quantization in quantum field theory is the name given to the construction of the Heisenberg commutators of the quantum field. Canonical quantization of the electromagnetic potential field, for example, runs into difficulties [8] in the contemporary standard model because of the assumption of a massless electromagnetic field with infinite range, and an identically zero photon mass. In special relativity this means that the special relativistic potential field A_μ can have only two physical components [8] and these are taken as the transverse components. However, A_μ must have four physical components to be manifestly covariant, and this is also a fundamental requirement of general relativity. Therefore there is a basic contradiction in the standard model, a contradiction that leads to well known difficulties [1]- [15]. It is shown in Section 3.2 that this contradiction is removed straightforwardly in ECE theory, in which the photon mass is identically non-zero as required. Without photon mass there can be no explanation of the Eddington experiment, contradicting the well known tests of the Einstein Hilbert (EH) field theory of gravitation, tests which are now known from NASA Cassini to be accurate to one part in one hundred thousand for light grazing the sun. So the existence of identically non-zero photon mass has been tested to this accuracy, because the Eddington experiment is explained in EH theory by considering the mass of the photon and the mass of the sun. If the photon mass is identically zero, the EH explanation makes no sense. Thus, photon mass is known very accurately to be identically non-zero. In Maxwell Heaviside field theory however, it is identically zero. This diametric contradiction is inherent in the standard model because gravitation is in that model a theory of general relativity and electromagnetism is a theory of special relativity (the Maxwell Heaviside (MH) field theory). Electromagnetism in the standard model is developed [8] [9] in terms of gauge theory. The simplest example of gauge theory is illustrated as follows. In MH theory the electromagnetic field in tensor notation is an anti-symmetric rank two tensor independent of gravitation:

$$F_{\mu\nu} = \partial_\mu A_\nu - \partial_\nu A_\mu. \tag{3.1}$$

This field tensor is unchanged under the mathematical transform:

$$A_\mu \rightarrow A_\mu + \partial_\mu \Lambda \tag{3.2}$$

which is a simple example of a gauge transformation [8]- [15]. Here Λ is any scalar function and the invariance of the field under the gauge transformation is an example of the Poincaré Lemma. The field is said to be gauge invariant, and in gauge field theory this is a central hypothesis which leads to a debate concerning the nature of $F_{\mu\nu}$ and A_μ. One or the other is regarded as fundamental, there being proponents of both views. ECE theory has shown that this debate is superfluous, and ECE has replaced gauge theory with a generally covariant unified field theory [1]– [7] in which the fundamental transformation is the general coordinate transformation of general relativity. The debate in the gauge theory of the standard model was initiated to a large extent by the discovery of the magnetic Aharonov Bohm effect by Chambers [8] [9].It was shown experimentally by Chambers that in regions where is zero, A_μ has a physical effect. The standard classical interpretation had been that $F_{\mu\nu}$ is physical but A_μ is unphysical. This interpretation was based directly on classical gauge theory, first proposed by Weyl and his contemporaries. In the ensuing forty year debate concerning the Aharonov Bohm (AB) effects some proponents have held the view that they are purely quantum effects, and in quantum theory the minimal prescription applies:

$$p_\mu \rightarrow p_\mu + eA_\mu \tag{3.3}$$

Here p_μ is the four-momentum of special relativity and $-e$ is the charge on the electron. In this view, used for example in the Dirac equation to explain the Stern Gerlach effect, A_μ is a physical property. In gauge theory it is held that A_μ can be transformed to $A_\mu + \partial_\mu \Lambda$ without affecting the field $F_{\mu\nu}$, so in consequence A_μ is unphysical. The philosophical basis of gauge theory is therefore questionable, the assumption that A_μ is unphysical and that $F_{\mu\nu}$ is physical is untenable and confusing even to the experts, thus a protracted forty year debate that reveals this confusion. In the standard model of the late twentieth century, gauge invariance was elevated to a central hypothesis of the electromagnetic, weak and strong fields and has been applied to the gravitational field, where it is clearly superfluous by Ockham's Razor. The general coordinate transformation is already sufficient for gravitational field theory and there is no need for the further postulate of gauge invariance. ECE theory has shown [1]– [7] that this is also true for unified field theory. The difficulties of using gauge theory outweigh any of its advantages, the latter being increasingly difficult to find as ECE is increasingly developed and accepted [16]. It has been shown [1]– [7] that ECE theory leads to a unified field theory in which the gravitational, electromagnetic, weak and strong fields are represented by the tangent spacetime at point P to the base manifold in standard Cartan geometry. The various fields are represented by various representation spaces in the tangent spacetime [1]– [7]. Gauge theory is superfluous in this context.

The difficulty inherent in the fundamental assumption of gauge invariance can be illustrated as follows using differential form notation [1]– [7] [17], where Eqs. (3.1) and (3.2) become:

$$A \rightarrow A + d\Lambda \tag{3.4}$$

$$F = d \wedge A = d \wedge (A + d\Lambda) \tag{3.5}$$

because:

$$d \wedge d\Lambda := 0. \tag{3.6}$$

Eq.(3.6) is the Poincaré Lemma in differential form notation. However, ECE theory now shows that the generally covariant foundation of electrodynamics unifies the latter with the other fields, notably the gravitational field. The standard model is unable to do this despite many attempts throughout the twentieth century. In ECE theory the relevant differential form equations of the electromagnetic sector are [1]- [7]:

$$F^a = d \wedge A^a + \omega^a{}_b \wedge A^b \tag{3.7}$$

$$d \wedge F^a = \mu_0 j^a \tag{3.8}$$

$$d \wedge \widetilde{F}^a = \mu_0 J^a \tag{3.9}$$

Here $\omega^a{}_b$ is the spin connection, which is identically non-zero because the electromagnetic field is always a spinning frame, not a static frame as in the standard model. The homogeneous current j^a is in general non-zero, it vanishes if and only if the electromagnetic and gravitational fields become independent and do not influence each other [1]- [7]. The presence of j^a (however tiny in magnitude) is of key importance, because it may be amplified by resonance [1]- [7] producing easily measurable electric power from ECE space-time, a new source of energy. In the standard model j^a does not exist, there is no concept of j^a in the standard model because electromagnetism there is a concept of special relativity superimposed on a flat Minkowski frame. Eq.(3.9) is the Hodge dual of Eq.(3.8) and here μ_0 is the vacuum permeability in S.I. units. The electromagnetic potential field in ECE theory is

$$A^a = A^{(0)} q^a \tag{3.10}$$

where $cA^{(0)}$ is the primordial voltage and q^a the tetrad field. Thus F^a is constructed from A^a through the first Cartan structure equation (3.7), and F^a obeys the first Bianchi identity, Eq.(3.8), and its Hodge dual (3.9). These are the classical field equations. The vector valued one-form $A^a{}_\mu$ and the vector valued two-form $F^a{}_{\mu\nu}$ are covariant under the general coordinate transformation [1]- [7] [17] according to the well known rules of standard Cartan geometry. If we attempt a "gauge transformation":

$$A^a \rightarrow A^a + d\Lambda^a \tag{3.11}$$

then:

$$F^a \rightarrow d \wedge A^a + \omega^a{}_b \wedge A^b + \omega^a{}_b \wedge d\Lambda^b = F^a + \omega^a{}_b \wedge d\Lambda^b \tag{3.12}$$

and F^a is not invariant: it must be coordinate covariant, not gauge invariant. Another fundamental problem of gauge theory in the standard model is that it uses a hypothesis superfluous to general relativity, the indices a of gauge theory are abstract mathematical concepts, whereas in Cartan geometry a is the index of the tangent space-time and thus well defined by geometry as required by relativity theory. In ECE theory the space-time that defines the electromagnetic field is the same as the space-time that defines all fields, including the gravitational field, in four physical dimensions, ct, X, Y and Z. Thus ECE is preferred to gauge theory by Ockham's Razor of philosophy, which asserts the use of the minimum number of concepts. In ECE theory a is already inherent in Cartan geometry, in gauge theory the label a is introduced as an extra mathematical assumption, i.e. of Yang Mills theory. Similarly ECE theory is preferred to String Theory by Ockham's razor, because String Theory uses superfluous parameters which are asserted arbitrarily to be dimensions. There is no experimental evidence for String Theory, and for this reason String Theory has been described as pre-Baconian. The experimental evidence for ECE theory is given in ref. [1] to [7] in a representative cross section of contemporary physics.

In ECE theory the fundamental wave equation of electrodynamics is [1]- [7]:

$$(\square + kT)\, A^{a}{}_{\mu} = 0 \tag{3.13}$$

where k is the Einstein constant and T is the index contracted canonical energy-momentum density of the unified field. The Einstein Ansatz asserts that:

$$R = -kT \tag{3.14}$$

where R is the scalar curvature. Using the correspondence principle of Einstein, the Proca equation emerges from Eq.(3.13) in the well defined limit:

$$kT \rightarrow \left(\frac{mc}{\hbar}\right) \tag{3.15}$$

where m is the identically non-zero photon mass indicated by the Eddington experiment, and $\hbar$ is the Planck constant. Note carefully that the d'Alembert wave equation of the standard model does not emerge from ECE theory, indicating that the d'Alembert wave equation is incomplete because it asserts identically zero photon mass. In the standard model the interpretation of the Proca equation is self-contradictory [8] because the Lagrangian needed to derive it is not gauge invariant. So for this reason the Proca equation is not used for canonical quantization in the standard model. The basic weak point here is again the assumption of gauge invariance, which has so long been thought of as the strength of gauge theory. In ECE theory there is no problem with the Proca equation because as we have seen, gauge invariance has been replaced by coordinate covariance in the unified field. Therefore the photon mass is identically non-zero as required by general relativity (photon mass was first proposed by Einstein) and the electromagnetic field is manifestly covariant with four physical components: time-like and three space-like, two transverse and one longitudinal. In the standard model the time-like and longitudinal components are dubiously removed by the Gupta Bleuler method [1]- [15]. The latter procedure is incorrect in general relativity, which prohibits the existence of a massless field. A massless field would mean no curvature R, and nothing at all (no field, no particles). Thus $A^{a}{}_{b}$ is manifestly covariant in ECE theory and can be canonically quantized in a rigorously correct way.

3.2 Canonical quantization of the ECE field

The potential field $A^{a}{}_{b}$ in ECE theory can be non-zero in regions where the electromagnetic field $F^{a}{}_{\mu\nu}$ is zero [7]. In this case the equations defining the potential in form notation are:

$$d \wedge A^1 = gA^2 \wedge A^3 \tag{3.16}$$

$$d \wedge A^2 = gA^3 \wedge A^1 \tag{3.17}$$

$$d \wedge A^3 = gA^1 \wedge A^2 \tag{3.18}$$

$$d \wedge A^0 = -d \wedge A^3 \tag{3.19}$$

In deriving these equations it has been assumed, for the sake of simplicity of argument only, that the gravitational and electromagnetic fields have become independent. This assumption is:

$$j^a = 0 \tag{3.20}$$

It is seen that Eqs.(3.16) to (3.19) all contain commutators of potentials. This commutator is precisely the quantity needed to construct a Heisenberg equation in the potential. This is the required canonical quantization. The quantity g is [9]- [14]:

$$g = \frac{\kappa}{A^{(0)}} = \frac{e}{\hbar} \tag{3.21}$$

so the quantized field is

$$A^2 \wedge A^3 = \frac{\hbar}{e} d \wedge A^1 \quad \text{et cyclicum.} \tag{3.22}$$

In tensor notation, Eq.(3.22) is:

$$\left[A^2{}_\mu, A^3{}_\nu\right] = \frac{\hbar}{e}\left[\partial_\mu, A^1{}_\nu\right] \quad \text{et cyclicum} \tag{3.23}$$

This is the Heisenberg type equation needed in canonical quantization, but it is important to realize that Eq.(3.23) is an equation of general relativity. The original Heisenberg equation is one of non-relativistic quantum mechanics as is well known. In regions where both $F^a{}_{\mu\nu}$ and $A^a{}_\mu$ are non-zero (which is the case in general):

$$F^1 = d \wedge A^1 - gA^2 \wedge A^3, \quad \text{et cyclicum} \tag{3.24}$$

and the commutators in tensor notation are:

$$\left[A^2{}_\mu, A^3{}_\nu\right] = \frac{\hbar}{e}\left(\left[\partial_\mu, A^1{}_\nu\right] - F^1{}_{\mu\nu}\right) \quad \text{et cyclicum} \tag{3.25}$$

Therefore it is seen that the commutators of potential needed for canonical quantization of the electromagnetic field, for example, are derived directly from Cartan geometry through the fundamental postulate (3.21). All four components of $A^a{}_\mu$ are non-zero in general and physical. The fundamental wave equation of the electro-dynamical sector of ECE theory is Eq.(3.13), which reduces to the Proca equation (3.15) using the correspondence principle. The principle of gauge invariance in special relativity is discarded in favor of the older principle of coordinate covariance in general relativity. Canonical quantization is then inherent in the method.

Finite electromagnetic fields in the approximation (3.20) are given by:

$$F^1 = d \wedge A^1 - gA^2 \wedge A^3 \tag{3.26}$$

$$F^2 = d \wedge A^2 - gA^3 \wedge A^1 \tag{3.27}$$

$$F^3 = d \wedge A^3 - gA^1 \wedge A^2 \tag{3.28}$$

Translating into the complex circular basis [1]- [7], [9]- [14]:

$$F^{(1)*} = d \wedge A^{(1)*} + igA^{(2)} \wedge A^{(3)} \quad \text{et cyclicum} \tag{3.29}$$

and in tensor notation:

$$F^{(1)*}_{\mu\nu} = \partial_\mu A^{(1)*}_\nu - \partial_\nu A^{(1)*}_\mu + ig\left(A^{(2)}_\mu A^{(3)}_\nu - A^{(2)}_\nu A^{(3)}_\mu\right). \quad \text{et cyclicum} \tag{3.30}$$

The magnetic components in the complex circular basis for example are:

$$\mathbf{B}^{(1)*} = \mathbf{\nabla} \times \mathbf{A}^{(1)*} + ig\mathbf{A}^{(2)} \times \mathbf{A}^{(3)}. \quad \text{et cyclicum} \tag{3.31}$$

A possible mathematical solution of Eqs.(3.26) to (3.31) is:

$$\mathbf{A}^{(1)} = \frac{A^{(0)}}{\sqrt{2}}(i\mathbf{i} + \mathbf{j})\, e^{i\phi} \tag{3.32}$$

$$\mathbf{A}^{(2)} = \mathbf{A}^{(1)*} = \frac{A^{(0)}}{\sqrt{2}}(-i\mathbf{i} + \mathbf{j})\, e^{-i\phi} \tag{3.33}$$

$$\mathbf{A}^{(3)} = -A^{(0)}\mathbf{k} = \mathbf{A}^{(3)*}, \quad B^{(0)} = \kappa A^{(0)} \tag{3.34}$$

If the electromagnetic phase is denoted:

$$\phi = \omega t - \kappa Z \tag{3.35}$$

it is found that the magnetic fields are:

$$\mathbf{B}^{(1)*} = \mathbf{B}^{(2)} = \frac{2B^{(0)}}{\sqrt{2}}(-i\mathbf{i} + \mathbf{j})\, e^{-i\phi} \tag{3.36}$$

$$\mathbf{B}^{(2)*} = \mathbf{B}^{(1)} = \frac{2B^{(0)}}{\sqrt{2}}(i\mathbf{i} + \mathbf{j})\, e^{i\phi} \tag{3.37}$$

$$\mathbf{B}^{(3)*} = \mathbf{B}^{(3)} = -B^{(0)}\mathbf{k}, \quad \mathbf{\nabla} \times \mathbf{A}^{(2)} = k\mathbf{A}^{(2)} \tag{3.38}$$

These are the results of $O(3)$ electrodynamics [1]- [7], [9]- [14], to which ECE theory reduces when gravitation has no influence on electromagnetism. The Evans (later ECE) spin field $\mathbf{B}^{(3)}$ is identically non-zero, and is the direct result of the spin connection in Eq. (3.7), i.e. the direct result of general relativity applied to electrodynamics self-consistently with gravitation. Quantization occurs from Eq.(3.21) using the de Broglie wave/particle dualism in combination with the minimal prescription

$$p = \hbar k \tag{3.39}$$

i.e.

$$p = eA^{(0)} \tag{3.40}$$

to obtain the Heisenberg commutators of the canonically quantized field:

$$\left[A^{(2)}{}_{\mu}, A^{(3)}{}_{\nu}\right] = \frac{-i}{g}\left(F^{(1)*}{}_{\mu\nu} - \left[\partial_{\mu}, A^{(1)*}{}_{\nu}\right]\right). \tag{3.41}$$
et cyclicum

Note that $A_{\mu}^{(2)}$ (and so on) is an energy-momentum within the factor e, via the minimal prescription, so energy-momentum Heisenberg commutators are given by Eq.(3.41):

$$\left[A^{a}{}_{\mu}, A^{b}{}_{\nu}\right] = A^{(0)2}\left[q^{a}{}_{\mu}, q^{b}{}_{\nu}\right] \tag{3.42}$$

where $q^{a}{}_{\mu}$ and $q^{a}{}_{\nu}$ are tetrad field/wave- functions. In ECE theory the tetrad is both the field and the wave-function, so due to this fact of Cartan geometry, ECE theory is a rigorous quantum field theory as required. Heisenberg-type commutators emerge naturally from the same geometry (i.e. from general relativity). These generally covariant commutators cannot be constructed in the standard model, which uses special relativity and intellectual gymnastics such as those of Gupta and Bleuler [8]to "remove" the time-like and longitudinal components of the four-potential. This removal is necessitated in turn by having to work with the fact that in special relativity (Poincaré group), only two components of a massless potential field can be physical, and the consequent but arbitrary assumption that these must be the transverse components of a plane wave. This flawed procedure is necessitated by the flawed assumption of a massless electromagnetic field, and the correct Proca equation cannot be used in the standard model because of the flawed assumption of gauge invariance. These flaws are all consequences of pathological science in the twentieth century, the pathology, or anthropomorphism, being that the Maxwell Heaviside theory"cannot be wrong". It was first realized as late as 1991 [18] that the $\mathbf{B}^{(3)}$ spin field was missing entirely from Maxwell Heaviside field theory [10]- [14] and ECE theory [1]- [7] is a direct consequence of that realization. ECE theory is a logical and self-consistent application of general relativity to all fields, including the electromagnetic field, and ECE theory is now generally accepted [16]. ECE replaces the convoluted, special relativistic, arguments of the standard model with a straightforward canonical quantization inherent in the method of general relativity. ECE theory can therefore be regarded as a benchmark theory or "new standard model" of physics.

Before proceeding to second quantization in Section 3.3 some discussion is given of single particle quantization [1]- [9] in ECE theory. In single particle quantization there is present a wave-function, for example the Dirac spinor, but on the traditional single particle level in the standard model [8] this wave-function has not yet been interpreted as a field. This means that multi particle phenomena such as transmutation cannot be given a satisfactory interpretation without "second quantization". In the standard model the latter procedure is the name given to the construction of Heisenberg commutators of the wave-function itself, which is thus interpreted as a quantum field. In ECE theory second quantization is automatic, (inherent in the method), because the tetrad is both the fundamental field and the wave-function on all levels. So to set the scene for the development of the fully rigorous ECE quantum field theory Section 3.3, the tetrad is first used here to obtain the fundamental single particle wave equations of special relativistic quantum mechanics [1]- [7] in well defined limits of the ECE wave equation via the correspondence principle of Einstein. These include the Klein-Gordon, Dirac and Proca equations.

The basic equations of generally relativistic quantum mechanics in ECE theory [1]- [7] are all obtained from basic equations of standard Cartan geometry [17]. The wave-function is always the tetrad $q^a{}_\mu$ as we have argued already. Two of these basic geometrical equations are the normalization condition:

$$q^a{}_\mu q^\mu{}_a = 1 \tag{3.43}$$

and the tetrad postulate:

$$D_\mu q^a{}_\nu = 0 \tag{3.44}$$

The ECE Lemma is obtained directly from the tetrad postulate, from the identity:

$$D^\mu \left(D_\mu q^a{}_\nu \right) := 0 \tag{3.45}$$

It follows straightforwardly [1]- [7] from Eq. (3.45) that:

$$\Box q^a{}_\mu = R q^a{}_\mu \tag{3.46}$$

where the scalar curvature of Eq.(3.46) must always be defined by:

$$R := q^\lambda{}_b \partial^\mu \left(\Gamma^\nu{}_{\mu\lambda} q^a{}_\nu - \omega^a{}_{\mu b} q^b{}_\lambda \right) \tag{3.47}$$

Here $\Gamma^\nu{}_{\mu\lambda}$ is the gamma connection. The latter becomes the Christoffel connection of Riemann geometry if and only if the torsion tensor vanishes:

$$T^\kappa{}_{\mu\nu} = \Gamma^\kappa{}_{\mu\nu} - \Gamma^\kappa{}_{\nu\mu} = 0 \tag{3.48}$$

The fundamental wave equation of ECE theory follows directly from the Lemma or subsidiary proposition (3.46) of Cartan geometry using the Einstein Ansatz (3.14). So the fundamental wave equation is always:

$$\left(\Box + kT \right) q^a{}_\mu = 0 \tag{3.49}$$

for all field/wave-functions $q^a{}_\mu$. Eq.(3.49) is seen to be the direct result of Cartan geometry and the Einstein Ansatz. Different representation spaces of the tetrad [1]- [7] then give the different fundamental fields of physics: the gravitational, electromagnetic, weak and strong, together with matter fields such as the fermions and bosons. Note that the wave-function has already been interpreted as the field, the fundamental tetrad field of general relativity [17]. So the required multi particle interpretation is inherent in the ECE theory from the outset. The Pauli exclusion principle, for example, or Fermi-Dirac and Bose-Einstein statistics, then emerge from the multi-particle interpretation.

It is always possible to write Eq.(3.43) as the classical:

$$q^a{}_\mu R q^\mu{}_a = R \tag{3.50}$$

By using Eq.(3.46) in Eq.(3.50) we obtain:

$$R = q^\mu{}_a \Box q^a{}_\mu \tag{3.51}$$

which is an operator equation indicating that R is the expectation value of the d'Alembertian operator $\Box$. Here $q^\mu{}_a$ is the inverse tetrad. Thus:

$$\Box q^a{}_\mu = R q^a{}_\mu \tag{3.52}$$

$$\Box q^{\mu}{}_{a} = R_1 q^{\mu}{}_{a} \tag{3.53}$$

where R_1 is to be defined. It follows from Eq.(3.53) that:

$$R = q^{\mu}{}_{a} \Box q^{a}{}_{\mu} = q^{\mu}{}_{a} R q^{a}{}_{\mu} \tag{3.54}$$

$$R_1 = q^{a}{}_{\mu} \Box q^{\mu}{}_{a} = q^{a}{}_{\mu} R_1 q^{\mu}{}_{a} \tag{3.55}$$

In quantum mechanics [19], Hermitian operators are used, because their eigenvalues are real-valued and thus physical. In Dirac bracket notation [19] a Hermitian system is defined by:

$$\langle m|\Omega|n\rangle = \langle n|\Omega|m\rangle^{*} \tag{3.56}$$

where $*$ denotes complex conjugate. A Hermitian matrix [20] is a square matrix unchanged by taking the transpose of its complex conjugate, e.g. if:

$$A = \begin{bmatrix} 1 & 1+i \\ 1-i & 3 \end{bmatrix}, \quad A^{*} = \begin{bmatrix} 1 & 1-i \\ 1+i & 3 \end{bmatrix} \tag{3.57}$$

then

$$\widetilde{A}^{*} = A \tag{3.58}$$

Eq.(3.8) translates into:

$$q^{a}{}_{\mu} \Box q^{\mu}{}_{a} = \left(q^{\mu}{}_{a} \Box q^{a}{}_{\mu}\right)^{*} \tag{3.59}$$

which implies:

$$R_1 = R^{*} \tag{3.60}$$

This means that the tetrad must be a Hermitian matrix. The eigenvalues of the tetrad eigenfunction are real-valued and physical. Now denote:

$$R = R^{'} + iR^{"} \tag{3.61}$$

$$R^{*} = R^{'} - iR^{"} \tag{3.62}$$

to obtain:

$$R^{'} = \frac{1}{2}\left(q^{\mu}{}_{a} \Box q^{a}{}_{\mu} + q^{a}{}_{\mu} \Box q^{\mu}{}_{a}\right) \tag{3.63}$$

$$R^{"} = \frac{-i}{2}\left(q^{\mu}{}_{a} \Box q^{a}{}_{\mu} - q^{a}{}_{\mu} \Box q^{\mu}{}_{a}\right) \tag{3.64}$$

Single particle relativistic equations in the limit of special relativity are regained using the correspondence principle of Einstein as follows:

$$|R^{'}| \rightarrow \left(\frac{mc}{\hbar}\right)^{2} \tag{3.65}$$

$$|R^{"}| \rightarrow 0 \tag{3.66}$$

In this limit it is self-consistently apparent that:

$$R = \frac{1}{2}\left(q^{\mu}{}_{a} \Box q^{a}{}_{\mu} + q^{a}{}_{\mu} \Box q^{\mu}{}_{a}\right) = \frac{1}{2}(R + R) = R' = -\left(\frac{mc}{\hbar}\right)^{2} \tag{3.67}$$

The Klein Gordon equation [8] is regained in the zero spin limit:

$$\phi = q^{0}{}_{0} = q^{1}{}_{1} = q^{2}{}_{2} = q^{3}{}_{3} \rightarrow 1 \tag{3.68}$$

so that the ECE wave equation reduces to:

$$\left(\square + \left(\frac{mc}{\hbar}\right)^2\right)\phi = 0 \tag{3.69}$$

where ϕ is a scalar eigenfunction without spin. Eq.(3.69) is:

$$\left(\left(\frac{1}{c^2}\frac{\partial^2}{\partial t^2} - \nabla^2\right) + \left(\frac{mc}{\hbar}\right)^2\right)\phi = 0 \tag{3.70}$$

The contra-variant and covariant four-momenta are defined as:

$$p^\mu = \left(\frac{E}{c}, \mathbf{p}\right), \quad p_\mu = \left(\frac{E}{c}, -\mathbf{p}\right) \tag{3.71}$$

Eq.(3.70) becomes the classical Einstein equation of special relativity:

$$p^\mu p_\mu = m^2c^2 \tag{3.72}$$

if the operator equivalence of quantum mechanics is used:

$$E \rightarrow i\hbar\frac{\partial}{\partial t}, \quad \mathbf{p} \rightarrow -i\hbar\boldsymbol{\nabla} \tag{3.73}$$

In the non-relativistic limit Eq.(3.72) becomes the Newtonian kinetic energy:

$$E = \frac{1}{2}\frac{p^2}{m} = \frac{1}{2}mv^2 \tag{3.74}$$

Using Eq.(3.73) in Eq.(3.74) gives the time-dependent Schrödinger equation of non-relativistic quantum mechanics:

$$\frac{\hbar^2}{2m}\nabla^2\phi = -i\hbar\frac{\partial\phi}{\partial t} \tag{3.75}$$

This is a single particle equation because the kinetic energy (3.74) is that of one free particle. The Klein Gordon equation is therefore a single particle equation of special relativistic quantum mechanics. In this case the free particle has zero spin and is defined by the properties (3.68) of the tetrad.

The Klein Gordon, Schrödinger and Newton equations have been obtained from Cartan geometry using the correspondence principle and operator equivalence (3.68):

$$p^\mu = i\hbar\partial^\mu \tag{3.76}$$

The minus sign in Eq.(3.14) is a convention. In the Born interpretation [19] the probability density of the Schrödinger equation is proportional to:

$$\rho = \phi^*\phi \tag{3.77}$$

where ϕ is a complex-valued quantity. The probability current [8] of the Schrödinger equation is defined as:

$$\mathsf{j} = -\frac{i\hbar}{2m}\left(\phi^*\boldsymbol{\nabla}\phi - \phi\boldsymbol{\nabla}\phi^*\right) \tag{3.78}$$

In order to make these definitions relativistic, define the four-current:

$$j^\mu = (\rho, \mathbf{j}) \tag{3.79}$$

By Noether's Theorem, this four-current must be conserved in a continuity equation:

$$\partial_\mu j^\mu = 0 \tag{3.80}$$

This is true as follows for the Schrödinger equation:

$$\begin{aligned}\frac{\partial \rho}{\partial t} + \boldsymbol{\nabla} \cdot \mathbf{j} &= \frac{\partial}{\partial t}(\phi^* \phi) - \frac{i\hbar}{2m}\left(\phi^* \boldsymbol{\nabla}^2 \phi - \phi \boldsymbol{\nabla}^2 \phi^*\right) \\ &= \phi^* \left(\frac{\partial \phi}{\partial t} - \frac{i\hbar}{2m}\nabla^2 \phi\right) + \phi\left(\frac{\partial \phi^*}{\partial t} + \frac{i\hbar}{2m}\nabla^2 \phi^*\right) = 0\end{aligned} \tag{3.81}$$

In the Klein-Gordon equation however, ρ must be defined as the time-like component of the four-current j^μ:

$$\rho = \frac{i\hbar}{2m}\left(\phi^* \frac{\partial \phi}{\partial t} - \phi \frac{\partial \phi^*}{\partial t}\right) \tag{3.82}$$

Thus [8]:

$$\partial_\mu j^\mu = \frac{i\hbar}{2m}\left(\phi^* \Box \phi - \phi \Box \phi^*\right) = 0 \tag{3.83}$$

In ECE theory Eq.(3.83) becomes:

$$R - R_1 = q^\mu{}_a \Box q^a{}_\mu - q^a{}_\mu \Box q^\mu{}_a \tag{3.84}$$

For a Hermitian $q^a{}_\mu$:

$$R_1 = R^* \tag{3.85}$$

and the continuity equation of ECE theory becomes:

$$\partial_\mu q^\mu = \frac{i\hbar}{2m}\left(q^\mu{}_a \Box q^a{}_\mu - q^a{}_\mu \Box q^\mu{}_a\right) \tag{3.86}$$

For real and physical scalar curvature:

$$R = R^* \tag{3.87}$$

and

$$\partial_\mu q^\mu = R - R^* = 0 \tag{3.88}$$

as required by Noether's Theorem.

The Klein Gordon equation in this single particle interpretation was abandoned in favor of the single particle Dirac equation as is well known. The Klein Gordon equation gives a negative ρ from Eq.(3.82), a negative probability makes no sense. In order to make sense of the Klein Gordon equation ϕ must become a field [8], leading in the standard model to second quantization. Therefore in ECE theory the multi-particle interpretation automatically gives a physically sensible probability current for a spin-less particle because the tetrad is both the wave-function and the quantized field whose commutators are always well defined. This is illustrated for the electromagnetic field in Section 3.3, and in the remainder of this Section, the Dirac spinor wave-function is introduced on the single particle level.

The Dirac equation is a limit [1]– [7] of the ECE wave equation using Eq.(3.15), but the Dirac wave-function has a half integral spin whose single particle interpretation defines the fermion as is well known [8]. The Dirac spinor is a tetrad in $SU(2)$ representation space [1]– [7]. The Dirac equation is therefore:

$$\left(\square + \left(\frac{mc}{\hbar}\right)^2\right) q^a{}_\mu = 0 \tag{3.89}$$

where the a and μ labels represent the four components of the Dirac spinor: $q^R{}_1$, $q^R{}_2$, $q^L{}_1$ and $q^L{}_2$. The Pauli spinors are:

$$\phi^R = \begin{bmatrix} q^R{}_1 \\ q^R{}_2 \end{bmatrix}, \quad \phi^L = \begin{bmatrix} q^L{}_1 \\ q^L{}_2 \end{bmatrix} \tag{3.90}$$

and the Dirac spinor is:

$$q^a{}_\mu = \begin{bmatrix} \phi^R \\ \phi^L \end{bmatrix}. \tag{3.91}$$

In standard quantum field theory [8] [9] the Dirac spinor is denoted by ψ and the convention $c = \hbar = 1$ is used. So Eq.(3.89) becomes:

$$\left(\square + m^2\right)\psi = 0 \tag{3.92}$$

which may be written [8] as:

$$\left(i\gamma^\mu \partial_\mu - m\right)\psi = 0 \tag{3.93}$$

where γ^μ is the Dirac matrix and where:

$$\gamma^\mu \partial_\mu = \gamma^0 \partial_0 + \gamma^i \partial_i \tag{3.94}$$

Written out in full, Eq.(3.93) is:

$$\left(i\gamma^\mu \partial_\mu - \left(\frac{mc}{\hbar}\right)^2\right) q^a{}_\mu = 0 \tag{3.95}$$

Using the shorthand notation (3.93) the Hermitian conjugate equation is defined [8]:

$$\psi^+ \left(-i\gamma^0 \overleftarrow{\partial_0} + i\gamma^i \overleftarrow{\partial_i} - m\right) = 0 \tag{3.96}$$

and the adjoint spinor is defined as:

$$\overline{\psi} = \psi^+ \gamma^0 \tag{3.97}$$

Here:

$$\gamma^i \gamma^0 = -\gamma^0 \gamma^i \tag{3.98}$$

so that:

$$\overline{\psi}\left(i\gamma^\mu \overleftarrow{\partial_\mu} + m\right) = 0 \tag{3.99}$$

The conserved four-current is thus defined naturally as:

$$j^\mu = \overline{\psi}\gamma^\mu \psi \tag{3.100}$$

and is conserved as follows:

$$\begin{aligned}\partial_\mu j^\mu &= \left(\partial_\mu \overline{\psi}\right)\gamma^\mu\psi + \overline{\psi}\gamma^\mu\left(\partial_\mu\psi\right)\\ &= \left(im\overline{\psi}\right)\psi + \overline{\psi}\left(-im\psi\right) = 0\end{aligned} \tag{3.101}$$

The probability density of the Dirac equation is thus rigorously positive-valued:

$$j^0 = \overline{\psi}\gamma^0\psi = \psi^+\psi = |\psi_1|^2 + |\psi_2|^2 + |\psi_3|^2 + |\psi_4|^2 > 0 \tag{3.102}$$

and is the Born probability of the free fermion or free Dirac particle. Only one fermion is being considered here so the Pauli exclusion principle and Fermi-Dirac statistics have not yet been deduced. To achieve that in the standard model requires second quantization [8].

In ECE theory:

$$\psi^+ \rightarrow q^{\mu+}{}_a, \quad \overline{\psi} \rightarrow q^{\mu+}{}_a\gamma^0 = \overline{q}^\mu{}_b, \quad j^\mu = \overline{q}^\nu{}_a\gamma^\mu q^a{}_\nu \tag{3.103}$$

and the conserved current is generally covariant as required. Second quantization of the Dirac spinor is inherent in ECE theory from the outset, as argued already, so ECE automatically theory gives the Pauli exclusion principle and Fermi-Dirac statistics from the required multi-particle interpretation of the tetrad field/wave-function. When applied to bosons, ECE theory automatically gives the Bose Einstein statistics for the same reason. The photon is well known to be an integral spin boson, and electrons and quarks are examples of half-integral spin fermions. ECE theory also gives the required multi particle interpretation of gravitons for the same reason, and so forth for any quantized field, the generally covariant, unified and quantized field of any spin, including zero spin. All other fields are limits of this field.

The negative energy states of the Dirac equation come from the fact that the spin $\frac{1}{2}$ particles obey the Pauli exclusion principle. Negative energy states are completely filled, so the exclusion principle prevents any more electrons entering the Dirac sea, made up of negative energy states. The Dirac sea is the vacuum in this picture, a vacuum made up of negative energy electrons, protons, neutrinos, neutrons and other fermions. If there is a vacancy or "hole" in the fermion sea, with energy $-|E|$, an electron with energy E fills it, emitting energy $2E$:

$$e^- + \text{hole} \rightarrow \text{energy} \tag{3.104}$$

The hole has charge e^+ and positive energy, and is called a positron. The energy difference is:

$$\Delta E = E - (-|E|) = 2E \tag{3.105}$$

Eq. (3.104) in particle theory is:

$$e^- + e^+ = 2\gamma \tag{3.106}$$

and this means that an electron and positron mutually annihilate to give two photons. In ECE theory the Dirac sea is defined by tetrads obeying the Pauli exclusion principle. The latter is understood by regarding the tetrad as a field and wave-function as argued already.

Quantities such as $p^\mu p_\mu$ are understood in ECE theory through the momentum tetrad [1]- [7]:

$$p^a{}_\mu = p^{(0)}q^a{}_\mu \tag{3.107}$$

Planck/Einstein/de Broglie quantization is given by:

$$p^a{}_\mu = \hbar\kappa^a{}_\mu = eA^a{}_\mu = eA^{(0)}q^a{}_\mu \tag{3.108}$$

The Heisenberg commutator equation is understood using the structure invariants of Cartan geometry (pp. 140 ff. of ref [1]):

$$x^a = \int_s T^a \tag{3.109}$$

$$\theta^a{}_b = \int_s R^a{}_b \tag{3.110}$$

and the Heisenberg equation is:

$$[x^a, p_b] = iJ^a{}_b = \hbar\theta^a{}_b \tag{3.111}$$

where

$$\theta^a{}_b = \frac{i}{\hbar}J^a{}_b \tag{3.112}$$

Thus:

$$[x^a, p_b] = \hbar\int_s R^a{}_b \tag{3.113}$$

where

$$R^a{}_b = -\frac{\kappa}{2}\epsilon^a{}_{bc}T^c \tag{3.114}$$

appropriate to spinning spacetime unaffected by curvature. Here T^c is Cartan torsion.

3.3 Second quantization of the tetrad field

As argued, the tetrad in ECE field theory is both the field and the wave-function. Therefore the generally covariant quantum field is defined by commutators inherent in Cartan geometry. In this section the position and conjugate momentum tetrads are defined from the appropriate Lagrangian density. Then the field tetrad is expanded in a Fourier series in order to define the creation and annihilation operators of the field. The number operator of the field is defined from the creation and annihilation operators and a multi-particle interpretation developed.

Eq.(3.46) may be derived [1]– [3] from an Euler Lagrange equation. For example if:

$$\frac{\partial\mathcal{L}}{\partial q^\nu{}_a} = -\partial^\mu\left(\frac{\partial\mathcal{L}}{\partial\left(\partial^\mu q^\nu{}_a\right)}\right) \tag{3.115}$$

and

$$\mathcal{L} = -\frac{c^2}{\kappa}\left(\frac{1}{2}\left(\partial_\mu q^a{}_\nu\right)\left(\partial^\mu q^\nu{}_a\right) + \frac{R}{2}q^a{}_\nu q^\nu{}_a\right) \tag{3.116}$$

the Eq.(3.46) follows. There is a freedom of choice in the Lagrangian density, it is chosen to give the ECE Lemma (3.46) through the Euler Lagrange equation (3.115). Define the position and conjugate momentum tetrads by:

$$x^a{}_\nu = x^{(0)}q^a{}_\mu \tag{3.117}$$

$$p^a{}_\mu = p^{(0)} q^a{}_\mu \tag{3.118}$$

The conjugate momentum is related to the position by:

$$p^a{}_\mu = \frac{1}{c} \frac{\partial \mathcal{L}}{\partial \left(\partial_\nu x^a{}_\mu\right)} \tag{3.119}$$

Eq.(3.119) is a canonical equation in the sense that it generalizes the well known classical result [20]:

$$p_i = \frac{\partial \mathcal{L}}{\partial \dot{x}_i} \tag{3.120}$$

Classical dynamics [20] can be developed with the Lagrange equation of motion:

$$\frac{d}{dt}\left(\frac{\partial L}{\partial \dot{q}_j}\right) = \frac{\partial L}{\partial q_j} \tag{3.121}$$

and the Hamilton equations of motion, the canonical equations:

$$\dot{q}_k = \frac{\partial H}{\partial p_k}, \quad -\dot{p}_k = \frac{\partial H}{\partial q_k} \tag{3.122}$$

where H is the Hamiltonian. These well known equations are found in any textbook on classical dynamics. Note that Eq.(3.119) is an equation of general relativity, whereas Eqs.(3.120) to (3.122) are non-relativistic.

The Cartan torsion is defined [1]- [7], as argued already in this paper, by:

$$T^a = d \wedge q^a + \omega^a{}_b \wedge q^b \tag{3.123}$$

and if

$$\omega^a{}_b = -\frac{\kappa}{2} \epsilon^a{}_{bc} q^c \tag{3.124}$$

then:

$$T^a = d \wedge q^a - \kappa q^b \wedge q^c \tag{3.125}$$

i.e.

$$T^1 = d \wedge q^1 - \kappa q^2 \wedge q^3 \quad \text{et cyclicum} \tag{3.126}$$

$$T^0 = d \wedge q^0 - \kappa q^2 \wedge q^1 \tag{3.127}$$

Eqs.(3.125) and (3.127) indicate the existence of commutators (wedge products). These appear on the classical level therefore in generally covariant unified field theory. In regions where the fundamental field is non-zero but where the torsion is zero:

$$d \wedge q^a = \kappa q^b \wedge q^c \tag{3.128}$$

In tensor notation, Eq.(3.128) is a commutator equation:

$$\left[\partial_\mu, q^a{}_\nu\right] = \kappa \left[q^b{}_\mu, q^c{}_\nu\right] \tag{3.129}$$

For the position tetrad:

$$\left[\partial_\mu, x^a{}_\nu\right] = \kappa \left[q^b{}_\mu, x^c{}_\nu\right] \tag{3.130}$$

Multiply both sides of Eq.(3.130) by $p^{(0)}$ to obtain:

$$p^{(0)} \left[\partial_\mu, x^a{}_\nu\right] = \kappa \left[p^b{}_\mu, x^c{}_\nu\right] \tag{3.131}$$

This is an equation of classical general relativity but it contains the commutator of position and conjugate momentum. This commutator is fundamentally important to quantum mechanics as is well known. Second quantization [8] [9] proceeds by setting up the equal time commutators for position and momentum: in the Heisenberg equation the position and conjugate momentum are defined at the same instant in time. Eq.(3.131) shows that the momentum tetrad $p^b{}_\mu$, and the partial derivative operator ∂_μ play the same role. This observation leads to the fundamental operator equivalence (3.76) of quantum mechanics. In ECE theory this operator equivalence is derived as in Eq.(3.131) from Cartan geometry in a generally covariant context.

In the de Broglie limit:

$$p^{(0)} = \hbar\kappa \tag{3.132}$$

and

$$\left[p^b{}_\mu, x^c{}_\nu\right] = \hbar\left[\partial_\mu, x^a{}_\nu\right] \tag{3.133}$$

Eq.(3.133) in the complex circular basis [1]– [3] is:

$$\left[p^{(b)}{}_\mu, x^{(c)}{}_\nu\right] = i\hbar\left[\partial_\mu, x^{(a)*}{}_\nu\right] \tag{3.134}$$

and is equivalent to the angular momentum commutator relations [1]– [14]:

$$\left[J^{(b)}{}_\mu, J^{(c)}{}_\nu\right] = i\hbar J^{(a)*}{}_{\mu\nu} \tag{3.135}$$

where:

$$J^{(a)}{}_{\mu\nu} = iJ^{(0)}\left[\partial_\mu, x^{(a)*}{}_\nu\right] \tag{3.136}$$

In vector notation:

$$\begin{gathered}\mathbf{J}^{(1)} \times \mathbf{J}^{(2)} = i\hbar\mathbf{J}^{(3)*}\\ \text{et cyclicum}\end{gathered} \tag{3.137}$$

In general relativity, Eq.(3.131) gives:

$$p^{(0)} = \left(\left[p^b{}_\mu, x^c{}_\nu\right] / \left[\partial_\mu, x^a{}_\nu\right]\right)\kappa \tag{3.138}$$

so the Planck constant has been identified as the limit:

$$\left[p^b{}_\mu, x^c{}_\nu\right] \rightarrow \hbar\left[\partial_\mu, x^a{}_\nu\right] \tag{3.139}$$

or, in the complex circular basis:

$$\left[p^{(b)}{}_\mu, x^{(c)}{}_\nu\right] \rightarrow i\hbar\left[\partial_\mu, x^{(a)*}{}_\nu\right] \tag{3.140}$$

The Planck constant is therefore defined by a particular type of Cartan geometry (Eq.(3.114)) in a well defined limit form the correspondence principle. Having recognized that Eq. (3.131) leads to quantization, its fully quantized form is:

$$\left[p^b{}_\mu, x^c{}_\nu\right] q^d{}_\rho = \left(\frac{p^{(0)}}{\kappa}\right)\left[\partial_\mu, x^a{}_\nu\right] q^d{}_\rho \tag{3.141}$$

where it has been recognized that the commutators act on the tetrad field/wavefunction $q^d{}_\rho$. We may write:

$$\delta^a_{\mu\nu} = \left[\partial_\mu, x^a{}_\nu\right] \tag{3.142}$$

In shorthand notation, Eq.(3.142) is:

$$(p \wedge x)\, q = \left(\frac{p^{(0)}}{\kappa}\right)(d \wedge x)\, q \tag{3.143}$$

where it is emphasized that q is both the wave-function and the field. So in this sense a rigorous quantum field theory emerges automatically from Cartan geometry. The commutators of Eq.(3.141) are between tetrads, i.e. field commutators as required. This conclusion is true for all fields of physics, including for example the electromagnetic potential field:

$$\left[A^a{}_\mu, a^b{}_\nu\right] = \frac{A^{(0)2}}{\kappa}\left[\partial_\mu, \kappa^a{}_\nu\right] \tag{3.144}$$

Eq.(3.137) is a cyclic relation between tetrads:

$$\begin{gathered}\mathbf{q}^{(1)} \times \mathbf{q}^{(2)} = i\mathbf{q}^{(3)*} \\ \text{et cyclicum}\end{gathered} \tag{3.145}$$

Following the usual development of quantum field theory [8] [9] Eqs.(3.32) can be expressed as:

$$\begin{gathered}\left[q_x, q_y\right] = iq_z \\ \text{et cyclicum}\end{gathered} \tag{3.146}$$

The raising and lowering or creation and annihilation operators [19] are:

$$q^+ = q_x + iq_y \tag{3.147}$$

$$q^- = q_x - iq_y \tag{3.148}$$

These are similar to the complex circular tetrads:

$$q^{(1)} = \frac{1}{\sqrt{2}}\left(q_x + iq_y\right) \tag{3.149}$$

$$q^{(2)} = \frac{1}{\sqrt{2}}\left(q_x - iq_y\right) \tag{3.150}$$

The commutator properties of Eqs.(3.147) and (3.148) are [19]:

$$\left[q^+, q_z\right] = -q^+ \tag{3.151}$$

$$\left[q^-, q_z\right] = q^- \tag{3.152}$$

$$\left[q^+, q^-\right] = 2q_z \tag{3.153}$$

The creation tetrad operator is then defined by:

$$q^+|n\rangle = c_n^+|n+1\rangle \tag{3.154}$$

and the annihilation tetrad operator by:

$$q^-|n\rangle = c_n^-|n-1\rangle \tag{3.155}$$

In quantum electrodynamics for example, q^+ increases the state $|n >$ to $|n+1 >$.The electromagnetic field is described by an infinite number of harmonic oscillators, one for every point in space. In this case:

$$q^-|n\rangle = n^{1/2}|n-1\rangle \tag{3.156}$$

$$q^+|n\rangle = (n+1)^{1/2}\,|n+1\rangle \tag{3.157}$$

The number operator is:

$$N = q^+q^- \tag{3.158}$$

so that:

$$N|n\rangle = n|n\rangle \tag{3.159}$$

The hamiltonian is:

$$\left.\begin{aligned} H &= \hbar\omega\left(N + \frac{1}{2}\right) \\ &= \hbar\omega\left(q^+q^- + \frac{1}{2}\right) \\ &= \hbar\omega\left(n + \frac{1}{2}\right) \end{aligned}\right\} \tag{3.160}$$

giving the zero point energy:

$$H_0 = \frac{1}{2}\hbar\omega \tag{3.161}$$

The fundamental tetrad field is expanded in a Fourier series [8]:

$$q = \sum_{\kappa}\left({q^+}_{\kappa}e^{-i\kappa\cdot\mathbf{r}} + {q^-}_{\kappa}e^{i\kappa\cdot\mathbf{r}}\right) \tag{3.162}$$

which may be developed into an integral in a volume V:

$$q = V\int\left({q^+}_{\kappa}e^{-i\kappa\cdot\mathbf{r}} + {q^-}_{\kappa}e^{i\kappa\cdot\mathbf{r}}\right)d^3\kappa \tag{3.163}$$

3.4 Electromagnetic Aharonov Bohm effect and field commutators

The electromagnetic Aharonov Bohm (EAB) effect is defined [1]- [7] by:

$$d\wedge A^a + {\omega^a}_b\wedge A^b = 0, \tag{3.164}$$

$${\omega^a}_b \neq 0 \tag{3.165}$$

it is important to note that the spin connection ${\omega^a}_b$ must be non-zero, this is a fundamental requirement of general relativity because the electromagnetic field is a spinning frame, the gravitational field a curving frame. The ECE spin field shows this to be true experimentally [1]- [7]. In order to obtain solutions of Eq.(3.164) the spin connection must be given an analytical form. The Faraday law of induction is contained in the approximation [1]- [7]:

$$d\wedge F^a = 0 \tag{3.166}$$

The Faraday law of induction is accurate under most laboratory applications, so in this approximation:

$$j^a = 0, \tag{3.167}$$

implying Eq.(3.124). The proportionality constant $-\kappa/2$ in Eq. (3.124) [1]- [7] has been assumed to be a scalar, the minus sign and the factor half have been chosen as a convention. More generally, the proportionality factor in Eq.(3.124) can be a tensor with different components. Its units are inverse meters, those of wave-number. Use the ECE Ansatz (3.10) to find that:

$$\omega^a{}_b = -\frac{g}{2}\epsilon^a{}_{bc}A^c \tag{3.168}$$

where:

$$g = \frac{\kappa}{A^{(0)}} \tag{3.169}$$

Therefore the EAB effect is described by wedge products or commutators of potential, those used on canonical quantization. Eq.(3.167) is defined to an excellent approximation in most applications by the Faraday Law of induction. Under resonance conditions however [1]- [7] the homogeneous current j^a may be amplified by many orders of magnitude [1]- [7]. Eq. (3.164) may be expanded in terms of its components as [1]- [7]:

$$\begin{gathered} d \wedge A^1 = gA^2 \wedge A^3 \\ \text{et cyclicum} \end{gathered} \tag{3.170}$$

$$d \wedge A^0 = -d \wedge A^3 \tag{3.171}$$

and the EAB effect is described by solutions to these four simultaneous equations.

Writing out Eq.(3.170) in tensor notation gives:

$$\begin{gathered} \partial_\mu A^3{}_\nu - \partial_\nu A^3{}_\mu = g\left(A^1{}_\mu A^2{}_\nu - A^1{}_\nu A^2{}_\mu\right) \\ \text{et cyclicum} \end{gathered} \tag{3.172}$$

or in the complex circular basis:

$$\begin{gathered} \partial_\mu A^{(3)}{}_\nu - \partial_\nu A^{(3)}{}_\mu = -ig\left(A^{(1)}{}_\mu A^{(2)}{}_\nu - A^{(1)}{}_\nu A^{(2)}{}_\mu\right) \\ \text{et cyclicum} \end{gathered} \tag{3.173}$$

Eq.(3.173) is for example:

$$\partial_x A^{(3)}{}_y - \partial_y A^{(3)}{}_x = \kappa A^{(0)} \tag{3.174}$$

However:

$$A^{(3)}{}_y = A^{(3)}{}_x = 0, \quad A^{(0)} \neq 0 \tag{3.175}$$

so the only solution is:

$$\kappa = 0 \tag{3.176}$$

Similarly, Eq.(3.173) gives:

$$\partial_0 A^{(3)}{}_z - \partial_z A^{(3)}{}_0 = -ig\left(A^{(1)}{}_0 A^{(2)}{}_z - A^{(1)}{}_z A^{(2)}{}_0\right) \tag{3.177}$$

and again Eq.(3.176) is the only solution because:

$$A^{(2)}{}_z = A^{(1)}{}_z = 0, \quad A^{(0)} \neq 0 \tag{3.178}$$

In order to obtain a self consistent solution to the simultaneous equations (3.170) to (3.171) it must be assumed that κ_i is a tensor:

$$\kappa_i = \begin{bmatrix} \kappa & 0 & 0 \\ 0 & \kappa & 0 \\ 0 & 0 & \kappa \end{bmatrix} \tag{3.179}$$

so Eqs.(3.170) to (3.171) become:

$$d \wedge A^1 = g A^2 \wedge A^3 \tag{3.180}$$

$$d \wedge A^2 = g A^3 \wedge A^1 \tag{3.181}$$

$$d \wedge A^3 = -d \wedge A^0 = 0, \quad A^1 \wedge A^2 \neq 0 \tag{3.182}$$

The solutions are:

$$\mathbf{A}^{(1)} = \frac{A^{(0)}}{\sqrt{2}} (\mathbf{i} - i\mathbf{j})\, e^{i(\omega t - \kappa z)} \tag{3.183}$$

$$\mathbf{A}^{(2)} = \frac{A^{(0)}}{\sqrt{2}} (\mathbf{i} + i\mathbf{j})\, e^{-i(\omega t - \kappa z)} \tag{3.184}$$

Thus:

$$\boldsymbol{\nabla} \times \mathbf{A}^{(1)} = \kappa \mathbf{A}^{(1)} \tag{3.185}$$

$$\mathbf{A}^{(1)} \times \mathbf{A}^{(2)} = i \mathbf{A}^{(0)2} \mathbf{k} \tag{3.186}$$

$$\boldsymbol{\nabla} \times \mathbf{A}^{(1)*} = -ig \mathbf{A}^{(2)} \times \mathbf{A}^{(3)} \tag{3.187}$$

$$\boldsymbol{\nabla} \times \mathbf{A}^{(2)*} = -ig \mathbf{A}^{(3)} \times \mathbf{A}^{(1)} \tag{3.188}$$

$$\nabla \times \mathbf{A}^{(3)*} = \mathbf{0} \tag{3.189}$$

Eqs.(3.183) to (3.189) are tetrad equations, with:

$$\mathbf{A}^{(3)} = A^{(0)} \mathbf{k} \tag{3.190}$$

Thus:

$$A^0 = -A^{(0)} \tag{3.191}$$

The complex circular basis is defined by the tetrad equations:

$$\mathbf{q}^{(1)} \times \mathbf{q}^{(2)} = i \mathbf{q}^{(3)*} \quad \text{et cyclicum} \tag{3.192}$$

where

$$\mathbf{q}^{(1)} = \mathbf{q}^{(2)*} = \frac{1}{\sqrt{2}} (\mathbf{i} - i\mathbf{j})\, e^{i(\omega t - \kappa t)} \tag{3.193}$$

$$\mathbf{q}^{(3)} = \mathbf{q}^{(3)*} = \mathbf{k} \tag{3.194}$$

These tetrads are the mechanism for defining the Cartan torsion and spinning space-time responsible for electromagnetic potential fields.

The EAB effect is caused by $\mathbf{A}^{(1)}$, $\mathbf{A}^{(2)}$ and $\mathbf{A}^{(3)}$ in regions where

$$\mathbf{E}^a = \mathbf{0}, \quad \mathbf{B}^a = \mathbf{0} \tag{3.195}$$

The EAB effect occurs in regions outside a laser or radar beam. The $\mathbf{A}^{(1)}$ and $\mathbf{A}^{(2)}$ components are rapidly oscillating, so on average:

$$\langle \mathbf{A}^{(1)} \rangle = \langle \mathbf{A}^{(2)} \rangle = \mathbf{0} \tag{3.196}$$

but

$$\mathbf{A}^{(1)} \times \mathbf{A}^{(2)} = iA^{(0)2}\mathbf{k} \tag{3.197}$$

and is non-zero on average. The electromagnetic field components in these regions are zero, for example:

$$\mathbf{B}^{(1)*} = \nabla \times \mathbf{A}^{(1)*} + ig\mathbf{A}^{(2)} \times \mathbf{A}^{(3)} = \mathbf{0} \tag{3.198}$$

$$\mathbf{B}^{(2)*} = \nabla \times \mathbf{A}^{(2)*} + ig\mathbf{A}^{(3)} \times \mathbf{A}^{(1)} = \mathbf{0} \tag{3.199}$$

Therefore the beam intensity or power density (watts per square meter) is also zero:

$$I = -\frac{ic}{\mu_0} |\mathbf{B}^{(1)} \times \mathbf{B}^{(2)}| = 0 \tag{3.200}$$

The power density I of the radar or laser beam is non-zero if and only if the oscillating electric and magnetic fields making up the beam are non-zero. Obviously, outside the beam there is no power density or beam intensity measurable experimentally. However the conjugate product or commutator $\mathbf{A}^{(1)} \times \mathbf{A}^{(2)}$ still exists outside the beam because in these regions:

$$\mathbf{A}^{(1)} = \mathbf{A}^{(2)*} \neq \mathbf{0} \tag{3.201}$$

Similarly, the color of the laser beam is due to its non-zero electric and magnetic fields. Outside the laser beam there is no color visible, but Eq.(3.201) is still true. For a static magnetic field (Chambers experiment) $\mathbf{B}$ is non-zero inside the iron whisker or solenoid, but outside $\mathbf{B}$ is zero and $\mathbf{A}$ non-zero [1]- [7], causing the well known magnetic (original) Aharonov Bohm effect. So the experiment to detect the EAB for the first time must be set up to observe the inverse Faraday effect [1]- [7] in regions outside the laser or radio frequency beam, i.e. in regions where I is zero as discussed already. As described in Appendix F of Vol. 3 of ref [12] the inverse Faraday effect in a gas of N electrons occupying a volume V produces in the sample a magnetic flux density (in Tesla):

$$\mathbf{B}^{(3)}_{\text{sample}} = \frac{N}{V} \frac{\mu_0 e^3 c^2}{2m\omega^2} \left(\frac{B^{(0)}}{\sqrt{m^2\omega^2 + e^2 B^{(0)2}}} \right) \mathbf{B}^{(3)}_{\text{beam}} \tag{3.202}$$

where $-e$ is the charge on the electron, m is its mass, μ_0 is the vacuum permeability in S.I., ω is the angular frequency of the beam, $B^{(0)}$ is its magnetic flux density magnitude and $\boldsymbol{B}^{(3)}_{\text{beam}}$ is its free space ECE spin field. Eq.(3.202) is the result of the Hamilton Jacobi equation [12] in the limit of special relativity and where the electromagnetic field has been assumed independent of the gravitational field. In off-resonance conditions in the laboratory this is true to an excellent approximation. At visible frequencies (laser beam):

$$|\mathbf{B}^{(3)}_{\text{sample}}| \to \frac{N}{V} \left(\frac{\mu_0 e^3 c^2}{2m^2\omega^3} \right) B^{(0)2} \tag{3.203}$$

and at radar frequencies:

$$|\mathbf{B}^{(3)}_{\text{sample}}| \rightarrow \frac{N}{V}\left(\frac{\mu_0 e^2 c^2}{2m\omega^2}\right) B^{(0)} \tag{3.204}$$

In terms of intensity I Eq.(3.203) is:

$$|\mathbf{B}^{(3)}_{\text{sample}}| \rightarrow \frac{N}{V}\left(\frac{\mu_0^2 e^3 c}{2m^2}\right) \frac{I}{\omega^3} \tag{3.205}$$

For an intensity $I = 5.5 \times 10^{12} Wm^{-2}$ and a $Nd - YaG$ frequency of $1.77 \times 10^{16} rads^{1}$ then $\left|\mathbf{B}^{(3)}\right| \sim 10^{-9} Tesla$ assuming the Avogadro number of electrons in the volume V of one cubic meter, i.e. This calculation is in excellent agreement with experimental data on the inverse Faraday effect [12].

The EAB effect at laser frequencies is, from Eq.(3.205), the magnetization of an electron gas due to $\mathbf{A}^{(1)} \times \mathbf{A}^{(2)}$ in regions where I is zero experimentally, but where the commutator $\mathbf{A}^{(1)} \times \mathbf{A}^{(2)}$ is not zero. The EAB is given by the same equation (3.205) but the interpretation of I is different. It is the intensity of the laser beam transmitted to regions outside the beam by the spin connection, i.e. by the spinning of space-time itself. Similarly the $\mathbf{B}$ field of an iron whisker is transmitted to regions outside by the spinning of space-time [1]– [7]. All AB effects are therefore experimental evidence for the spinning of space-time and thus for ECE theory. The AB effects are all effects of the generally covariant electromagnetic field and its commutators used in canonical quantization as already argued. The latter is always defined (indexless shorthand notation [1]–[7]) by:

$$F = D \wedge A \tag{3.206}$$

The EAB experiment therefore has to accurately reproduce the conditions:

$$F = D \wedge A = 0, \quad A \neq 0 \tag{3.207}$$

This is done for the static magnetic field in the Chambers experiment, and in various other designs that reproduce this experiment. The electric and gravitational AB effects have also been observed, but the EAB has not yet been observed. The AB experiments define what is meant by a field in generally covariant electrodynamics. It is a field of force accompanied by kinetic energy. When the field of force is zero the potential energy may still be non-zero, and there is a potential for the creation of a field of force. In the magnetic AB effect for example there is no magnetic force field outside the iron whisker, but there is still a potential for the creation of a magnetic field of force. This potential is generated by the spinning of space-time itself. The gravitational AB effect is due to the potential for force generated by the curving of space-time. In regions of zero gravity, i.e. locally zero Riemann curvature:

$$R = D \wedge \omega = 0, \quad \omega \neq 0 \tag{3.208}$$

and the spin connection still exists. So the usual properties of an electromagnetic beam for example are due to non-zero force fields (in this case oscillating electric and magnetic force fields). These properties include intensity (heat), color (spectrum), transmission of signals form transmitter to receiver, and so on. In ECE theory there exists a property of an electromagnetic field hitherto unmeasured, the EAB effect due to the potential for the creation of a field of force in regions where the field itself is zero. The potential is the tetrad, the properties that create the EAB effect are the spin connection and the commutator needed for canonical quantization.

Acknowledgements The British Parliament, Prime Minister and Head of State are thanked for the award of a Civil List Pension in recognition of distinguished service to Britain in science. The staff of AIAS and others are thanked for many interesting discussions.

Bibliography

[1] M. W. Evans, Generally Covariant Unified Field Theory (Abramis, 2005), volume 1.

[2] ibid., vol. 2. (Abramis, 2006).

[3] ibid., vol 3 (Abramis, 2006).

[4] L. Felker, The Evans Equations of Unified Field Theory (preprint on www.aias.us and www.atomicprecision.com).

[5] H. Eckardt and L. Felker, popular article on www.aias.us and www.atomicprecision.com

[6] M. W. Evans, Generally Covariant Dynamics, first paper (Chapter 1) of vol. 4 of ref. (1) (www.aias.us and www.atomicprecision.com).

[7] M. W. Evans, Geodesics and the Aharonov Bohm Effects in ECE Theory, second paper (chapter 2) of vol. 4 of ref. (1) (www.aias.us and www.atomicprecision.com).

[8] L. H. Ryder, Quantum Field Theory (Cambridge Univ. Press, 1996, 2nd ed.).

[9] S. Weinberg, The Quantum Theory of Fields, (Cambridge Univ. Press, 2005).

[10] M. W. Evans and L. B. Crowell, Classical and Quantum Electrodynamics and the $\boldsymbol{B}^{(3)}$ Field, (World Scientific, Singapore, 2001).

[11] M. W. Evans (ed.). Modern Non-linear Optics, a special topical issue in three parts of I. Prigogine and A. A. Rice (series eds.), 'Advances in Chemical Physics' (Wiley Inter-science, New York, 2001, 2nd. ed.), vols. 119(1) to 119(3).

[12] M. W. Evans and J.-P. Vigier, The Enigmatic Photon (Kluwer, Dordrecht, 1994 to 2002 hardback and softback), in five volumes.

[13] M. W. Evans and A. A. Hasanein, The Photomagneton in Quantum Field Theory (World Scientific, Singapore, 1994).

[14] M. W. Evans and S. Kielich (eds.), first edition of ref. (11), vols 85(1) 85(3), (1992, reprinted 1993, softback 1997).

[15] M. W. Evans, papers in Found. Phys and Found. Phys. Lett., 1994 to present.

[16] Feedback sites for www.aias.us and www.atomicprecision.com, introductions of refs. (1) to (3).

[17] S. P. Carroll. Lecture Notes in General Relativity, (graduate courses at Harvard, UCSB and Chicago, public domain) arXiv : gr -gc 973019 v1 1997.

[18] M. W. Evans, Physica B, 182, 227, 237 (1992).

[19] P. W. Atkins, Molecular Quantum Mechanics, (Oxford Univ. Press, 1983, 2nd ed.).

[20] G. Stephenson, Mathematical Methods for Science Students (Longmans, London, 1968).

[21] J. B. Marion and S. T. Thornton, Classical Dynamics of Particles and Systems, (HBC, New York, 1988, 3rd. ed.).

Chapter 4

The effect of torsion on the Schwarzschild Metric and light deflection due to gravitation

(Paper 58)
by
Myron W. Evans,
Alpha Institute for Advanced Study (AIAS).
(emyrone@aol.com, www.aias.us, www.atomicprecision.com)

Abstract

The effect of torsion on the Schwarzschild metric and light deflection due to gravitation is calculated straightforwardly using the tetrad method at the root of Einstein Cartan Evans (ECE) unified field theory. Consideration of torsion changes several of the assumptions at the root of standard model cosmologies such as Big Bang, and torsion is shown to affect the deflection of light due to gravitation. Thus, any deviations from Einstein Hilbert theory may be explained by the presence of torsion.

Keywords: Einstein Cartan Evans (ECE) unified field theory, Schwarzschild metric, light deflection due to gravitation, effect of torsion on standard model cosmologies.

4.1 Introduction

Light deflection due to gravitation is a famous prediction of gravitational general relativity, and is based on the Einstein Hilbert (EH) field equation published independently by Einstein and Hilbert in 1916 as is well known. The phenomenon of light deflection by the sun can now be measured to an accuracy of one part in one hundred thousand (NASA Cassini) and even more accurate tests are being prepared by NASA. It is shown in Section 4.3 that any small deviations from the EH result that may become observable can be understood straightforwardly as being due to space-time torsion in general relativity. The Cartan torsion is of key importance to the recently inferred [1]– [16] Einstein Cartan Evans (ECE) unified field theory because the electromagnetic field is Cartan torsion within a factor $cA^{(0)}$ with the units of volts and thus referred to as the primordial voltage. The EH equation is well known to produce twice the Newtonian result for the deflection angle of light grazing a mass, such as the mass of the sun. In Section 4.2 this result is derived straightforwardly using the tetrads appropriate to the Schwarzschild metric (SM). The latter was used in the original and famous test by Eddington and co-workers and is used here to illustrate the effect of torsion. More generally in ECE field theory metrics must be calculated in the presence of Cartan torsion, which changes many of the basic assumptions of standard model cosmology. In the presence of Cartan torsion the Ricci cyclic equation is no longer true, the Riemann tensor is no longer anti-symmetric in its first two indices, the symmetric metric and symmetric Ricci tensor are true only if the central part of torsion affected motion is considered, and the symmetric Christoffel symbol must be replaced by a more general and asymmetric gamma connection. The neglect of Cartan torsion in cosmologies such as Big Bang is arbitrary. Without Cartan torsion the gravitational field cannot be unified with the electromagnetic field, which as originally inferred by Cartan himself, is the Cartan torsion within $cA^{(0)}$ [1]– [16]. Attempts to interpret astronomical data in terms of a purely central cosmology such as Big Bang are therefore purposeless because torsion is likely to pervade all cosmologies. There is no reason to assert that Cartan curvature is always large in magnitude in comparison with Cartan torsion. This EH assumption appears to be true for the sun, but may not be true for other cosmological objects.

4.2 Calculation of gravitational light deflection using the tetrad method

The SM is well known to be the first solution to the Einstein Hilbert field equation, and was inferred in 1916. The SM metric is the static solution for a spherically symmetric space-time and produces a deflection of light twice that expected from Newtonian theory. For light deflection from the sun this result of the SM has been verified by NASA Cassini to one part in one hundred thousand. So for the sun, EH theory is adequate to this accuracy. For other systems however, this may not be the case at all, because there is no reason to assume that Cartan torsion is small in magnitude compared with Cartan curvature for all cosmological objects [1]– [16]. The SM $g_{\mu\nu}$ is necessarily symmetric in its indices:

$$g_{\mu\nu} = g_{\nu\mu} \tag{4.1}$$

and defines the square of the line element:

$$ds^2 = g_{\mu\nu} dx^\mu dx^\nu \tag{4.2}$$

where x^μ is the four-coordinate:

$$x^\mu = (ct, x, y, z) \,. \tag{4.3}$$

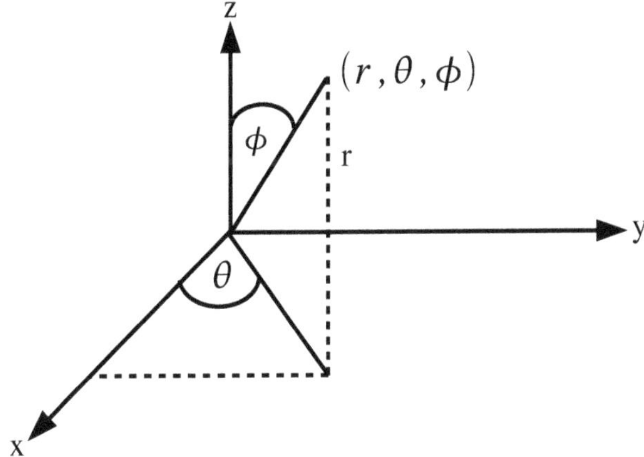

Figure 4.1: Spherical polar system

This symmetric metric is defined in terms of the tetrad of ECE theory [1]- [16] by:

$$g_{\mu\nu} = q^a{}_\mu q^b{}_\nu \eta_{ab} \tag{4.4}$$

where η_{ab} is the Minkowski metric of flat space-time. The latter is defined by:

$$\eta_{ab} = \begin{bmatrix} -1 & 0 & 0 & 0 \\ 0 & 1 & 0 & 0 \\ 0 & 0 & 1 & 0 \\ 0 & 0 & 0 & 1 \end{bmatrix}. \tag{4.5}$$

In spherical polar coordinates the line element is:

$$ds^2 = -c^2 dt^2 + dr^2 + r^2 d\Omega^2 \tag{4.6}$$

where

$$d\Omega^2 = d\theta^2 + \sin^2\theta d\phi^2 \tag{4.7}$$

and the SM in spherical polar coordinates and complete S.I. units is well known to be:

$$ds^2 = -\left(1 - \frac{2GM}{c^2 r}\right) c^2 dt^2 + \left(1 - \frac{2GM}{c^2 r}\right)^{-1} dr^2 + r^2 d\Omega^2. \tag{4.8}$$

Here G is the Newton gravitational constant, M is the mass of the object responsible for the light deflection (e.g. the sun), c is the speed of light and where r is the radial coordinate of the spherical polar system defined in Fig. 4.1: The SM reduces to the Minkowski result in the limit of large r or small M as is well known. In Cartesian coordinates the Minkowski metric is found from:

$$ds^2 = -c^2 dt^2 + dx^2 + dy^2 + dz^2 \tag{4.9}$$

and in spherical polar coordinates it is:

$$\eta_{ab} = \begin{bmatrix} -1 & 0 & 0 & 0 \\ 0 & 1 & 0 & 0 \\ 0 & 0 & r^2 & 0 \\ 0 & 0 & 0 & r^2\sin^2\phi \end{bmatrix}. \tag{4.10}$$

The SM in spherical polar coordinates is:

$$g_{\mu\nu} = \begin{bmatrix} -\left(1-\frac{2GM}{c^2r}\right) & 0 & 0 & 0 \\ 0 & \left(1-\frac{2GM}{c^2r}\right)^{-1} & 0 & 0 \\ 0 & 0 & r^2 & 0 \\ 0 & 0 & 0 & r^2\sin^2\phi \end{bmatrix}. \tag{4.11}$$

Therefore from a comparison of the diagonal elements in Eqs.(4.10) and (4.11) the tetrads of the SM may be found straightforwardly. The non-zero Minkowski elements in spherical polar coordinates are:

$$\eta_{00} = -1, \quad \eta_{11} = 1, \quad \eta_{22} = r^2, \quad \eta_{33} = r^2\sin^2\phi \tag{4.12}$$

and the non-zero SM elements in the same coordinates are:

$$g_{00} = -\left(1-\frac{2GM}{c^2r}\right), \quad g_{11} = \left(1-\frac{2GM}{c^2r}\right)^{-1}, \quad g_{22} = r^2, \quad g_{33} = r^2\sin^2\phi \tag{4.13}$$

where in general:

$$g_{00} = q^a{}_0 q^b{}_0 \eta_{ab} \\ \vdots \tag{4.14}$$

$$g_{33} = q^a{}_3 q^b{}_3 \eta_{ab} \tag{4.15}$$

Considering only the diagonal elements Eqs.(4.14) to (4.15) simplify to:

$$g_{00} = g^0{}_0 q^0{}_0 \eta_{00} \\ \vdots \tag{4.16}$$

$$g_{33} = g^3{}_3 q^3{}_3 \eta_{33} \tag{4.17}$$

Therefore the required tetrad elements of the SM are:

$$q^0{}_0 = \left(1-\frac{2GM}{c^2r}\right)^{1/2} \tag{4.18}$$

$$q^1{}_1 = \left(1-\frac{2GM}{c^2r}\right)^{1/2} \tag{4.19}$$

$$q^2{}_2 = 1 \tag{4.20}$$

$$q^3{}_3 = 1. \tag{4.21}$$

In the limit of large r or small M these reduce to the correct Minkowski elements:

$$g_{00} \to \eta_{00} \quad \text{etc.} \tag{4.22}$$

so Eqs.(4.16) – (4.17) are correctly compatible with this limit. According to the ECE Lemma [1]- [16]

$$\Box q^0{}_0 = R_0 q^0{}_0 \tag{4.23}$$

$$\Box q^1{}_1 = R_1 q^1{}_1 \tag{4.24}$$

so scalar curvatures R_0 and R_1 are generated by two of the tetrad elements of the SM. There are no ECE scalar curvatures produced by the Minkowski metric, and this result is compatible with the fact that that metric describes a flat space-time with no curvature. The four tetrads of the Minkowski metric are all unity. In spherical polar coordinates:

$$r = \left(x^2 + y^2 + z^2\right)^{1/2} \tag{4.25}$$

so Eqs.(4.23) and (4.24) reduce to:

$$\nabla^2 q^0{}_0 = -R_0 q^0{}_0 \tag{4.26}$$

$$\nabla^2 q^1{}_1 = -R_1 q^1{}_1 \tag{4.27}$$

compatible with the fact that the SM is a static solution of the EH field equation for a spherically symmetric spacetime.

The spherical polar coordinates and Cartesian coordinates are related by:

$$\left.\begin{aligned} x &= r\sin\phi\cos\theta \\ y &= r\sin\phi\sin\theta \\ z &= r\cos\phi \end{aligned}\right\} \tag{4.28}$$

so:

$$x^2 + y^2 + z^2 = r^2. \tag{4.29}$$

The infinitesimal elements are defined [17] by:

$$\left.\begin{aligned} dx &= -r\sin\phi\sin\theta d\theta + r\cos\phi\cos\theta d\phi + \sin\phi\cos\theta dr \\ dy &= r\sin\phi\cos\theta d\theta + r\cos\phi\sin\theta d\phi + \sin\phi\sin\theta dr \\ dz &= -r\sin\phi d\theta + \cos\phi dr \end{aligned}\right\} \tag{4.30}$$

so the square of the line element is:

$$ds^2 = dx^2 + dy^2 + dz^2 = dr^2 + r^2 d\phi^2 + r^2\sin^2\phi d\theta^2. \tag{4.31}$$

The space-like metric elements in curvilinear coordinates are the squares of the scale factors [17]:

$$g_{11} = h^2{}_1, \quad g_{22} = h^2{}_2, \quad g_{33} = h^2{}_3. \tag{4.32}$$

The scale factors in spherical polar coordinates [17] are:

$$h_1 = h_r = 1, \quad h_2 = h_\phi = r, \quad h_3 = h_\theta = r\sin\phi \tag{4.33}$$

in Euclidean space-time. The surface of a sphere is:

$$S = \int_0^{2\pi} d\theta \int_0^{\pi} r^2 \sin\phi d\phi = 4\pi r^2 \tag{4.34}$$

and the volume of a sphere is:

$$V = \int_0^r S dr = \frac{4}{3}\pi r^3. \tag{4.35}$$

The Euclidean unit vectors of the spherical polar coordinate system are [17]:

$$\left.\begin{aligned} \mathbf{e}_r &= \sin\phi\cos\theta\mathbf{i} + \sin\phi\sin\theta\mathbf{j} + \cos\phi\mathbf{k} \\ \mathbf{e}_\phi &= \cos\phi\cos\theta\mathbf{i} + \cos\phi\sin\theta\mathbf{j} - \sin\phi\mathbf{k}. \\ \mathbf{e}_\theta &= -\sin\theta\mathbf{i} + \cos\theta\mathbf{j} \end{aligned}\right\} \tag{4.36}$$

where $\mathbf{i}$, $\mathbf{j}$ and $\mathbf{k}$ are the unit vectors of the Cartesian system. The Euclidean vector field in spherical polar coordinates is therefore:

$$\begin{aligned} \mathbf{V} &= V_r\mathbf{e}_r + V_\phi\mathbf{e}_\phi + V_\theta\mathbf{e}_\theta \\ &= V_x\mathbf{i} + V_y\mathbf{j} + V_z\mathbf{k}. \end{aligned} \tag{4.37}$$

In Cartan geometry [1]- [16] [18], the governing equations of the EH equation and the SM are torsion-less:

$$T^a = d\wedge q^a + \omega^a{}_b \wedge q^b = 0 \tag{4.38}$$

$$R^a{}_b = d\wedge\omega^a{}_b + \omega^a{}_c\wedge\omega^c{}_b \tag{4.39}$$

$$R^a{}_b \wedge q^b = 0 \tag{4.40}$$

$$d\wedge R^a{}_b + \omega^a{}_c \wedge R^c{}_b - R^a{}_c\wedge\omega^c{}_b = 0. \tag{4.41}$$

Here T^a is the Cartan torsion form, q^a is the Cartan tetrad form, $\omega^a{}_b$ is the spin connection, and $\omega^a{}_b$ is the curvature or Riemann form of Cartan geometry. The elements of the tetrad of the SM are diagonal as shown already, and the non-vanishing elements of the Riemann tensor of the SM are:

$$R^0{}_{101}, R^0{}_{202}, R^0{}_{303}, R^0{}_{212}, R^0{}_{313}, R^1{}_{212}, R^1{}_{313}, R^2{}_{323}. \tag{4.42}$$

The Riemann form and Riemann tensor are related by [1]- [16] [18]:

$$R^a{}_{b\mu\nu} = q^a{}_\rho q^a{}_b R^\rho{}_{\sigma\mu\nu}. \tag{4.43}$$

In the presence of the Cartan torsion, equations (4.38) to 4.41 become:

$$T^a = d\wedge q^a + \omega^a{}_b\wedge q^b \tag{4.44}$$

$$R^a{}_b = d\wedge\omega^a{}_b + \omega^a{}_c\wedge\omega^c{}_b \tag{4.45}$$

$$d\wedge T^a + \omega^a{}_b\wedge T^b := R^a{}_b\wedge q^b \tag{4.46}$$

$$d\wedge R^a{}_b + \omega^a{}_c\wedge R^c{}_b - R^a{}_c\wedge\omega^c{}_b := 0. \tag{4.47}$$

Eqs.(4.44) and (4.45) are the two Cartan structure equations, and Eqs.(4.46) and (4.47) are the two Bianchi identities. These are well known equations of standard Cartan geometry and form the basis of ECE theory [1]- [16] through the ansatzen:

$$A^a = A^{(0)}q^a \tag{4.48}$$

$$F^a = A^{(0)}T^a \tag{4.49}$$

first proposed by Cartan himself in well known correspondence with Einstein. Here $A^{(0)}$ is the electromagnetic potential form, and $F^{(0)}$ is the electromagnetic field form. In the EH equation and SM there is no consideration given to the interaction of gravitation with other fields such as electromagnetism. In the presence of torsion the familiar Ricci cyclic equation (4.40) of EH theory and the SM is no longer obeyed. In tensor notation the Ricci cyclic equation is:

$$R_{\sigma\mu\nu\rho} + R_{\sigma\rho\mu\nu} + R_{\sigma\nu\rho\mu} = 0 \tag{4.50}$$

but this is not the case in the presence of torsion. The latter means therefore that the Riemann tensor is no longer anti-symmetric in its first two indices, and that the Christoffel connection becomes the general gamma connection no longer symmetric in its lower two indices. Cartan torsion fundamentally changes cosmologies based on the EH equation, for example Big Bang.

Restricting attention in this section to the EH field theory, the spin connection of the SM may be obtained from the tetrad of the SM using:

$$d \wedge q^a + \omega^a{}_b \wedge q^b = 0. \tag{4.51}$$

The Riemann form and the spin connection are related by the second Cartan structure equation 4.45:

$$R^a{}_b = d \wedge \omega^a{}_b + \omega^a{}_c \wedge \omega^c{}_b. \tag{4.52}$$

In Section 4.3 the equation (4.51) will be perturbed by a small torsion δT^a, to give:

$$d \wedge q^a + \omega^a{}_b \wedge q^b = \delta T^a \tag{4.53}$$

while in the rest of Section 4.2 the light deflection of the SM will be calculated by the tetrad method. This is shown to be much simpler and easier to use and understand than the conventional metric method [18] [19]. Use of the tetrad method also allows the effect of torsion to be calculated via equation (4.53).

The SM written out in spherical polar coordinates (Fig. 4.1) is:

$$ds^2 = -\left(1 - \frac{2GM}{c^2 r}\right) c^2 dt^2 + \left(1 - \frac{2GM}{c^2 r}\right)^{-1} dr^2 + r^2 d\phi^2 + r^2 \sin^2 \phi d\theta^2. \tag{4.54}$$

Light travels along null paths:

$$ds^2 = 0. \tag{4.55}$$

Now restrict consideration to a single plane through the center of mass:

$$\theta = 0. \tag{4.56}$$

Therefore Eq.(4.54) becomes:

$$c^2 dt^2 = \left(1 - \frac{2GM}{c^2 r}\right)^{-2} dr^2 + r^2 \left(1 - \frac{2GM}{c^2 r}\right)^{-1} d\phi^2. \tag{4.57}$$

The metric corresponding to this equation is:

$$g_{\mu\nu} = \begin{bmatrix} \left(1 - \frac{2GM}{c^2 r}\right)^{-2} & 0 \\ 0 & r^2 \left(1 - \frac{2GM}{c^2 r}\right)^{-1} \end{bmatrix} \tag{4.58}$$

which reduces to the Minkowski metric for large r or small M:

$$\eta_{ab} = \begin{bmatrix} 1 & 0 \\ 0 & r^2 \end{bmatrix}. \tag{4.59}$$

Therefore using Eq.(4.4) the tetrads are:

$$q_{rr} = \left(1 - \frac{2GM}{c^2 r}\right)^{-1}, \tag{4.60}$$

$$q_{\phi\phi} = \left(1 - \frac{2GM}{c^2 r}\right)^{-1/2}. \tag{4.61}$$

If:

$$c^2 r \gg 2GM \tag{4.62}$$

then:

$$q_{rr} \rightarrow 1 + \frac{2GM}{c^2 r} + \cdots \tag{4.63}$$

$$q_{\phi\phi} \rightarrow 1 + \frac{GM}{c^2 r} + \cdots \tag{4.64}$$

The tetrad element q_{rr} means that r is not a straight line, it is a curve:

$$\zeta\left(r\right) = \zeta^{(0)} q_{rr} \tag{4.65}$$

where $\zeta^{(0)}$ is a scalar proportionality factor. By differentiation with respect to r:

$$c^2 \frac{\partial q_{rr}}{\partial r} = -\frac{2GM}{r^2}. \tag{4.66}$$

The Newtonian force between a photon of mass m and the sun of mass M is:

$$F = -\frac{GmM}{r^2}. \tag{4.67}$$

The force from Eq.(4.66) is:

$$F = -mc^2 \frac{\partial q_{rr}}{\partial r} = -\frac{2GmM}{r^2}. \tag{4.68}$$

This is twice the Newtonian force and is $\partial q_{rr}/\partial r$ multiplied by the photon rest energy:

$$E_0 = mc^2 = \hbar\omega_0. \tag{4.69}$$

Eq.4.69 is the Planck / Einstein / de Broglie equation. Using the equivalence of inertial and gravitational mass, the force from Eq.(4.68) is:

$$F = mg = -mc^2 \frac{\partial q_{rr}}{\partial r} \tag{4.70}$$

so the acceleration due to gravity is due to the r derivative of the radial tetrad within a factor c^2:

$$g = -c^2 \frac{\partial q_{rr}}{\partial r}. \tag{4.71}$$

The angle of deflection in the Eddington experiment is defined by Fig. 4.2:

The Newtonian result is:

$$\delta\left(\text{Newton}\right) = \frac{2MG}{c^2 r_0} \tag{4.72}$$

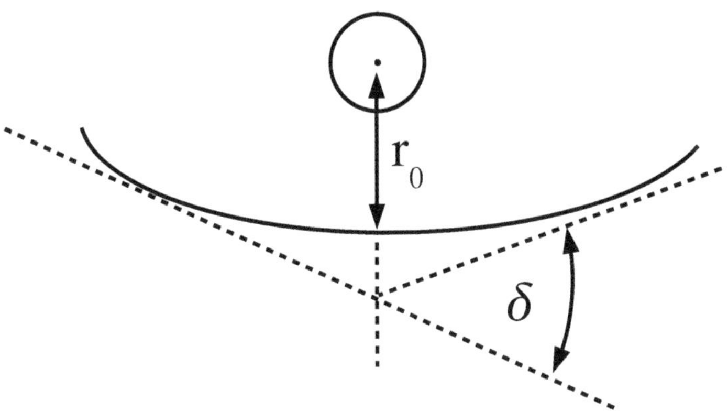

Figure 4.2: The angle of deflection in the Eddington experiment

where r_0 is the distance of closest approach. So the result from the EH theory is twice this from Eq.(4.68):

$$\delta\,(\text{Schwarzschild}) = \frac{4MG}{c^2 r_0} \tag{4.73}$$

Using the tetrad method the effect of Cartan torsion on this result will be calculated in Section 4.3. The tetrad method developed in this Section for the first time, is straightforward, and is ideally suited to calculate the effect of torsion from Eq.(4.53) from standard Cartan geometry. The metric method of calculating the Eddington deflection is much more complicated.

4.3 Torsional perturbation of light deflection due to gravity

The angle of deflection in the absence of torsional perturbation is given from the result in Eq.(4.73) by:

$$\delta = 2\,(q_r r - 1)_{r=r_0} \tag{4.74}$$

In the absence of torsion, Eq.(4.51) gives:

$$d \wedge q_{rr,0} = -\omega_0 \wedge q_{rr,0} \tag{4.75}$$

where ω_0 is the spin connection in the absence of torsion. In the presence of a small torsional perturbation Eq.(4.75) becomes:

$$d \wedge q_{rr,T} = -\omega \wedge q_{rr,T} + \delta T \tag{4.76}$$

In a first approximation:

$$\omega \sim \omega_0 \tag{4.77}$$

and:

$$\omega \wedge q_{rr,T} \sim \omega_0 \wedge q_{rr,0} \tag{4.78}$$

so

$$d \wedge q_{rr,T} - d \wedge q_{rr,0} \sim \delta T \tag{4.79}$$

or

$$d \wedge (\delta q_{rr}) \sim \delta T \tag{4.80}$$

and:

$$\Delta\delta \sim 2\,(\delta q_{rr})_{r=r_0} \tag{4.81}$$

From Eqs.(4.80) and (4.81) it is clear that the torsional perturbation δT will change the angle of deflection by $\Delta\delta$. In Cartesian coordinates introduce a perturbation of the type:

$$\delta q_{rr} = \frac{1}{\sqrt{2}}\,(1 - i)\,e^{i\phi} \sim \frac{1}{\sqrt{2}}\phi \tag{4.82}$$

for $\phi \ll 1$. Thus:

$$\Delta\delta \sim \frac{2}{\sqrt{2}}\phi \tag{4.83}$$

This is a simple illustration of the effect of torsion on the angle of deflection of light due to gravitation. From experimental data (NASA Cassini) on gravitational lensing within the solar system it is known that ϕ must be very small for the sun photon system because the EH result (torsionless or baseline result) is accurate to one part in one hundred thousand. For other cosmological objects such as rotating pulsars of great mass, the effect of torsion could be much larger. In this illustration the SM has been assumed to be approximately true in the presence of a torsional perturbation. Metrics in a generally covariant unified field theory must however be calculated from the second Bianchi identity of Cartan geometry. The torsionless SM is calculated as a solution of the second Bianchi identity of Riemann geometry, in which torsion is zero.

4.4 Discussion

Naive unification of the gravitational and electromagnetic fields was first attempted by Reissner [20] and independently by Nordstrom [21], shortly after the discovery of the Schwarzschild metric. Naive unification takes place without any consideration of the Cartan torsion, using the minimal substitution rule:

$$\partial_\mu \to D_\mu \tag{4.84}$$

The electromagnetic field in naive unification cannot therefore be the Cartan torsion and the effect of electromagnetism is introduced through the addition of an electromagnetic term to the canonical energy momentum of EH field theory. Einstein was dissatisfied with naive unification, and the idea that the electromagnetic field is the Cartan torsion was first suggested by Cartan himself in well known correspondence with Einstein during the twenties and thirties of the last century. Einstein then worked on unification until 1955, as is well known, but did not develop a satisfactory theory. The minimal substitution rule does not produce uniquely defined results [18] [19] and still uses the Christoffel symbol of torsionless EH theory. It was not until the inference of the experimentally observable ECE spin field ($\boldsymbol{B}^{(3)}$) in 1992 [1]- [16] that the general covariance of electromagnetism began to be correctly developed and it was not until 2003 to present that the correct mathematical structure for ECE unification finally emerged from $\boldsymbol{B}^{(3)}$ theory and gauge theory ($O(3)$ electrodynamics [1]- [16]). Naive unification does not produce an ECE spin field, which requires the use of Cartan torsion. The spin field is the direct result of the spinning space-time necessary to describe generally covariant electromagnetism unified in a self-consistent and rigorous geometrical manner with gravitation and the other fundamental fields. With the Christoffel connection of naive unification there is no spinning space-time, only a curving space-time. It is self-inconsistent to add an electromagnetic term to the canonical energy-momentum tensor without spinning space-time. This internal inconsistency is present in all naive unification schemes, such as that of Newman et al. [22] for the Kerr metric. There are several other phenomena [1]- [16] now known to be explicable with ECE but not by naive unification. Misner [23] for example, has used the tetrad method in a gravitational context, but again does not consider Cartan torsion in any relevant detail. Newman and Penrose [24] developed the tetrad method for use with spinors, but again in a restricted gravitational context using the null tetrad. Spinors were discovered by Cartan himself in 1913 [25]. There is some discussion of the method of Newman and Penrose by Barrett [26] but this does not provide even the basis for a generally covariant unified field theory. Throughout the twentieth century, there was difficulty in the development of a generally covariant unified field theory because the ECE spin field was not known. The spin field was inferred only in 1992 [1]- [16]. In the twentieth century, undue reliance continued to be placed on the Maxwell Heaviside (MH) field theory inferred in the nineteenth century. The MH theory is not generally covariant [1]- [16], it is special relativity, and therefore can only be Lorentz covariant. The MH theory does not use a spinning space-time, required for self consistent unification, and for this reason cannot produce an ECE spin field $\boldsymbol{B}^{(3)}$. MH theory must be made generally covariant before it can be unified with gravitation. This is an obvious point, but one which was overlooked for a hundred years or more. In the twentieth century, considerable confusion was caused by the Copenhagen School, especially by the Heisenberg Uncertainty Principle. The latter has no place in physics, which must be an objective and causal subject as recognised by Bacon in the seventeenth century. Several independent experimental refutations of the Uncertainty Principle are now available [1]- [16]. ECE theory produces a rigorous and generally covariant quantum field theory [1]- [16] without using the Copenhagen assertions. In a new twenty first century perspective, Copenhagen is little more than subjective assertion, or pathological science where an idea is not evaluated critically. The Uncertainty Principle deliberately introduces obscure anthropomorphism into science, and for this reason was immediately rejected by the causal realist school of thought led by Einstein, Schrödinger, de Broglie and followers.

The neglect of Cartan torsion restricted twentieth century cosmology to models such as the Big Bang. This model was immediately rejected by Hoyle and followers, as is well known. ECE theory has thrown considerable new light on this twentieth century debate [1]- [16]. An oscillatory cosmological model is favored by ECE theory [1]- [16]. This point may be illustrated in a simple manner as follows. For the general metric in spherically symmetric space-time:

$$ds^2 = -e^{2\alpha(t,r)}c^2dt^2 + e^{2\beta(t,r)}dr^2 + r^2d\Omega^2 \tag{4.85}$$

and therefore the tetrads are:

$$q^0{}_0 = e^\alpha, \quad q^1{}_1 = e^\beta, \quad q^2{}_2 = q^3{}_3 = 1 \tag{4.86}$$

The ECE Lemmas are therefore:

$$\Box e^\alpha = R_0 e^\alpha \tag{4.87}$$

$$\Box e^\beta = R_1 e^\beta \tag{4.88}$$

The differentiations in Eqs.(4.87) and (4.88) are therefore defined by:

$$\Box e^\alpha = \left(\frac{1}{c^2}\frac{\partial^2}{\partial t^2} - \frac{\partial^2}{\partial x^2} - \frac{\partial^2}{\partial y^2} - \frac{\partial^2}{\partial z^2}\right) e^\alpha \tag{4.89}$$

where from the Leibnitz Theorem:

$$\frac{\partial^2}{\partial t^2}e^\alpha = \frac{\partial}{\partial t}\left(\frac{\partial \alpha}{\partial t}e^\alpha\right) = \left(\frac{\partial^2 \alpha}{\partial t^2} + \left(\frac{\partial \alpha}{\partial t}\right)^2\right) e^\alpha \tag{4.90}$$

etc.

Thus:

$$\Box e^\alpha = \left(\Box\alpha + \frac{1}{c^2}\left(\frac{\partial \alpha}{\partial t}\right)^2 - \left(\frac{\partial \alpha}{\partial x}\right)^2 - \left(\frac{\partial \alpha}{\partial y}\right)^2 - \left(\frac{\partial \alpha}{\partial z}\right)^2\right) e^\alpha \tag{4.91}$$

and the scalar curvatures are:

$$R_0 = \Box\alpha + \frac{1}{c^2}\left(\frac{\partial \alpha}{\partial t}\right)^2 - \left(\frac{\partial \alpha}{\partial x}\right)^2 - \left(\frac{\partial \alpha}{\partial y}\right)^2 - \left(\frac{\partial \alpha}{\partial z}\right)^2 \tag{4.92}$$

$$R_1 = \Box\beta + \frac{1}{c^2}\left(\frac{\partial \beta}{\partial t}\right)^2 - \left(\frac{\partial \beta}{\partial x}\right)^2 - \left(\frac{\partial \beta}{\partial y}\right)^2 - \left(\frac{\partial \beta}{\partial z}\right)^2 \tag{4.93}$$

All spherically symmetric space-time solutions of the ECE theory obey this result. Eqs. (4.87) and (4.88) are equations of classical and causal physics. If it were possible to find complex valued solutions:

$$\alpha = \alpha^{'} + i\alpha^{"} \tag{4.94}$$

$$\beta = \beta^{'} + i\beta^{"} \tag{4.95}$$

then Eqs.(4.87) and (4.88) would become eigen-equations via the imaginary components $i\alpha^{"}$ and $i\beta^{"}$. For real valued α and β however there is only one R_0 and only one R_0. In standard model (twentieth century) cosmology, the existence of the ECE Lemma is not known, and Big Bang for example depends on torsionless solutions of the EH equation. The latter is only a limit of ECE cosmology. Wave cosmologies for example can be developed in ECE theory by considering the tetrads to be defined by:

$$q^0{}_0 q^0{}_0 = e^{2\alpha} \tag{4.96}$$

$$q^1{}_1 q^1{}_1 = e^{2\beta} \tag{4.97}$$

If these tetrads are complex valued:

$$q^0{}_0 = e^{\alpha' + i\alpha''}, \quad q^0{}_0{}^* = e^{\alpha' - i\alpha''}, \tag{4.98}$$

$$q^1{}_1 = e^{\beta' + i\beta''}, \quad q^1{}_1{}^* = e^{\beta' - i\beta''} \tag{4.99}$$

then:

$$q^0{}_0 q^0{}_0{}^* = e^2\alpha \tag{4.100}$$

$$q^1{}_1 q^1{}_1{}^* = e^2\beta \tag{4.101}$$

where $*$ denotes complex conjugate. Eqs.(4.98) to (4.101) have solutions:

$$\alpha = \alpha', \quad \beta = \beta' \tag{4.102}$$

for all α''. Therefore:

$$q^0{}_0 = e^{\alpha} e^{i\alpha''}, \quad q^1{}_1 = e^{\beta} e^{i\beta''} \tag{4.103}$$

and the transformations:

$$q^0{}_0 \rightarrow e^{i\alpha''} q^0{}_0 \tag{4.104}$$

$$q^1{}_1 \rightarrow e^{i\beta''} q^1{}_1 \tag{4.105}$$

leave the metric elements unchanged:

$$g_{00} \rightarrow e^{-i\alpha''} g_{00} e^{i\alpha''} \tag{4.106}$$

$$g_{11} \rightarrow e^{-i\beta''} g_{00} e^{i\beta''} \tag{4.107}$$

The oscillatory or wave cosmologies of ECE can therefore be defined by the eigenequations:

$$\Box e^{i\alpha''} = R_0'' e^{i\alpha''} \tag{4.108}$$

$$\Box e^{i\beta''} = R_1'' e^{i\beta''} \tag{4.109}$$

where for one eigenfunction there are many eigenvalues R_0'' and R_1''. Big Bang severely restricts what is actually available in cosmology to a uniformly expanding universe represented by Eqs.(4.87) and (4.88). More generally, ECE theory gives wave cosmologies described by Eqs.(4.108) and (4.109). Most generally, ECE gives cosmologies in which torsion and curvature play an equal role.

The interaction of electromagnetism and gravitation (i.e. of torsion and curvature) is of key importance also on the microscopic scale, as well as the macroscopic scale represented by cosmology. ECE theory now allows this fact to be much better defined. The hydrogen (H) atom on a microscopic scale, for example, is made up of an electron bound to a proton. The mass of the H atom is less [18] than the sum of the mass of a proton and an electron. The reason is that there is a negative binding energy. To separate the electron from the proton energy has to be used. There is interaction of electromagnetism with gravitation inside the H atom and this interaction produces the mass deficit referred to already. A standard model text such as ref. [18] deduces that gravitation must interact with all forms of energy and momentum. This is another way of stating the Einstein equivalence principle [18]. The latter means that the equations of general relativity must reduce to those of special relativity in the absence of gravitation. In special relativity and in the non-relativistic limit, the sum of the proton and electron masses would be the same as the mass of the H atom.

So the H atom in ECE theory is described by:

$$\Box q^a{}_\mu = R q^a{}_\mu \tag{4.110}$$

$$R = -kT \tag{4.111}$$

$$R = q^\lambda{}_a \partial^\mu \left(\Gamma^\nu{}_{\mu\lambda} q^a{}_\nu - \omega^a{}_{\mu b} q^b{}_\lambda \right) \tag{4.112}$$

Here $q^a{}_b$ is the wavefunction and also the field [1]– [16], R is the ECE scalar curvature, k is the Einstein constant, T is the index contracted canonical energy-momentum tensor, $\Gamma^\nu{}_{\mu\lambda}$ is the general gamma connection and $\omega^a{}_{\mu b}$ is the spin connection. The Einstein equivalence principle means that:

$$kT \rightarrow \left(\frac{mc}{\hbar} \right)^2 \tag{4.113}$$

in the limit of no gravitation (special relativity). Here m is the mass of the H atom. So it is seen that in the presence of gravitation (Eq.(4.110)) the mass of the H atom is changed from the value given by Eq.(4.113), which is the Dirac equation of the H atom:

$$\left(\Box + \left(\frac{mc}{\hbar} \right)^2 \right) q^a{}_\mu = 0 \tag{4.114}$$

The electromagnetic interaction between the electron and proton in the H atom is described by the ECE field equations:

$$d \wedge F^a = \mu_0 j^a \tag{4.115}$$

$$d \wedge \widetilde{F}^a = \mu_0 J^a \tag{4.116}$$

$$F^a = d \wedge A^a + \omega^a{}_b \wedge A^b \tag{4.117}$$

and thus by a linear inhomogeneous differential equation [1]– [16]:

$$d \wedge \left(d \wedge A^a + \omega^a{}_b \wedge A^b \right) = \mu_0 j^a \tag{4.118}$$

$$j^a = \frac{A^{(0)}}{\mu_0} \left(R^a{}_b \wedge q^b - q^a{}_b \wedge T^b \right) \tag{4.119}$$

At resonance, j^a can be amplified by many orders of magnitude, giving rise to a new source of electric power. This energy is tapped from the H atom, and has recently been observed experimentally [1]– [16]. The H orbitals given by the Dirac equation give no hint of the existence of this energy. The Schrödinger equation is the non-relativistic limit of the Dirac equation and gives even less information about the generally covariant nature of the H atom. The standard model of the H atom is described by the Schrödinger equation with the Coulomb Law. The latter is given by the:

$$j^a = 0 \tag{4.120}$$

limit of the ECE field equations (4.115) to (4.117).

The essence of the ECE theory is the use of the tetrad, which is both the fundamental unified field and also the unified wave-function. In this sense classical and quantum mechanics are unified, they are both manifestations of Cartan geometry, and the needless mysteries of the Copenhagen school are removed from physics. The tetrad may also be used to give a deeper meaning to the Eddington experiment. The relativistic result (4.73) is twice the Newtonian result, and at first sight does not reduce to the Newtonian result. The reason for this is that Eq.(4.73) is derived using Eq.(4.55) for motion infinitesimally close to the speed of light of the photon of mass m, (the lightest particle known in nature). Newtonian dynamics deals with particles moving at $v \ll c$. In this limit the light-like condition (4.55) no longer holds, and the radial metric must be calculated from Eq.(4.8). The relevant tetrad to consider is:

$$q^1{}_1 = \left(1 - \frac{2GM}{c^2 r}\right)^{-1/2} \tag{4.121}$$

and when:

$$2GM \ll c^2 r \tag{4.122}$$

this is:

$$q^1{}_1 \rightarrow 1 + \frac{GM}{c^2 r} \tag{4.123}$$

Therefore Eq.(4.65) is replaced by:

$$\zeta\,(r) = \zeta^{(0)} q^1{}_1 \tag{4.124}$$

so we obtain:

$$c^2 \frac{\partial q^1{}_1}{\partial r} = -\frac{GM}{r^2} \tag{4.125}$$

and the force:

$$F = -\frac{GmM}{r^2} \tag{4.126}$$

This result is the same as the Newtonian force governing the orbit of a mass m around a mass M, so the deflection is Eq.(4.72). This describes the Kepler laws and the orbit of a planet around the sun. However, the interpretation of Eq.(4.126) is different from that of Newton, who derived his inverse square law using an Euclidean space. Time was considered by Newton as a distinct from space. The ECE result (4.126) is derived by considering space and time to be unified into a spacetime with in general curvature and torsion.

Acknowledgements The British Parliament, Prime Minister and Head of State are thanked for a Civil List Pension and the staff and environment of AIAS for many interesting discussions.

Bibliography

[1] M. W. Evans, Generally Covariant Unified Field Theory (Abramis, 2005), volume one.

[2] ibid., volumes two and three (Abramis 2006, preprints on www.aias.us and www.atomicprecision.com).

[3] L. Felker, The Evans Equations of Unified Field Theory (preprint on www.aias.us and www.atomicprecision.com).

[4] H. Eckardt and L. Felker, articles on www.aias.us and www.atomicprecision.com.

[5] P. H. Pinter, What is Life? (Abramis, 2006).

[6] M. W. Evans (ed.), Modern Non-Linear Optics a special topical issue of I. Prigogine and S. A. Rice, Advances in Chemical Physics (Wiley-Interscience, New York, 2001, 2nd. Edition), vols. 119(1) to 119(3).

[7] M. W. Evans and L. B. Crowell, Classical and Quantum Electrodynamics and the $\boldsymbol{B}^{(3)}$ Field (World Scientific, Singapore, 2001).

[8] M. W. Evans and J.-P. Vigier, The Enigmatic Photon (Kluwer, Dordrecht, 1994 to 2002, hardback and softback), in five volumes.

[9] M. W. Evans and A. A. Hasanein, The Photomagneton in Quantum Field Theory (World Scientific, Singapore, 1994).

[10] M. W. Evans and S. Kielich. (Eds.), first Edition of ref. (6) (Wiley-Interscience, New York, 1992, 1993 and 1997, hardback and softback), vols. 85(1) to 85(3).

[11] M. W. Evans, The Photon's Magnetic Field, Optical NMR Spectroscopy (World Scientific, Singapore, 1992).

[12] M. W. Evans, Physica B, 182, 227 and 237 (1992), the original $\boldsymbol{B}^{(3)}$ papers.

[13] M. W. Evans, papers and letters in Foundations of Physics and Foundations of Physics Letters, 1994 onwards.

[14] M. W. Evans, Generally Covariant Dynamics (paper 55 of the ECE series, preprint on www.aias.us and www.atomicprecision.com).

[15] M. W. Evans, Geodesics and the Aharonov Bohm Effects in ECE Theory (paper 56 of the ECE series, preprint on www.aias.us and www.atomicprecision.com).

[16] M. W. Evans, Canonical and Second Quantization in Generally Covariant Quantum Field Theory (paper 57 of the ECE series, preprint on www.aias.us and www.atomicprecsiion.com).

[17] E. G. Milewski (Chief Ed.), The Vector Analysis Problem Solver (Research and Education Association, New York City, 1987).

[18] S. P. Carroll, Lecture Notes in General Relativity, (graduate courses at Harvard, UCSB and Chicago, public domain) arXiv : gr - gc 973019 v1 1997.

[19] R. M. Wald, General Relativity (Univ Chicago Press, 1984).

[20] H. Reissner, Ann. Phys., 50, 106 (1916).

[21] G. Nordstrom, Proc. Kon. Ned. Acad. Wet., 20, 1238 (1918).

[22] E. T. Newman, E. Couch, K. Chiannapared, A. Exton, A. Prakash and R. Torrence, J. Math. Phys., 6, 918 (1965).

[23] C. W. Misner, J. Math. Phys., 4, 924 (1963).

[24] E. T. Newman and R. Penrose, J. Math. Phys., 3, 566 (1962).

[25] E. Cartan, Bull. Soc. Math. de France, 41, 53 (1913); E. Cartan, The Theory of Spinors (Dover, New York, 1981).

[26] T. W. Barrett in T. W. Barrett and D. M. Grimes (eds.), Advanced Electromagnetism (World Scientific, Singapore, 1995), pp. 278 ff.

Chapter 5

The resonant Coulomb Law from ECE Theory: Application to the hydrogen atom

(Paper 59)
by
Myron W. Evans,
Alpha Institute for Advanced Study (AIAS).
(emyrone@aol.com, www.aias.us, www.atomicprecision.com)

Abstract

The Einstein Cartan Evans (ECE) generally covariant unified field theory is applied to the Coulomb law by defining the electric field in terms of a scalar potential and a spin connection vector. The resulting equation modifies the Poisson equation into a linear inhomogeneous differential equation with resonant solutions. When applied to the hydrogen (H) atom as a model, the resonances thus produced by the spin connection can cause amplified oscillations, ionization and the production of a free electron. The latter may be used in a circuit to produce electric power.

Keywords: Einstein Cartan Evans (ECE) unified field theory, resonant Coulomb law, hydrogen atom, electric power from space-time.

5.1 Introduction

The Einstein Cartan Evans (ECE) field theory [1]– [16] is a generally covariant unified field theory based on standard Cartan geometry. As first suggested by Cartan himself, the electromagnetic field form (F^a) is the Cartan torsion form (T^a) within a factor $A^{(0)}$. Here $cA^{(0)}$ is the scalar potential in volts. The electromagnetic potential form (A^a) is the tetrad form (q^a) within the same factor $A^{(0)}$. The ECE field theory has been applied to many of the major areas of contemporary physics [1]– [16] and has been accepted as mainstream physics [17]. This is unsurprising in view of the fact that it is a logical extension to all fields of the theory of general relativity first applied by Einstein and Hilbert to the gravitational field. ECE theory has been tested in a number of ways [1]– [16] both for mathematical self consistency and against experimental data and has been shown to be a rigorous quantum field theory in which the tetrad is both the fundamental field and the wave-function.

In Section 5.2 the ECE theory is applied straightforwardly to the Coulomb law, and the Poisson equation is developed into a linear inhomogeneous differential equation with resonant solutions due to the spin connection of ECE theory [1]– [16]. In the simplest approximation, and considering only the scalar potential, the spin connection modifies the relation between the electric field strength (E in volt/m) and the scalar potential (ϕ in volts) to:

$$\mathbf{E} = -\left(\boldsymbol{\nabla} + \boldsymbol{\omega}\right)\phi \tag{5.1}$$

where $\boldsymbol{\omega}$ is a vector valued component of the complete spin connection form $\omega^a{}_{\mu b}$. In the standard model (Maxwell Heaviside field theory of special relativity) the equivalent of Eq.(5.1) is well known to be:

$$\mathbf{E} = -\boldsymbol{\nabla}\phi \tag{5.2}$$

so the term $\boldsymbol{\omega}$ introduces an extra potential from the ECE space-time of general relativity. In ECE theory the Coulomb law in the simplest approximation has the same form as the Coulomb law in the standard model:

$$\boldsymbol{\nabla} \cdot \mathbf{E} = \frac{\rho}{\epsilon_0} \tag{5.3}$$

but the relation between $\mathbf{E}$ and ϕ in ECE theory is always given by Eq.(5.1). Therefore it follows that:

$$\nabla^2\phi + \boldsymbol{\omega} \cdot \boldsymbol{\nabla}\phi + \left(\boldsymbol{\nabla} \cdot \boldsymbol{\omega}\right)\phi = -\frac{\rho}{\epsilon_0} \tag{5.4}$$

This is a linear inhomogeneous differential equation with the well known resonant solutions first inferred by the two Bernoulli's in 1739 and reported by Euler in 1743. Thus ECE gives the well known results of the Coulomb law, known to be a very precise law of physics, but also gives the possibility of extracting electric field strength from space-time (represented by the spin connection vector $\boldsymbol{\omega}$). Repeatable experimental results [18] indicate that E can be extracted at resonance from Eq.(5.4).

In Section 5.3 this phenomenon [18] is modeled with the H atom, used only as a simple thought model. It is shown that at resonance form Eq.(5.4) amplified oscillations develop and the H atom is ionized into a free electron and proton. The free electron can be used in a circuit to provide electric power from space-time. This model illustrates theoretically what has already been shown experimentally [18]to be possible. There is no standard model explanation for the results of ref. [18] because in the standard model:

$$\nabla^2\phi = -\frac{\rho}{\epsilon_0} \tag{5.5}$$

and there are no resonant solutions of this (Poisson) equation.

5.2 Resonant Coulomb Law

This law is deduced [1]– [16] from the first Cartan structure equation [19]:

$$T^a = d \wedge q^a + \omega^a{}_b \wedge q^b \tag{5.6}$$

and the first Bianchi identity:

$$d \wedge T^a := R^a{}_b \wedge q^b - \omega^a{}_b \wedge T^b \tag{5.7}$$

where $R^a{}_b$ is the Riemann form of Cartan geometry. The fundamental ECE assumptions are:

$$A^a = A^{(0)} q^a, \quad F^a = A^{(0)} T^a, \tag{5.8}$$

so Eq.(5.6) produces the relation between F^a and A^a:

$$F^a = d \wedge A^a + \omega^a{}_b \wedge A^b \tag{5.9}$$

and Eq.(5.7) gives the homogeneous field equation [1]– [16]:

$$d \wedge F^a = \mu_0 j^a = A^{(0)} \left(R^a{}_b \wedge q^b - \omega^a{}_b \wedge T^b \right) \tag{5.10}$$

where μ_0 is the vacuum S.I, permeability and j^a is the homogeneous current. The Hodge dual of Eq.(5.10) is:

$$d \wedge \widetilde{F}^a = \mu_0 J^a = A^{(0)} \left(\widetilde{R}^a{}_b \wedge q^b - \omega^a{}_b \wedge \widetilde{T}^b \right) \tag{5.11}$$

and defines the inhomogeneous field equation where J^a is the inhomogeneous current.

The Coulomb Law is part of Eq.(5.11), which is a linear, inhomogeneous differential equation of Cartan geometry with resonant solutions. In vector notation Eq.(5.9) is developed into two equations:

$$\mathbf{E}^a = -\frac{\partial \mathbf{A}^a}{\partial t} - c \boldsymbol{\nabla} A^{0a} - c\omega^{0a}{}_b \mathbf{A}^b + c\boldsymbol{\omega}^a{}_b A^{0b}, \tag{5.12}$$

$$\mathbf{B}^a = \boldsymbol{\nabla} \times \mathbf{A}^a - \boldsymbol{\omega}^a{}_b \times \mathbf{A}^b, \tag{5.13}$$

where the superscripts a and b derive from Cartan geometry [1]– [16]. Here $\mathbf{E}^a$ is the electric field strength (volt/m) and $\mathbf{B}^a$ the magnetic flux density (tesla). The scalar potential is:

$$\phi^a = cA^{0a} \tag{5.14}$$

and the vector potential is A^a. The spin connection form is a four vector with scalar $(\omega^{0a}{}_b)$ and vector $(\boldsymbol{\omega}^a{}_b)$ components [1]– [16]. In vector notation Eqs. (5.10) and (5.11) become four field equations of ECE theory. These are the generally covariant Gauss law of magnetism :

$$\nabla \cdot \mathbf{B}^a = \mu_0 \widetilde{j}^{0a}, \tag{5.15}$$

the generally covariant Faraday law of induction:

$$\boldsymbol{\nabla} \times \mathbf{E}^a + \frac{\partial \mathbf{B}^a}{\partial t} = \mu_0 \widetilde{\mathbf{j}}^a, \tag{5.16}$$

the generally covariant Coulomb law:

$$\nabla \cdot \mathbf{E}^a = c\mu_0 \widetilde{J}^{0a} = \frac{\rho^a}{\epsilon_0}, \tag{5.17}$$

and the generally covariant Amperé-Maxwell law:

$$\nabla \times \mathbf{B}^a - \frac{1}{c^2}\frac{\partial \mathbf{E}^a}{\partial t} = \frac{\mu_0}{c}\widetilde{\mathbf{J}}^a \tag{5.18}$$

The various resonance equations of ECE field theory [1]– [16] are found by substituting Eqs. (5.12) and (5.13) into Eqs.(5.15) to (5.18). The four currents are defined by:

$$\widetilde{j}^{a\nu} = \left(\frac{1}{c}\widetilde{j}^{a0}, \widetilde{\mathbf{j}}^a\right) \tag{5.19}$$

and

$$\widetilde{J}^{a\nu} = \left(\frac{1}{c}\widetilde{J}^{a0}, \widetilde{\mathbf{J}}^a\right) \tag{5.20}$$

The corresponding equations of the standard model (the Maxwell Heaviside field theory of special relativity) are well known to be:

$$\mathbf{E} = -\frac{\partial \mathbf{A}}{\partial t} - \nabla\phi \tag{5.21}$$

and:

$$\mathbf{B} = \nabla \times \mathbf{A} \tag{5.22}$$

The Gauss law of magnetism of the standard model is:

$$\nabla \cdot \mathbf{B} = 0 \tag{5.23}$$

The Faraday law of induction is:

$$\nabla \times \mathbf{E} + \frac{\partial \mathbf{B}}{\partial t} = \mathbf{0} \tag{5.24}$$

The Coulomb law is:

$$\nabla \cdot \mathbf{E} = \frac{\rho}{\epsilon_0} \tag{5.25}$$

and the Ampére Maxwell law is:

$$\nabla \times \mathbf{B} - \frac{1}{c^2}\frac{\partial \mathbf{E}}{\partial t} = \frac{\mu_0}{c}\widetilde{\mathbf{J}} \tag{5.26}$$

Eqs.(5.12) to (5.18) are equations of a generally covariant unified field theory, Eqs.(5.21) to (5.26) are equations of special relativity, where the electromagnetic field is not unified with other fields such as the gravitational field.

The ECE Coulomb law is obtained by substituting Eq.(5.12) into Eq.(5.17):

$$\nabla^2 A^{a0} + \frac{1}{c}\frac{\partial}{\partial t}(\nabla \cdot \mathbf{A}^a) + \nabla \cdot \left(\omega^{0a}{}_b \mathbf{A}^b\right) - \nabla \cdot \left(\omega^a{}_b A^{0b}\right) = -\mu_0 \widetilde{J}^{0a} \tag{5.27}$$

Eq.(5.27) is a driven damped oscillator equation [19] with resonance solutions. On the right hand side appears a small "driving force". On the left hand side there are Hooke's law and damping terms. At resonance the scalar and vector potentials can be greatly amplified for an initially small driving charge current density, a process which is well known to obey Noether's Theorem under all circumstances and thus to conserve energy/momentum and charge/current density.

The standard model equivalent of Eq.(5.27) is the Poisson equation:

$$\nabla \cdot \left(\frac{\partial \mathbf{A}}{\partial t} + \nabla \phi \right) = -\frac{\rho}{\epsilon_0} \tag{5.28}$$

which does not give resonant solutions because the spin connection terms are missing in the flat or Minkowski space-time of Maxwell Heaviside field theory. Resonance is due to the spin connection, and so resonance is due to space-time itself. The simplest type of resonance equation is [19]:

$$\ddot{x} + 2\beta\dot{x} + \omega_0^2 x = \alpha \cos \omega t \tag{5.29}$$

where there is a damping term proportional to $\dot{x}$, a Hooke's law term proportional to x and a driving force on the right hand side. The particular integral [19] is:

$$x_p(t) = \frac{\alpha \cos(\omega t - \delta)}{\left(\left(\omega_0^2 - \omega^2\right)^2 + 4\omega^2\beta^2 \right)^{1/2}} \tag{5.30}$$

where:

$$\delta = \tan^{-1}\left(\frac{2\omega\beta}{\omega_0^2 - \omega^2} \right) \tag{5.31}$$

and the resonance frequency is:

$$\omega_0 = \omega. \tag{5.32}$$

At resonance:

$$\frac{d}{d\omega}\left(\frac{\alpha}{\left(\left(\omega_0^2 - \omega^2\right)^2 + 4\omega^2\beta^2 \right)^{1/2}} \right)\Bigg|_{\omega=\omega_r} = 0. \tag{5.33}$$

The quality factor of resonance is defined by:

$$Q = \frac{\omega_R}{2\beta} \tag{5.34}$$

and is infinite for an undamped oscillator. On the left hand side of Eq.(5.29) there is a double derivative in time, a single derivative in time, and a term linear in x. The Coulomb law (5.27) can be rewritten as:

$$\begin{aligned} \nabla^2 A^{a0} + \frac{1}{c}\nabla \cdot \frac{\partial \mathbf{A}^a}{\partial t} + \left(\omega^{0a}{}_b \nabla \cdot \mathbf{A}^b - \omega^a{}_b \cdot \nabla A^{0b} \right) + \\ \left(\nabla \omega^{0a}{}_b \cdot \mathbf{A}^b - A^{0b} \nabla \cdot \omega^a{}_b \right) = -\mu_0 \tilde{J}^{0a} \end{aligned} \tag{5.35}$$

On the left hand side there is a double derivative in distance $\nabla^2 A^{a0}$, single derivatives in distance, $(\omega^{0a}{}_b \nabla \cdot - \omega^a{}_b \cdot \nabla A^{0b})$, and terms linear in the potential, $\left(\nabla \omega^{0a}{}_b\right) \cdot \mathbf{A}^b - A^{0b} \nabla \cdot \omega^a{}_b$. The driving term appears on the right hand side of Eq.(5.34). There is also a time dependent term $\frac{1}{c}\nabla \cdot \frac{\partial \mathbf{A}^a}{\partial t}$ on the left hand side.

In the rest of this section Eq.(5.34) is simplified with approximations to the point where it becomes analytically soluble. The latter solution is applied in Section 5.3 to the H atom. To proceed, a quasi electrostatic situation is used in which the vector potential is assumed to be very small. Eq.(5.34) the reduces to an equation in the scalar potential ϕ^a:

$$\nabla^2 \phi^a - \boldsymbol{\omega}^a{}_b \cdot \nabla \phi^b - \left(\nabla \cdot \boldsymbol{\omega}^a{}_b \right) \phi^b = -\frac{\rho^a}{\epsilon_0} \tag{5.36}$$

In order to produce resonance amplification, it must be assumed that the charge density is oscillatory, for example:

$$\rho^a = -\rho^a (0) \cos (\kappa Z) \tag{5.37}$$

The meaning of the index on the scalar potential is clarified if we consider it to be complex valued, whereby for example:

$$\phi^{(1)} = \phi^{(2)*} = \frac{1}{\sqrt{2}} (1 - i) e^{i(\omega t - \kappa Z)} \tag{5.38}$$

If the scalar potential is real-valued the indices can be omitted. For simplicity of argument define the vector spin connection through only one of its elements, for example:

$$\boldsymbol{\omega} = -\boldsymbol{\omega}^0{}_0 \tag{5.39}$$

Eq.(5.12) then reduces to its simplest form:

$$\mathbf{E} = -\left(\nabla + \boldsymbol{\omega} \right) \phi. \tag{5.40}$$

The term $\boldsymbol{\omega}\phi$ adds a term to the electric field as described in the introduction to this paper. The linear inhomogeneous differential equation in its simplest form is therefore:

$$\nabla^2 \phi + \boldsymbol{\omega} \cdot \nabla \phi + (\nabla \cdot \boldsymbol{\omega}) \phi = \frac{\rho}{\epsilon_0} \tag{5.41}$$

Its particular integral is

$$\phi_p (Z) = \frac{\rho^{(0)}}{\epsilon_0} \frac{\cos (\kappa Z - \delta)}{\left(\left(\frac{\partial \omega}{\partial Z} - \kappa^2 \right)^2 + \omega^2 \kappa^2 \right)^{1/2}} \tag{5.42}$$

where:

$$\delta = \tan^{-1} \left(\frac{\omega \kappa}{\frac{\partial \omega}{\partial Z} - \kappa^2} \right) \tag{5.43}$$

At resonance:

$$\kappa_0^2 = \frac{\partial \omega}{\partial Z} \tag{5.44}$$

and the scalar potential ϕ becomes highly oscillatory. Eq.(5.41) may be rewritten as

$$\phi_p (Z) = \frac{e}{4\pi\epsilon_0 V_0} \frac{\cos (\kappa Z - \delta)}{\left(\left(\frac{\partial \omega}{\partial Z} - \kappa^2 \right)^2 + \omega^2 \kappa^2 \right)^{1/2}} \tag{5.45}$$

where e is charge and V_0 is a volume. The potential energy from Eq.(5.44) is:

$$V_E = -e\phi_p (Z) \tag{5.46}$$

When:

$$\kappa = 0 \tag{5.47}$$

Eq.(5.45) reduces to:

$$V_E = -\frac{e^2}{4\pi\epsilon_0 r} \tag{5.48}$$

if:

$$r := V_0 \frac{\partial \omega}{\partial Z} \tag{5.49}$$

5.3 Application to the H atom

It is well known that the H atom is accurately described in the vast majority of applications [20] by the Schrödinger equation with the Coulomb law limit (5.47):

$$\hat{H}\psi = E\psi \tag{5.50}$$

where the hamiltonian operator is:

$$\hat{H} = -\frac{\hbar^2}{2\mu}\nabla^2 + V_E \tag{5.51}$$

Here μ is the reduced mass:

$$\mu = \frac{m_e m_p}{m_e + m_p} \tag{5.52}$$

where m_p and m_e are masses of proton and electron respectively. These equations are written in S.I. units. The Schrödinger equation of the H atom is therefore:

$$-\frac{\hbar^2}{2\mu}\nabla^2\psi - \frac{e^2}{4\pi\epsilon_0 r}\psi = E\psi \tag{5.53}$$

In spherical polar coordinates [20]:

$$\frac{1}{r}\frac{\partial^2}{\partial r^2}r\psi + \frac{1}{r^2}\Lambda^2\psi + \frac{\mu e^2}{2\pi\epsilon_0\hbar^2 r}\psi = -\frac{2\mu E}{\hbar^2}\psi \tag{5.54}$$

and the wave-function is the product of radial and angular parts:

$$\psi(r,\theta,\phi) = R(r)\,Y(\theta,\phi) \tag{5.55}$$

These are respectively the associated Laguerre polynomials and spherical harmonics with an effective potential:

$$V_{\text{eff}} = -\frac{e^2}{4\pi\epsilon_0 r} + \frac{e(l+1)\hbar^2}{2\mu r^2} \tag{5.56}$$

These equations define the energy levels of the H atom in terms of the principal quantum number n:

$$E = -\frac{\mu e^4}{32\pi^2\epsilon_0^2\hbar^2}\cdot\frac{1}{n^2} \tag{5.57}$$

The complete atomic orbitals of H are:

$$\psi_{nlm_e}(r,\theta,\phi) = R_{ne}(r)\,Y_{em_e}(\theta,\phi) \tag{5.58}$$

These are the s, p, d, f, ... atomic orbitals defined respectively by $l = 0, 1, 2, 3$, ... In the vast majority of applications these are well known to be defined by the limit (5.47) of the resonant Coulomb law. The complete resonant Coulomb law will change all these orbitals, and at resonance will ionize the H atom by resonance, breaking it into a free electron and a free proton. The energy required to do this is well known [20] to be 13.6 eV:

$$\Delta E = \frac{\mu e^4}{32\pi^2\epsilon_0^2\hbar^2} = 13.6eV = 2.2 \times 10^{-18}J \tag{5.59}$$

This is the potential energy needed for a bound electron to be removed to infinity. If an electron is given more than this energy by tuning as in Eq.(5.43), its state becomes an unbound state, an unquantized continuum state [20]. Another effect of tuning to resonance through Eq.(5.43) is to lift the well known n - fold degeneracy of the H atom. So the spectral lines of the H atom will be affected by resonance.

In order to produce the spectrum the potential energy (5.47) must be used in Eq.(5.50) and the resulting Schrödinger equation solved numerically. Experimental methods must be found to produce the resonance (5.43) in the H atom. This thought model is intended only as a plausible mechanism for explaining the results of ref. [18] in the simplest possible approximation. The small driving force oscillations in the H atom can be modeled on the well known phenomenon [20] of zitterbewegung (jitterbugging), known to exist from quantum electrodynamics. If a molecule such as hydrogen deuteride (HD) is considered, the HD bond vibration may be used as the driving force. In more complicated molecules there are several vibrational and rotational frequencies as is well known. It is also well known that a molecule can be ionized by tuning a laser to one of these frequencies. Instead of using a laser, the spin connection of space-time has been used in this paper. The end result is a surge in the scalar potential for a small oscillatory driving charge/current density within the molecule. Thus the output power is much greater than the input power without any violation of the Noether Theorem. In the experimental work of ref. [18] this ratio has been observed in a scientifically repeatable manner to be of the order 100,000 or more. There is no explanation for this observation in the standard model but there is a straightforward explanation in ECE theory. The work in this paper must be made more precise, and applied to engineering situations of practical interest, but this paper illustrates the principle with a thought model based on the H atom.

Acknowledgements The British Parliament, Prime Minister and Head of State are thanked for the award of a Civil List Pension in recognition of distinguished service to Britain in science, and the staff and environment of AIAS are thanked for many interesting discussions.

Bibliography

[1] M. W. Evans, Generally Covariant Unified Field Theory (Abramis, 2005, softback), vol. One.

[2] M. W. Evans, ibid., vols. 2 - 4 (Abramis, 2006 and 2007, preprints on www.aias.us and www.atomicprecision.com).

[3] L. Felker, The Evans Equations of Unified Field Theory (preprints on www.aias.us and www.atomicprecision.com). ; H. Eckardt and L. Felker, papers on these websites.

[4] M. W. Evans, Generally Covariant Dynamics (paper 55 of the ECE series, preprints on www.aias.us and www.atomicprecision.com).

[5] M. W. Evans, Geodesics and the Aharonov Bohm Effects in ECE Theory (paper 56 of the ECE series, preprints on www.aias.us and www.atomicprecision.com).

[6] M. W. Evans, Canonical and Second Quantization in Generally Covariant Quantum Field Theory (paper 57 of the ECE series, preprints on www.aias.us and www.atomicprecision.com).

[7] M. W. Evans, The Effect of Torsion on the Schwarzschild Metric and Light Deflection due to Gravitation (paper 58 of the ECE series, preprints on www.aias.us and www.atomicprecision.com.).

[8] M. W. Evans et al., papers and letters in Found. Phys. Lett. and Found Phys., 1994 to present.

[9] M. W. Evans (ed.), Modern Non-linear Optics a special topical issue in three parts of I. Prigogine and S. A. Rice (series eds.), Advances in Chemical Physics (Wiley-Interscience, New York, 2001, 2nd. Edition), vols. 119(1) to 119(3).

[10] M. W. Evans and L. B. Crowell, Classical and Quantum Electrodynamics and the $\boldsymbol{B}^{(3)}$ Field (World Scientific, Singapore, 2001).

[11] M. W. Evans and J.-P. Vigier, The Enigmatic Photon (Kluwer, Dordrecht, 1994 to 2002, hardback and softback), vols. 1 - 5.

[12] M. W. Evans and A. A. Hasanein, The Photomagneton in Quantum Field Theory (World Scientific, Singapore, 1994).

[13] M. W. Evans and S. Kielich (eds.), first edition of ref. (9) (Wiley Interscience, New York, 1992, reprinted 1993 and 1997), vols. 85(1) to 85(3).

[14] P. H. Pinter, What is Life? (Abramis, 2006).

[15] M. W. Evans, The Photon's Magnetic Field, Optical NMR Spectroscopy (World Scientific, 1992).

[16] M. W. Evans, Physica B, 182, 227, 237 (1992).

[17] Feedback sites for www.aias.us and www.atomicprecision.com, indicating essentially unanimous acceptance of the ECE theory worldwide.

[18] Mexican group work associated with A.I.A.S, communications and independent observations 2000 to 2006.

[19] J. B. Marion and S. T. Thornton, Classical Dynamics of Particles and Systems (HB Publishers, New York, 1998, 3rd edition) chapter three.

[20] P. W. Atkins, Molecular Quantum Mechanics (Oxford Univ. Press, 1983, 2nd edition).

Chapter 6

Application of Einstein-Cartan-Evans (ECE) Theory to atoms and molecules: Free electrons at resonance

(Paper 60)
by
Myron W. Evans Alpha Institute for Advanced Study (AIAS).
(emyrone@aol.com www.aias.us, www.atomicprecision.com)

Abstract

The ECE theory is applied to atoms and molecules, using the hydrogen (H) atom as an example. The ECE wave equation is used in the non-relativistic quantum limit (Schrödinger equation) but the generally covariant Coulomb law is used, incorporating the spin connection of ECE space-time. All Coulombic interactions in atoms and molecules must be generally covariant, meaning that the spin connection must be used. This realizations opens up many new possibilities, in particular the ionization of the atom or molecule by tuning the spin connection. This is not possible in the standard model, even on a conceptual level, because the latter in electrodynamics is a theory of special relativity with no spin connection.

Keywords: Einstein Cartan Evans (ECE) field theory; spin connection, atoms and molecules, H atom, space-time resonance, new energy.

6.1 Introduction

Recently [1]- [17] a generally covariant unified field theory has been proposed and accepted in the natural sciences [18] as being a mathematically correct and workable unified field theory, with major consequences. In this paper the ECE wave equation is applied in some detail to the well known quantum theory of atoms and molecules. The most important consequence for electrical engineering is that free electrons can be produced when the spin connection of ECE field theory is tuned to resonance. This means that there is a novel source of electric power available from materials and/or circuits of the right design, as proven experimentally recently [19]. This fact is illustrated in this paper for the hydrogen (H) atom. In Section 6.2 the Dirac and Schrödinger equations are recovered from the generally covariant ECE wave equation. In Section 6.3 the Coulomb law used in the Schrödinger equation of atoms and molecules is made generally covariant by incorporating the spin connection within the context of ECE theory. Finally in Section 6.4 it is shown how the H atom may be ionized by tuning the spin connection to resonance, at which resonant kinetic energy is inputted to the atom or molecule from space-time. If this kinetic energy exceeds the ionization energy the electrons break free from the protons of the nucleus. The free electrons thus produced may be used for the generation of electric power.

6.2 Dirac and Schrödinger equations from the ECE wave equation

The ECE wave equation [1]- [17] is:

$$(\Box + kT)q^a_\mu = 0 \tag{6.1}$$

where

$$R = -kT = q^\lambda_a \partial^\mu \left(\Gamma^\nu_{\mu\lambda} q^a_\nu - \omega^a{}_{\mu b} q^b_\lambda\right) \tag{6.2}$$

Here k is Einsteins constant, T is the index reduced canonical energy-momentum density, q^a_μ is the tetrad form, $\Gamma^\nu_{\mu\lambda}$ is the general connection of Riemann geometry, and $\omega^a{}_{\mu b}$ is the spin connection of Cartan geometry. Using Einstein's equivalence principle [20] Eq.(6.1) must reduce to equations of special relativity when there is no gravitational field present. The free particle Dirac equation is recovered from the ECE wave equation in the limit:

$$kT = \left(\frac{mc}{\hbar}\right)^2 = \frac{1}{\lambda^2} \tag{6.3}$$

where m is the mass of the free fermion (e.g. an electron), c is the speed of light, $\hbar$ is the reduced Planck constant and λ is the Compton wavelength. The free fermion Dirac equation is therefore:

$$\left(\Box + \frac{m^2c^2}{\hbar^2}\right) q^a_\mu = 0 \tag{6.4}$$

The Dirac spinor is therefore the tetrad q^a_μ in the limit of special relativity defined by Einstein's equivalence principle. The Dirac spinor is therefore defined by the equation that defines the tetrad in $SU(2)$ representation space:

$$\begin{bmatrix} V^R \\ V^L \end{bmatrix} = \begin{bmatrix} q_1^R & q_2^R \\ q_1^L & q_2^L \end{bmatrix} \begin{bmatrix} V^1 \\ V^2 \end{bmatrix} \tag{6.5}$$

Adopting the usual [21] field particle theory notation for the Dirac spinor:

$$\psi = \begin{bmatrix} \zeta^R \\ \zeta^L \end{bmatrix} \tag{6.6}$$

where the Pauli spinors are:

$$\zeta^R = \begin{bmatrix} q_1^R \\ q_2^R \end{bmatrix}, \quad \zeta^L = \begin{bmatrix} q_1^L \\ q_2^L \end{bmatrix} \tag{6.7}$$

So the Dirac equation is the familiar wave equation [21]:

$$\left(\Box + \frac{m^2c^2}{\hbar^2}\right)\psi = 0 \tag{6.8}$$

This can be factorized into the first order differential equation [1]- [17]:

$$\left(i\gamma^\mu \partial_\mu - \frac{mc}{\hbar}\right)\psi = 0 \tag{6.9}$$

where γ^μ is the Dirac matrix. In vector notation and in S.I. units equation (6.9) becomes two simultaneous equations:

$$(E + c\boldsymbol{\sigma}\cdot\mathbf{p})\zeta^L(\mathbf{p}) = mc^2\zeta^R(\mathbf{p}) \tag{6.10}$$

$$(E - c\boldsymbol{\sigma}\cdot\mathbf{p})\zeta^R(\mathbf{p}) = mc^2\zeta^L(\mathbf{p}) \tag{6.11}$$

where:

$$\zeta^L(\mathbf{0}) = \zeta^R(\mathbf{0}) \tag{6.12}$$

Here E is the total relativistic energy, $\boldsymbol{p}$ is the relativistic momentum, and $\boldsymbol{\sigma}$ denotes the Pauli matrix. These well known vector equations of relativistic quantum mechanics are therefore limits of the ECE wave equation in the absence of a gravitational interaction between fermions.

The Schrödinger equation is the non-relativistic limit of the Dirac equation. However the former equation is written in $O(3)$ representation space and the latter in $SU(2)$ representation space with half integral fermion spin and Fermi-Dirac statistics in thermodynamics. So the link between the two equations cannot be forged in a trivial manner. However, the Dirac equation (6.8) is a Klein Gordon equation [21] for each component of the tetrad, i.e. for $q_1^R, q_2^R, q_1^L, q_2^L$. Each Klein Gordon equation is formed from the Einstein equation of special relativity using the fundamental operator equivalence:

$$p^\mu = \left(\frac{E}{c}, \boldsymbol{p}\right) = i\hbar\partial_\mu = i\hbar\left(\frac{1}{c}\frac{\partial}{\partial t}, -\boldsymbol{\nabla}\right) \tag{6.13}$$

In this way, the classical limit of the Dirac equation is found through the Klein Gordon equation and is the Einstein equation:

$$E^2 = c^2p^2 + m^2c^4 \tag{6.14}$$

where:

$$E = \gamma mc^2, \quad E_0 = mc^2, \quad \gamma = \left(1 - \frac{v^2}{c^2}\right)^{-1/2} \tag{6.15}$$

Here E is the total relativistic energy, E_0 is the rest energy, and

$$\boldsymbol{p} = \gamma m \boldsymbol{v} \tag{6.16}$$

is the relativistic momentum. From Eq.(6.16) it is seen that:

$$\begin{aligned} p^2c^2 &= \gamma^2 m^2 c^4 \frac{v^2}{c^2} \\ &= \gamma^2 m^2 c^4 \left(1 - \frac{1}{\gamma^2}\right) \\ &= \gamma^2 m^2 c^4 - m^2 c^4 \end{aligned} \tag{6.17}$$

The relativistic kinetic energy is obtained from Eq.(6.16) and is:

$$\begin{aligned} T &= mc^2(lc\gamma - 1) \\ &= mc^2 \left(1 - \frac{v^2}{c^2}\right)^{-1/2} - mc^2 \\ &\sim mc^2 \left(1 + \frac{1}{2}\frac{v^2}{c^2} + \cdots\right) - mc^2 = \frac{p^2}{2m} \end{aligned} \tag{6.18}$$

when $v \ll c$. Eq.(6.18) is the Newtonian kinetic energy and in Eq.(6.18) p is the Newtonian momentum magnitude:

$$p = mv \tag{6.19}$$

The free particle Schrödinger equation is obtained from

$$T = \frac{p^2}{2m} \tag{6.20}$$

using Eq.(6.13). So:

$$-\frac{\hbar^2}{2m}\nabla^2\psi = E\psi = i\hbar\frac{\partial\psi}{\partial t} \tag{6.21}$$

The kinetic energy or hamiltonian operator is defined as:

$$\widehat{H} := -\frac{\hbar^2}{2m}\nabla^2 \tag{6.22}$$

so we obtain the familiar:

$$\widehat{H}\psi = E\psi \tag{6.23}$$

In the presence of potential energy V, the hamiltonian operator becomes:

$$\widehat{H} := -\frac{\hbar^2}{2m}\nabla^2 + V \tag{6.24}$$

Unlike the Dirac equation, the Schrödinger equation has no sense of helicity or half integral spin, and when using the Schrödinger equation the orbital angular momentum L is replaced by $L + 2S$ where S is the spin angular momentum.

The hydrogen atom in the standard model is described from Eq.(6.23) by adding a potential energy:

$$V = -\frac{e^2}{4\pi\epsilon_0} \cdot \frac{1}{r} \tag{6.25}$$

from the Coulomb law of the standard model [22]. Here ϵ_0 is the vacuum permittivity, e is the charge on the proton, $-e$ is the charge on the electron, and r is the radial component of the spherical polar coordinate system. In ECE theory the Coulomb law of the standard model is modified to include a spin connection [1]– [17], and this modification can result in resonance ionization of the H atom to give free electrons, as shown later in this paper.

The d'Alembertian operator in Eq.(6.4) is:

$$\begin{aligned} \square = \partial^\mu \partial_\mu &= -\frac{p^\mu p_\mu}{\hbar^2} \\ &= -\frac{1}{\hbar^2}\left(\frac{E^2}{c^2} - p^2\right) \\ &= \frac{1}{c^2}\frac{\partial^2}{\partial t^2} - \nabla^2 \end{aligned} \tag{6.26}$$

so

$$-\hbar^2 \nabla^2 q^a_\mu = \left(\frac{E^2}{c^2} - m^2c^2\right) q^a_\mu \tag{6.27}$$

This is a special relativistic form of Eq.(6.23). Multiplying Eq.(6.27) by c^2 and using Eq.(6.14) it is found that Eq.(6.27) is:

$$-c^2\hbar^2\nabla^2 q^a_\mu = c^2 p^2 q^a_\mu \tag{6.28}$$

i.e.

$$-\hbar^2\nabla^2 q^a_\mu = p^2 q^a_\mu \tag{6.29}$$

where:

$$p^2 = \frac{E^2}{c^2} - m^2c^2 \tag{6.30}$$

To describe the H atom with the standard Dirac equation, a Coulombic term is added to Eq.(6.27) to give:

$$-\left(\frac{\hbar\nabla^2}{2m} + \frac{e^2}{4\pi\epsilon_0 r}\right) q^a_\mu = \frac{1}{2m}\left(\frac{E^2}{c^2} - m^2c^2\right) q^a_\mu \tag{6.31}$$

In the relativistic quantum limit of ECE theory a resonance Coulomb law can be added to Eq.(6.31). The most rigorous method however is to use the ECE field equations [1]– [17] in the ECE wave equation.

6.3 The resonant laws of classical electrodynamics in general relativity and ECE theory

The Coulomb and Ampére Maxwell laws in ECE theory are:

$$\boldsymbol{\nabla} \cdot \boldsymbol{E}^a = \mu_0 c \widetilde{J}^{0a} \tag{6.32}$$

$$\nabla \times \boldsymbol{B}^a - \frac{1}{c^2}\frac{\partial \boldsymbol{E}^a}{\partial t} = \frac{\mu_0}{c}\widetilde{\boldsymbol{J}}^a \tag{6.33}$$

where

$$\boldsymbol{E}^a = -\frac{\partial \boldsymbol{A}^a}{\partial t} - \nabla\phi^a - c\omega^{0a}{}_b\boldsymbol{A}^b + \boldsymbol{\omega}^a{}_b\phi_b \tag{6.34}$$

$$\boldsymbol{B}^a = \nabla \times \boldsymbol{A}^a - \boldsymbol{\omega}^a{}_b \times \boldsymbol{A}^b \tag{6.35}$$

Here E^a is the electric field strength, $\boldsymbol{B}^a$ is the magnetic flux density, μ_0 is the vacuum permeability, ϕ^a is the scalar potential in volts, $\boldsymbol{A}^a$ is the vector potential, and the inhomogeneous four-current [1]– [17] is defined by:

$$\widetilde{J}^{\mu a} = (\widetilde{J}^{0a}, \widetilde{\boldsymbol{J}}^a) \tag{6.36}$$

For simplicity of notation and development, the polarization indices may be dropped. This process is equivalent to the use of a simplified form of the spin connection, which in general is the four-vector:

$$\omega^a{}_{\mu b} = (\omega^a{}_{0b}, -\boldsymbol{\omega}^a{}_b) \tag{6.37}$$

The spin connection is a fundamental feature of general relativity [1]– [17] and indicates that the field is the frame itself. The electromagnetic field is spinning space-time and the gravitational field is curving space-time. In the standard model the electromagnetic field is a separate entity superimposed on a flat or Minkowski space-time - the Lorentz covariant Maxwell Heaviside field theory of special relativity. In order to unify electrodynamics with gravitation and other fields the ECE theory is needed [1]– [17]. In this simplified notation Eqs.(6.32) to (6.36) become:

$$\nabla \cdot \boldsymbol{E} = \frac{\rho}{\epsilon_0} \tag{6.38}$$

$$\nabla \times \boldsymbol{B} - \frac{1}{c^2}\frac{\partial E}{\partial t} = \frac{\mu_0}{c}\widetilde{\boldsymbol{J}} \tag{6.39}$$

$$\boldsymbol{E} = -\frac{\partial A}{\partial t} - \nabla\phi - c\omega^0\boldsymbol{A} + \phi\boldsymbol{\omega} \tag{6.40}$$

$$\boldsymbol{B} = \nabla \times \boldsymbol{A} - \boldsymbol{\omega} \times \boldsymbol{A} \tag{6.41}$$

where ρ is the charge density in coulombs per cubic meter, and $\widetilde{\boldsymbol{J}}$ is the current density. The four current in this case is:

$$\widetilde{J}^\mu = (c\rho, \widetilde{\boldsymbol{J}}) \tag{6.42}$$

Here ω^0 is the time-like and $\boldsymbol{\omega}$ the space-like parts of the complete spin connection four-vector. Thus $\boldsymbol{\omega}$ is a vector in three dimensions, and ω^0 is a scalar.

The Coulomb law is a law of electro-statics described by the scalar potential and in ECE theory is described by the equations:

$$\nabla \cdot \boldsymbol{E} = \frac{\rho}{\epsilon_0} \tag{6.43}$$

$$\boldsymbol{E} = (-\nabla + \boldsymbol{\omega})\phi \tag{6.44}$$

These can be combined to give the resonant Poisson equation [1]– [17]:

$$\nabla^2\phi - \nabla \cdot (\boldsymbol{\omega}\phi) = -\frac{\rho}{\epsilon_0} \tag{6.45}$$

At resonance ϕ is greatly amplified and kinetic energy resonance can occur [23] and if this law is incorporated into the Schrödinger equation (6.23), it may be shown numerically that the atom ionizes if the resonant kinetic energy is greater than the ionization energy of H, 13.6 eV [22]. The free electrons from ionization are produced by energy form space-time, and the free electrons are a source of electric power for engineering applications. This type of resonance has recently been shown to exist in nature 6.24, using circuits and materials more complicated than H. The latter is used to illustrate the scientific principles at work.

If on the other hand the scalar potential is neglected and attention is confined to the vector potential only, then:

$$\boldsymbol{E} = -\frac{\partial A}{\partial t} - c\omega^0 \boldsymbol{A} \tag{6.46}$$

and we obtain the following resonance equation from the Ampére Maxwell law (6.39):

$$\begin{aligned} &\frac{1}{c^2}\frac{\partial^2 \boldsymbol{A}}{\partial t^2} + \frac{\omega^0}{c}\frac{\partial \boldsymbol{A}}{\partial t} + (\boldsymbol{\nabla}\cdot\boldsymbol{\omega})\boldsymbol{A} - (\boldsymbol{A}\cdot\boldsymbol{\nabla})\boldsymbol{\omega} \\ &- \nabla^2\boldsymbol{A} + \boldsymbol{\nabla}(\boldsymbol{\nabla}\cdot\boldsymbol{A}) - \boldsymbol{\omega}(\boldsymbol{\nabla}\cdot\boldsymbol{A}) + (\boldsymbol{\omega}\cdot\boldsymbol{\nabla})\boldsymbol{A} = \frac{\mu_0}{c}\widetilde{\boldsymbol{J}} \end{aligned} \tag{6.47}$$

If a quasi plane-wave approximation is used:

$$\boldsymbol{\nabla}\cdot\boldsymbol{A} \sim 0 \tag{6.48}$$

and if $\widetilde{\boldsymbol{J}}$ can be expressed as a sum of time dependent and r dependent parts:

$$\widetilde{\boldsymbol{J}} \sim \widetilde{\boldsymbol{J}}(t) - \widetilde{\boldsymbol{J}}(r) \tag{6.49}$$

two resonance equations in $\boldsymbol{A}$ are obtained:

$$\frac{1}{c^2}\frac{\partial^2 \boldsymbol{A}}{\partial t^2} + \frac{\omega^0}{c}\frac{\partial \boldsymbol{A}}{\partial t} + (\boldsymbol{\nabla}\cdot\boldsymbol{\omega})\boldsymbol{A} = \frac{\mu_0}{c}\widetilde{\boldsymbol{J}}(t) \tag{6.50}$$

and

$$\nabla^2\boldsymbol{A} - (\boldsymbol{\omega}\cdot\boldsymbol{\nabla})\boldsymbol{A} + (\boldsymbol{A}\cdot\nabla)\boldsymbol{\omega} = \frac{\mu_0}{c}\widetilde{\boldsymbol{J}}(r) \tag{6.51}$$

At resonance, $\boldsymbol{A}$ is greatly amplified and kinetic energy inputted from space-time.

Similarly, the magneto-static ECE equations are:

$$\boldsymbol{\nabla}\cdot\boldsymbol{B}^a = \mu_0\widetilde{j}^{0a} \tag{6.52}$$

$$\boldsymbol{\nabla}\times\boldsymbol{B}^a = \frac{\mu_0}{c}\widetilde{\boldsymbol{J}}^{0a} \tag{6.53}$$

$$\boldsymbol{B}^a = \boldsymbol{\nabla}\times\boldsymbol{A}^a - \boldsymbol{\omega}^a{}_b\times\boldsymbol{A}^b \tag{6.54}$$

which give the equations:

$$\boldsymbol{\nabla}\cdot(\boldsymbol{\omega}^a{}_b\times\boldsymbol{A}^b) = -\mu_0\widetilde{j}^{0a} \tag{6.55}$$

and

$$\boldsymbol{\nabla}\times(\boldsymbol{\nabla}\times\boldsymbol{A}^a - \boldsymbol{\omega}^a{}_b\times\boldsymbol{A}^b) = \frac{\mu_0}{c}\widetilde{\boldsymbol{J}}^a \tag{6.56}$$

Eq.(6.55) is the generally covariant Gauss law of magnetism, and Eq.(6.56) is the generally covariant Ampére law of magnetism. In these equations:

$$\omega^1{}_b \times \boldsymbol{A}^b = \boldsymbol{\omega}^1{}_1 \times \boldsymbol{A}^1 + \boldsymbol{\omega}^1{}_2 \times \boldsymbol{A}^2 + \boldsymbol{\omega}^1{}_3 \times \boldsymbol{A}^3 \\ \text{etc.,} \tag{6.57}$$

If we define:

$$\widetilde{j}^{0a} := \widetilde{j}_1^{0a} + \widetilde{j}_2^{0a} + \widetilde{j}_3^{0a} \tag{6.58}$$

Eq.(6.55) for example splits into three equations:

$$\boldsymbol{\nabla} \cdot (\boldsymbol{\omega}^1{}_1 \times \boldsymbol{A}^1) = -\mu_0 \widetilde{J}_1^{01} \\ \text{etc.,} \tag{6.59}$$

and the notation can be simplified again by considering each equation to be of the form:

$$\boldsymbol{\nabla} \cdot (\boldsymbol{\omega} \times \boldsymbol{A}) = -\mu_0 \widetilde{J}^0 \tag{6.60}$$

Therefore:

$$\boldsymbol{\omega} \times \boldsymbol{A} = -\mu_0 \int \widetilde{j}^0 d\boldsymbol{r} \tag{6.61}$$

Similarly, Eq.(6.56) can be split into three equations of the form:

$$\boldsymbol{\nabla} \times (\boldsymbol{\nabla} \times \boldsymbol{A} - \boldsymbol{\omega} \times \boldsymbol{A}) = \frac{\mu_0}{c} \widetilde{\boldsymbol{J}} \tag{6.62}$$

so we obtain

$$\boldsymbol{\nabla} \times (\boldsymbol{\nabla} \times \boldsymbol{A}) = \frac{\mu_0}{c} \widetilde{\boldsymbol{J}} - \mu_0 \boldsymbol{\nabla} \times \int \widetilde{j}^0 d\boldsymbol{r} \tag{6.63}$$

Using vector identities [25], Eq.(6.62) is the resonance equation:

$$\boldsymbol{\nabla}(\boldsymbol{\nabla} \cdot \boldsymbol{A}) - \nabla^2 \boldsymbol{A} - \boldsymbol{\omega}(\boldsymbol{\nabla} \cdot \boldsymbol{A}) + (\boldsymbol{\omega} \cdot \boldsymbol{\nabla})\boldsymbol{A} + (\boldsymbol{\nabla} \cdot \boldsymbol{\omega})\boldsymbol{A} - (\boldsymbol{A} \cdot \boldsymbol{\nabla})\boldsymbol{\omega} = \frac{\mu_0}{c} \widetilde{\boldsymbol{J}} \tag{6.64}$$

and again $\boldsymbol{A}$ is amplified at resonance, giving novel effects in magnetism. Considering the Z component of $\widetilde{\boldsymbol{J}}$ gives:

$$\frac{\partial^2 A_X}{\partial Z \partial X} + \frac{\partial^2 A_Y}{\partial Z \partial Y} + \omega_X \frac{\partial A_Z}{\partial X} + \omega_Y \frac{\partial A_Z}{\partial Y} - \omega_Z \frac{\partial A_X}{\partial X} - \omega_Z \frac{\partial A_Y}{\partial Y} + \\ \left(\frac{\partial \omega_X}{\partial X} + \frac{\partial \omega_Y}{\partial Y} \right) A_Z - \frac{\partial \omega_Z}{\partial X} A_X - \frac{\partial \omega_Z}{\partial Y} A_Y = \frac{\mu_0}{c} \widetilde{J}_Z \tag{6.65}$$

If it assumed for simplicity that only A_X is non-zero the resonant structure simplifies to:

$$\frac{\partial}{\partial Z} \left(\frac{\partial A_X}{\partial X} \right) - \omega_Z \frac{\partial A_X}{\partial X} - \left(\frac{\partial \omega_Z}{\partial X} \right) A_X = \frac{\mu_0}{c} \widetilde{J}_Z \tag{6.66}$$

producing amplification of A_X at resonance. In this simple example, A_X must be a mathematical function of both X and Z, and ω_Z must be a function of X. At resonance, amplification of the vector potential and the magnetic flux density occurs by tuning the spin connection of space-time. This is a rigorous result of general relativity (ECE theory) applied to the theory of magnetism. The latter is unified with the gravitational field, producing the magneto-gravitational field. These are some simple examples of many types of novel resonant equations of electrodynamics and gravitational theory available from ECE theory.

The generally covariant Coulomb law is of particular importance to the quantum theory of atoms and molecules, the theory of absorption, and to well developed numerical methods such as density functional theory. General covariance in atoms and molecules works its way into all aspects of the subject [22], for example transition electric dipole moment theory, where the interaction hamiltonian [22] is:

$$H^{(1)}(t) = -\boldsymbol{\mu} \cdot \boldsymbol{E}(t) \tag{6.67}$$

The electric field $\boldsymbol{E}$ in general relativity must always be defined through the spin connection, so the important new property of space-time resonance is present throughout the subject of atomic and molecular quantum mechanics. Similarly, the interaction of a magnetic dipole moment $\boldsymbol{m}$ and a magnetic flux density produces the interaction hamiltonian:

$$H^{(2)}(t) = -\boldsymbol{m} \cdot \boldsymbol{B}(t) \tag{6.68}$$

Here again, $\boldsymbol{B}$ must always be defined by Eq.(6.35) in general relativity, and therefore in terms of the spin connection. In the simplest instance, Eq.(6.67) is:

$$H^{(1)}(t) = ezE(t) \tag{6.69}$$

and the eE part of this hamiltonian comes from the resonant Coulomb law in general relativity. The Laporte selection rule [22] for example, is governed by the interaction hamiltonian (6.67). The transition dipole moment is zero unless it is totally symmetric under the symmetry operations of the system. For the Laporte selection rule, this operation is the parity inversion operation:

$$\widehat{P}(\boldsymbol{r}) = -\boldsymbol{r} \tag{6.70}$$

Electric dipole transitions are allowed only if they involve a change of parity. Big charge shifts long distances, giving intense absorption lines 6.22. Time dependent perturbation theory 6.22 shows that the transition rate and the spectral intensity of an absorption line depend on the square of the matrix element of the perturbation, i.e. the square of

$$\langle n_2 l_2 m_{l2} \mid \boldsymbol{\mu} \mid n_1 l_1 m_{l1} \rangle \tag{6.71}$$

in Dirac bracket notation. In ECE theory $H^{(1)}$ depends on the spin connection of space-time so at spin connection resonance or space-time resonance the spectral absorption band is greatly affected. Recent repeatable experiments [24] show beyond scientific doubt that such resonances can be induced in the laboratory. Inside an atom or molecule, the electron can be promoted by space-time resonance not only from the s to the p orbital, but to higher odd-parity orbitals such as the f orbital and so on up to the continuum state, where the electron breaks free of the nucleus, and may be used to generate electric power as a free electron (i.c. clectric) current in a circuit or cable. This novel phenomenon in the natural and life sciences is due to space-time itself, and not due to a device such as a tuned laser.

Therefore the process of absorption by an atom or molecule is governed by:

$$H = -\boldsymbol{\mu} \cdot \boldsymbol{E} \tag{6.72}$$

The electric field strength in this process is always governed by:

$$\boldsymbol{\nabla} \cdot \boldsymbol{E}^a = \frac{\rho^a}{\epsilon_0} \tag{6.73}$$

$$\boldsymbol{E}^a = -\frac{\partial A^a}{dt} - \nabla \phi^a - c\omega^{0a}{}_b \boldsymbol{A}^b + \boldsymbol{\omega}^a{}_b \phi^b \tag{6.74}$$

giving the most general type of resonance equation:

$$-\boldsymbol{\nabla} \cdot \frac{\partial A^a}{dt} - \nabla^2 \phi^a - c\boldsymbol{\nabla} \cdot (\omega^{0a}{}_b \boldsymbol{A}^b) + \boldsymbol{\nabla} \cdot (\boldsymbol{\omega}^a{}_b \phi^b) = \frac{\rho^a}{\epsilon_0} \tag{6.75}$$

The transition dipole moment $\boldsymbol{\mu}$ is defined by the charge density ρ^a of an electron in an orbital. For a plane wave in the standard model, Eq.(6.75) reduces to the Poisson equation:

$$\nabla^2 \phi^a = -\frac{\rho^a}{\epsilon_0} \tag{6.76}$$

and orbital angular momentum is imparted to the electron by an electromagnetic field (photon). This well known theory does not consider the spin connection of generally covariant unified field theory and loses much information. In Eq.(6.75):

$$\boldsymbol{\nabla} \cdot (\phi^b \boldsymbol{\omega}^a{}_b) = (\boldsymbol{\nabla} \cdot \boldsymbol{\omega}^a{}_b)\phi^b + \boldsymbol{\omega}^a{}_b \cdot \boldsymbol{\nabla}\phi^b \tag{6.77}$$

and

$$\boldsymbol{\nabla} \cdot (\omega^{0a}{}_b \boldsymbol{A}^b) = (\boldsymbol{\nabla} \cdot \boldsymbol{A}^b)\omega^{0a}{}_b + \boldsymbol{A}^b \cdot \boldsymbol{\nabla}\omega^{0a}{}_b \tag{6.78}$$

If the vector potential is regarded as being a plane wave in the first approximation, then $\boldsymbol{\omega}^a{}_b$ is dual [1]– [17] to $\boldsymbol{A}^c$, and in this first approximation:

$$\boldsymbol{\nabla} \cdot \frac{\partial \boldsymbol{A}^a}{dt} \sim \boldsymbol{\nabla} \cdot \boldsymbol{A}^b \sim \boldsymbol{\nabla} \cdot \boldsymbol{\omega}^a{}_b \sim 0 \tag{6.79}$$

So Eq.(6.75) becomes the resonance equation:

$$-\nabla^2 \phi^a - c(\boldsymbol{\nabla}\omega^{0a}{}_b) \cdot \boldsymbol{A}^b + \boldsymbol{\omega}^a{}_b \cdot \boldsymbol{\nabla}\phi^b = \frac{\rho^a}{\epsilon_0} \tag{6.80}$$

If for convenience of argument we assume that:

$$c\boldsymbol{A}^b = \boldsymbol{\phi}^b \tag{6.81}$$

and that $\boldsymbol{\omega}^a{}_b$ is negative in sign, then we obtain:

$$\nabla^2 \phi^a + (\boldsymbol{\nabla}\omega^{0a}{}_b) \cdot \boldsymbol{\phi}^b + \boldsymbol{\omega}^a{}_b \cdot \boldsymbol{\nabla}\phi^b = \frac{\rho^a}{\epsilon_0} \tag{6.82}$$

or

$$\nabla^2 \phi^a + g\boldsymbol{\phi}^c \cdot \boldsymbol{\nabla}\phi^b + g\boldsymbol{\nabla}\phi^c \cdot \boldsymbol{\phi}^b = \frac{\rho^a}{\epsilon_0} \tag{6.83}$$

where g is a constant. This again is a simple resonance equation [1]– [17] showing that resonant kinetic energy can be inputted into the atom, molecule or material from space-time itself. The latter gives an unlimited amount of electric power in theory. However there may be limitations of design and efficiency as usual in engineering. Designs must be optimized as usual by experimentation and practical experience [24].

An important application of the Coulombic term in the Schrödinger equation is the well used density functional theory and computational code. The Hartree term for example is

$$V_H = \int \frac{e^2 n_s(\boldsymbol{r}')}{\mid \boldsymbol{r} - \boldsymbol{r}' \mid} d^3 r' \tag{6.84}$$

and is used in the effective single particle potential V_S of the Kohn Sham [26] equations of the auxiliary system of density functional methods. The Hartree term describes the electron to electron Coulomb repulsion and so obeys a resonance equation in ECE theory. If for simplicity we omit the polarization superscripts in Eq.(6.75) this resonance equation is:

$$\nabla^2 V_H - \boldsymbol{\omega} \cdot \boldsymbol{\nabla} V_H - (\boldsymbol{\nabla} \cdot \boldsymbol{\omega}^a{}_b) V_H = -\frac{e\rho}{\epsilon_0} \tag{6.85}$$

so space-time resonance [1]– [17] amplifies the electron to electron repulsion, causing ionization if the kinetic energy inputted from space-time at resonance exceeds the binding energy of the atom, molecule or material. The task of the engineer is to devise a method to induce such resonance, and this has been shown to be possible 6.24. Space-time resonance may also be the explanation for the well known Tesla coil and the well known electron avalanche phenomena in electrical engineering. Therefore space-time resonance has been known to electrical engineers and inventors but has only now found an explanation from the rigorous application of general relativity to classical electrodynamics (ECE theory).

In the standard model of classical electrodynamics there is no space-time resonance because there is no spin connection, Maxwell-Heaviside theory being a theory of flat spacetime (special relativity and Lorentz covariant). The Hartree term is based on the solution to the standard model Poisson equation:

$$\nabla^2 \phi = -\frac{\rho}{\epsilon_0} \tag{6.86}$$

The solution is well known to be [27], in S.I units:

$$\phi = \frac{1}{4\pi\epsilon_0} \int \frac{\rho(\boldsymbol{r}')}{\mid \boldsymbol{r} - \boldsymbol{r}' \mid} d^3 r' \tag{6.87}$$

The Hartree term in S.I. units is therefore:

$$\phi_H = \frac{1}{4\pi\epsilon_0} \int \frac{e n_s(\boldsymbol{r})}{\mid \boldsymbol{r} - \boldsymbol{r}' \mid} d^3 r' \tag{6.88}$$

and the charge density is expressed in terms of a number density n_s with the units of inverse meters cubed. In ECE theory (general relativity), the Hartree potential obeys:

$$-\nabla^2 \phi_H + \boldsymbol{\omega} \cdot \boldsymbol{\nabla} \phi_H + (\boldsymbol{\nabla} \cdot \boldsymbol{\omega}) \phi_H = \frac{e n_s(\boldsymbol{r})}{\epsilon_0} \tag{6.89}$$

At space-time resonance ϕ_H and $n_s(\boldsymbol{r})$ are amplified. ECE theory may be applied to density functional code by using Eq.(6.89) instead of the standard model:

$$\nabla^2 \phi_H = -\frac{e n_s(\boldsymbol{r})}{\epsilon_0} \tag{6.90}$$

Some more discussion of how this process may be coded, graphed and animated is given in Section 6.4.

An H atom is described therefore in the standard model by a combination of the non-relativistic Schrödinger equation and the classical Coulomb law as summarized in the following equations:

$$-\frac{\hbar^2}{2m}\nabla^2\psi = (E-V)\psi, \quad V = -e\phi, \quad \phi = \frac{e}{4\pi\epsilon_0 r} \tag{6.91}$$

The electric field strength between the electron and proton inside the H atom is defined in the standard model by:

$$\boldsymbol{E} = -\boldsymbol{\nabla}\phi, \quad \boldsymbol{\nabla}\cdot\boldsymbol{E} = \frac{\rho}{\epsilon_0} \tag{6.92}$$

In the non-relativistic quantum limit of ECE theory [1]– [17] the Schrödinger equation is used with a potential defined by the resonance equation:

$$\nabla^2\phi - (\boldsymbol{\nabla}\cdot\boldsymbol{\omega})\phi - \boldsymbol{\omega}\cdot\boldsymbol{\nabla}\phi = -\frac{\rho}{\epsilon_0} \tag{6.93}$$

i.e. the equation:

$$\nabla^2\phi = -\frac{\rho}{\epsilon_0} + (\boldsymbol{\nabla}\cdot\boldsymbol{\omega}) + \boldsymbol{\omega}\cdot\boldsymbol{\nabla}\phi \tag{6.94}$$

In the standard model, Eq.(6.94) is the Poisson equation:

$$\nabla^2\phi = -\frac{\rho}{\epsilon_0} \tag{6.95}$$

So the effect of the spin connection $\boldsymbol{\omega}$ is to add a charge density:

$$\rho_1 = -\epsilon_0 \boldsymbol{\nabla}\cdot(\phi\boldsymbol{\omega}) \tag{6.96}$$

In the first approximation ϕ in Eq.(6.94) may be taken to be roughly the Coulomb potential:

$$\phi_C = \frac{1}{4\pi\epsilon_0}\int \frac{\rho(\boldsymbol{r}')}{\mid \boldsymbol{r}-\boldsymbol{r}'\mid} d^3r' \tag{6.97}$$

so

$$\rho_1 \sim -\boldsymbol{\nabla}\cdot\left(\frac{\boldsymbol{\omega}}{4\pi}\int \frac{\rho(\boldsymbol{r}')}{\mid \boldsymbol{r}-\boldsymbol{r}'\mid} d^3r'\right) \tag{6.98}$$

where $\rho(\boldsymbol{r}')$ is the Coulombic charge density used in density functional code for the H atom. Therefore, in this rough approximation:

$$\nabla^2\phi \sim -\frac{1}{\epsilon_0}(\rho + \rho_1) \tag{6.99}$$

and

$$\phi \sim \frac{e_{eff}}{4\pi\epsilon_0 r} \tag{6.100}$$

where e_{eff} is an effective charge perturbed by the spin connection. At resonance, ϕ becomes very large, so a perturbation approximation can only be used self-consistently in an off-resonant condition. The space-time resonance can be thought of as:

$$\boldsymbol{r} \rightarrow \boldsymbol{r}' \tag{6.101}$$

in Eq.(6.98), defining a "radius of resonance".

Space-time resonance affects all the well known [22] phenomena of molecular and material quantum mechanics used in physics, chemistry and the life sciences. To end this section this fact is illustrated with reference to exchange energy and its concomitant Fermi repulsion and with the theory of absorption of a photon. In ECE theory, absorption of a photon is again described by a resonance equation and is understood entirely in terms of geometry, as required by general relativity. The exchange energy occurs in the simplest instance in helium (He), which has two electrons and two protons. The Schrödinger equation of the He atom [22] is:

$$\widehat{H}\psi = E\psi \tag{6.102}$$

$$\widehat{H} = -\frac{\hbar^2}{2m_e}(\nabla_1^2 + \nabla_2^2) - \frac{2e^2}{4\pi\epsilon_0 r_1} - \frac{2e^2}{4\pi\epsilon_0 r_2} + \frac{e^2}{4\pi\epsilon_0 r_{12}} \tag{6.103}$$

The equation is a function of the coordinates of the two electrons, r_1 and r_2:

$$\widehat{H}(\boldsymbol{r}_1, \boldsymbol{r}_2)\psi = E(\boldsymbol{r}_1, \boldsymbol{r}_2)\psi \tag{6.104}$$

It is not possible to solve this equation analytically and various levels of approximation are used as is well known: for example perturbation theory, density functional theory, *ab initio* theory. In perturbation theory [22] a rough approximation is attempted, taking the perturbation to be the electron-electron repulsion. This approximation is expressed as:

$$H^{(0)} = H_1 + H_2, \quad H_i = -\frac{\hbar^2}{2m_e}\nabla_i^2 - \frac{2e^2}{4\pi\epsilon_0 r_i} \tag{6.105}$$

where

$$H_i\psi(\boldsymbol{r}_i) = E_i\psi(\boldsymbol{r}_i) \tag{6.106}$$

Thus

$$\begin{aligned}(H_1 + H_2)\psi(\boldsymbol{r}_1, \boldsymbol{r}_2) &= (H_1 + H_2)\psi(\boldsymbol{r}_1)\psi(\boldsymbol{r}_2)\\ &= H_1\psi(\boldsymbol{r}_1)\psi(\boldsymbol{r}_2) + \psi(\boldsymbol{r}_1)H_2\psi(\boldsymbol{r}_2)\\ &= E_1\psi(\boldsymbol{r}_1)\psi(\boldsymbol{r}_2) + E_2\psi(\boldsymbol{r}_1)\psi(\boldsymbol{r}_2)\\ &= (E_1 + E_2)\psi(\boldsymbol{r}_1, \boldsymbol{r}_2)\end{aligned} \tag{6.107}$$

The unperturbed wave-function of He is a product of two H type wave-functions:

$$\psi(\boldsymbol{r}_1, \boldsymbol{r}_2) = \psi_{n_1 l_1 m_{l1}}(\boldsymbol{r}_1)\psi_{n_2 l_2 m_{l2}}(\boldsymbol{r}_2) \tag{6.108}$$

with energies:

$$E = -4hcR_\infty\left(\frac{1}{n_1^2} + \frac{1}{n_2^2}\right) \tag{6.109}$$

The electron to electron repulsion in He introduces the first order correction [22]:

$$E^{(1)} = J = \langle n_1 l_1 m_{l_1}\,;\, n_2 l_2 m_{l_2} \mid \frac{e^2}{4\pi\epsilon_0 r_{12}} \mid n_1 l_1 m_{l_1}\,;\, n_2 l_2 m_{l_2}\rangle \tag{6.110}$$

defining the Coulomb integral:

$$J = \frac{e^2}{4\pi\epsilon_0 r_{12}} \int \mid \psi_1(\boldsymbol{r}_1)^2 \mid \frac{1}{r_{12}} \mid \psi_2(\boldsymbol{r}_2) \mid^2 d\tau_1 d\tau_2 \tag{6.111}$$

In ECE theory however the repulsion term is governed by a space-time resonance equation:

$$\nabla^2\phi(\boldsymbol{r}_{12}) - \boldsymbol{\omega}\cdot\boldsymbol{\nabla}\phi(\boldsymbol{r}_{12}) - (\boldsymbol{\nabla}\cdot\boldsymbol{\omega})\phi(\boldsymbol{r}_{12}) = -\frac{\rho}{\epsilon_0} \tag{6.112}$$

where ϕ is the repulsion potential, giving rise to a repulsion potential energy:

$$V(\boldsymbol{r}_{12}) = -e\phi(\boldsymbol{r}_{12}) \tag{6.113}$$

At space-time resonance ϕ and V are amplified, and this process causes ionization if the binding energy is exceeded by the repulsion energy. At resonance the perturbation approach is however invalidated, but these remarks illustrate the fact that input of energy from space-time can result in the ionization of He, releasing free electrons for use in electric power generation [24].

The exchange integral is a quantum phenomenon with no classical equivalent, as is well known. In He the wave-functions of the exchange integral are:

$$\psi_\pm(\boldsymbol{r}_1,\boldsymbol{r}_2) = \frac{1}{\sqrt{2}}(\psi_1(\boldsymbol{r}_1)\psi_2(\boldsymbol{r}_2) \pm \psi_2(\boldsymbol{r}_1)\psi_1(\boldsymbol{r}_2)) \tag{6.114}$$

where ψ_1 and ψ_2 are H like for $Z = 2$ [22]. In this case the Coulombic electron to electron repulsion terms in He lift the degeneracy of the two product functions, giving an energy separation $2K$ where the integral K is the quantum correction to the Coulomb integral J. In ECE theory, both the Coulomb and exchange integrals are affected by space-time resonance. The exchange integral is the basis [22] for Fermi hole theory and the electron hole theory used in semiconductors and related materials such as bio-materials. The amplitude vanishes in Fermi hole theory when

$$\boldsymbol{r}_1 = \boldsymbol{r}_2 \tag{6.115}$$

Therefore space-time resonance in ECE theory will affect the whole of semiconductor theory and related areas of physics, chemistry, materials science and engineering. As should be expected, these are all areas in which general relativity should be applied for self consistency in the natural, engineering and life sciences. In the standard model, general relativity is applied only to gravitational theory in physics.

Atomic absorption theory is another area in which general relativity should be applied with ECE theory. Atomic systems can be represented classically [28] as linear oscillators. When electromagnetic radiation falls on matter, atoms and molecules vibrate. A resonant frequency occurs at one of the spectral frequencies of the system, and if this is the same as the light frequency, electromagnetic energy is absorbed, causing the atom or molecule to vibrate with a large amplitude. Large electromagnetic fields are produced (radiated) by the oscillating electric charge inside the atom or molecule. This process is described in ECE theory by Eq.(21.26) of [1] through the minimal prescription. In the presence of an electromagnetic field the ECE Lemma [1]– [17]:

$$\Box q^a_\mu = R q^a_\mu \tag{6.116}$$

is changed to:

$$\Box q^a_\mu = (R + R_1 + R_2) q^a_\mu \tag{6.117}$$

In the Dirac limit the extra curvatures R_1 and R_2 are defined by:

$$| R_1 | = \frac{e^2 A^*_a A^a}{\hbar^2}, \quad | R_2 | = \frac{emc\gamma^a (e^2 A^*_a + A^a)}{\hbar^2} \tag{6.118}$$

This process is effectively a minimal prescription in which $\boldsymbol{\nabla}$ is changed to

$$\boldsymbol{\nabla} \rightarrow \boldsymbol{\nabla} - \frac{ie}{\hbar} \boldsymbol{A} \tag{6.119}$$

Therefore in this case the spin connection is:

$$\boldsymbol{\omega} = -\frac{ie}{\hbar} \boldsymbol{A} \tag{6.120}$$

The modulus of Eq.(6.119) gives:

$$\hbar \kappa = e A^{(0)} \tag{6.121}$$

where

$$\kappa = | \boldsymbol{\omega} | \tag{6.122}$$

is a wave-number and magnitude of the spin connection. Eq.(6.120) describes the absorption of a photon of energy $\hbar\omega$, momentum $\hbar\kappa$ and angular momentum $\pm\hbar$. It can be seen that the photon originates in the spin connection of space-time itself. In this process the kinetic energy operator of the Schrödinger equation is changed as follows:

$$-\frac{\hbar^2}{2m} \nabla^2 \rightarrow -\frac{\hbar^2}{2m} (\boldsymbol{\nabla} - \boldsymbol{\omega})^2 \tag{6.123}$$

Thus:

$$\nabla^2 \psi \rightarrow ((\boldsymbol{\nabla} - \boldsymbol{\omega}) \cdot (\boldsymbol{\nabla} - \boldsymbol{\omega}))\, \psi \tag{6.124}$$

and the process of photon absorption by an atom is described by the resonance equation:

$$-\frac{\hbar^2}{2m} (\nabla^2 - \boldsymbol{\omega} \cdot \boldsymbol{\nabla} - \boldsymbol{\nabla} \cdot \boldsymbol{\omega} + \omega^2) \psi = (E - V) \psi \tag{6.125}$$

where V is the resonant Coulombic term of ECE theory.

6.4 Suggested numerical methods in density functional code

Starting with the H atom, density functional code can be run with the standard Coulomb potential to give base line output. The latter describes the well known $2n^2$ - fold degeneracy of the H atom [22], where n is the principal quantum number. Electron spin produces $s = \pm 1/2$, doubling the degeneracy from n^2 - fold. This high level of degeneracy is not present if the potential energy V does not vary as $1/r$ [22], i.e. the degeneracy is caused by the use of the Coulomb potential. In ECE theory the degeneracy is removed at resonance, but remains in an off resonance condition. This should be demonstrated numerically by using the ECE resonance equation:

$$\nabla^2\phi - \boldsymbol{\nabla}\cdot(\boldsymbol{\omega}\phi) = \frac{\rho}{\epsilon_0} \tag{6.126}$$

at the point in the density functional code where the Coulomb potential is normally coded. The standard density functional code is then resumed and the output graphed and animated. This process can be repeated in any atom or molecule or material such as a crystal which can be described by contemporary density functional code. The spin connection enters into all areas of quantum mechanics in which the Coulomb potential is used, for example the electron to proton, electron to electron and proton to proton terms in the Dirac or Schrödinger equations. Use of the generally covariant ECE wave equation [1]–[17] will show the effect of gravitation on atomic and molecular quantum mechanics, an entirely new subject area in the natural and life sciences, for example DNA studies and genetics. ECE theory also applies [1] to the electro-weak field (radio activity) and the strong nuclear force between a neutron and a proton in a nucleus. The interaction of the strong, electromagnetic and gravitational fields inside a nucleus determines for example the natural abundance of the elements, and here again, ECE theory applies, with concomitant space-time resonance equations. At a sub nuclear level, ECE theory determines the science of quarks and gluons, and elementary particle theory.

Acknowledgements The British Parliament, Prime Minister and Head of State are thanked for the award of a Civil List Pension in recognition of distinguished service to science. The staff of AIAS are thanked for many interesting and formative e mail discussions, in particular Dr Horst Eckardt of the Siemens Company. Prof. W. O. George is thanked for running density functional and *ab initio* code on H and other molecules, and Dr Eckardt for graphical output.

Bibliography

[1] M. W. Evans, Generally Covariant Unified Field Theory (Abramis 2005) vol.1.

[2] M. W. Evans, ibid., vols. 2-4, (in press, 2006 - 2007, www.aias.us and www.atomicprecision.com).

[3] L. Felker, The Evans Equations of Unified Field Theory (preprint on www.aias.us and www.atomicprecision.com).

[4] H. Eckardt and L. Felker, paper on www.aias.us and www.atomicprecision.com, H. Eckardt, papers on these websites.

[5] M. W. Evans, Generally Covariant Dynamics(paper 55) of the ECE series, in volume 4 of ref. (1), preprints on www.aias.us and www.atomicprecision.com)

[6] M. W. Evans, Geodesics and the Aharonov Bohm Effects (paper 56 ECE series).

[7] M. W. Evans, Canonical and Second Quantization in Generally Covariant Quantum field Theory (paper 57 of ECE series).

[8] M. W. Evans, The Effect of Torsion on the Schwarzschild Metric and Light Deflection due to Gravitation (paper 58 of ECE series).

[9] M. W. Evans, The Resonance Coulomb Law from ECE Theory: Application to the Hydrogen Atom (paper 59 of ECE theory).

[10] M. W. Evans, papers and letters in Found. Phys. Lett., and Found. Phys., 1993 to present, developing O(3) electrodynamics and ECE theory.

[11] M. W. Evans (ed.), Modern Nonlinear Optics, a special topical issue of I. Prigogine and S. A. Rice (series eds.), Advances in Chemical Physics (Wiley Interscience, New York, 2001. 2nd ed.), vols. 119(1) to 119(3).

[12] M. W. Evans and S. Kielich (eds.), first edition of ref. (11), (Wiley Interscience, New York, 1992, reprinted 1993 and 1997), vols. 85(1) to 85(3).

[13] M. W. Evans and L. B. Crowell, Classical and Quantum Electrodynamics and the $\boldsymbol{B}^{(3)}$ Field (World Scientific, New Jersey, 2001).

[14] M. W. Evans and J.-P. Vigier, The Enigmatic Photon (Kluwer, Dordrecht, 1994 to 2002, hardback and softback), vols. 1 - 5.

[15] M. W. Evans and A. A. Hasanein, The Photomagneton in Quantum field Theory (World Scientific, New Jersey, 1994).

[16] M. W. Evans, The Photons Magnetic Field (World Scientific, New Jersey, 1992).

[17] M. W. Evans, Physica B, 182, 227, 237 (1992), the first $\boldsymbol{B}^{(3)}$ papers.

[18] Feedback sites to www.aias.us and www.atomicprecision.com, indicating intense worldwide interest for two and a half years in the natural and engineering sciences and related areas.

[19] Mexican Group results affiliated to AIAS, observed independently, National University of Mexico Conference, February 2006.

[20] S. P. Carroll, Space-time and Geometry, an Introduction to General Relativity (Addison Wesley, New York, 2004).

[21] L. H. Ryder, Quantum Field Theory (Cambridge Univ Press, 1996, 2nd. ed.).

[22] P. W. Atkins, Molecular Quantum Mechanics (Oxford Univ. Press., 1983, 2nd. ed.).

[23] J. B. Marion and S. T. Thornton, Classical Dynamics of Particles and Systems (HB College Publishers, New York, 1988, 3rd. Ed.), chapter 3.

[24] Indicated by several independent contributions and discussion on www.aias.us and www.atomicprecision.com.

[25] E. G. Milewski (Chied Ed.), The Vector Analysis Problem Solver (Research and Education Association, New York City, 1987).

[26] H. Eckardt, e mail communications, and density functional code packages.

[27] J. D. Jackson, Classical Electrodynamics (Wiley, New York, 1998, 3rd. ed.).

[28] see discussion on linear inhomogeneous differential equations and Q factors in ref. (23).

Chapter 7

Space-time resonance in the Coulomb Law

(Paper 61)
by
Myron W. Evans, and H. Eckardt
Alpha Institute for Advanced Study (AIAS).
(emyrone@aol.com, horsteck@aol.com, www.aias.us,
www.atomicprecision.com)

Abstract

The Coulomb law is derived from general relativity applied to classical electrodynamics within Einstein Cartan Evans (ECE) unified field theory. The radial component of the spin connection is modeled to be of the form $1/r$, where r is the radial component of the spherical polar coordinate system. The Coulomb potential so obtained may be amplified by space-time resonance. If this resonant Coulomb potential is used in a computation of the radial orbitals of the H atom, for example, the latter ionizes if the kinetic energy inputted from space-time at resonance exceeds the ionization potential energy (13.6 eV). The free electrons so released may be used as a novel source of electric power.

Keywords: Einstein Cartan Evans (ECE) field theory, resonant Coulomb law, radial orbitals of the H atom, free electrons from resonance, source of electric power from space-time.

7.1 Introduction

The theory of general relativity was developed for the gravitational field, as is well known, and has recently been tested in the solar system [1] to one part in one hundred thousand with the NASA Cassini experiments. It is therefore logical to extend general relativity to other areas of physics, notably classical electrodynamics, thereby developing a generally covariant unified field theory [2]- [18] for the natural, engineering and life sciences. In the standard model, classical electrodynamics is a theory of special relativity - the Maxwell Heaviside (MH) field theory [19]. The well known Coulomb law is part of the MH field theory and is usually regarded as one of the most precise laws in physics [19], [20]. The Coulomb law is the basis for the quantum theory of atomic and molecular spectra for example, and is used in many of the advanced computational techniques employed in this area of physics and chemistry. When the MH theory is extended from special to general relativity [2]– [18] with the Einstein Cartan Evans (ECE) theory, important new features develop in all the basic laws of classical electrodynamics, including the Coulomb law. These features emanate from the spin connection of ECE space-time. The Minkowski space-time of the MH theory is the well known flat space-time [19] of special relativity, but ECE space-time is characterized by the presence of both curvature and torsion [2]– [18]. In general relativity (ECE theory) the electromagnetic field is spinning space-time and the gravitational field is curving space-time. The spinning and curving may interact through standard Cartan geometry [21] and therefore the electromagnetic and gravitational fields may interact as verified experimentally in the well known bending of light by gravity. This phenomenon has been observed with great precision in the recent NASA Cassini experiments. ECE theory has been accepted [22] as the first classical explanation of this phenomenon [2]– [18]. The original well known inference of this effect by Einstein and others is based on a semi-classical approach, where the photon mass gravitates with the mass of the sun according to the Einstein Hilbert (EH) field theory of gravitation published in 1916. A classical explanation was not possible prior to ECE theory because standard model electrodynamics is special relativity un-unified with gravitational general relativity. ECE theory [2]– [18] provides a relatively simple and practical unified field theory based on the fundamental and well known principle of general covariance [21]. Unification occurs on both classical and quantum levels, and so ECE theory has been accepted as unifying general relativity with quantum mechanics, a major aim of physics throughout the twentieth century.

In Section 7.2 the Coulomb law is developed within the context of ECE field theory using a simple model of the spin connection, which is assumed to have a $1/r$ radial dependence, where (r, θ, ϕ) is the spherical polar coordinate system [23]. The result is that the Poisson equation is extended to a second order differential equation through which the scalar potential may be amplified at resonance according to well known mathematical principles [24]. This capacity for resonance is due to the presence of the spin connection of ECE space-time itself. Resonance of this type is not possible in a flat space-time, because in a flat space-time there is no spin connection. The latter indicates that the electromagnetic field is spinning space-time. The latter inference is indicated independently by several other phenomena [2]– [18], notably the magnetization of matter by electromagnetic radiation (the inverse Faraday effect) and the presence of the ECE spin field ($\boldsymbol{B}^{(3)}$ [25]) in all types of electromagnetic radiation. The inverse Faraday effect is magnetization due to the $\boldsymbol{B}^{(3)}$ spin field. The latter originates [2]– [18] in the spin connection, which works its way into other observable phenomena throughout the whole of the natural, engineering and life sciences.

In Section 7.3 some graphical results are given from the resonant Coulomb law, and it is shown how this produces free electrons from the H atom by ionizing the latter with kinetic energy inputted from space-time at resonance. The H atom is used here as a simple model material. The release of free electrons at space-time resonance has been observed recently [26] and shown to be a repeatable phenomenon. The material and circuit designs used in this series of experiments [26] are much more complicated than H, but the latter serves as a model to illustrate the theoretical principles at work - those of general relativity applied to classical electrodynamics with ECE theory.

7.2 The ECE resonance Coulomb Law

The law is given [2]- [18] from the first Cartan structure equation:

$$T^a = d \wedge q^a + \omega^a{}_b \wedge q^b \tag{7.1}$$

and the first Bianchi identity:

$$d \wedge T^a + \omega^a{}_b \wedge T^b = R^a{}_b \wedge q^b \tag{7.2}$$

with the ECE Ansatz:

$$A^a = A^{(0)} q^a, \quad F^a = A^{(0)} T^a \tag{7.3}$$

Here T^a is the torsion form, $R^a{}_b$ is the Riemann or curvature form, q^a is the tetrad form, is the spin connection form, A^a is the electromagnetic potential form, $cA^{(0)}$ is the primordial voltage, and F^a is the electromagnetic field form. The Ansatz was first proposed by Cartan in well known correspondence with Einstein in the first part of the twentieth century, but was not developed into ECE theory until the spring of 2003 [2]- [18]. Eqs. (7.1) to (7.3) lead to [2]- [18]:

$$\mathbf{E}^a = -\frac{\partial \mathbf{A}^a}{\partial t} - \nabla \phi^a - c\omega^{0a}{}_b \mathbf{A}^b + \phi^b \boldsymbol{\omega}^a{}_b \tag{7.4}$$

$$\nabla \cdot \mathbf{E}^a = c\mu_0 \widetilde{J}^{0a} \tag{7.5}$$

in vector notation. Here $\mathbf{E}^a$ is the electric field strength (volts per meter), μ_0 Is the vacuum S.I. permeability, $\widetilde{J}^{oa}$ is the time-like component of the inhomogeneous four-current of ECE theory, c is the vacuum speed of light, $\mathbf{A}^a$ is the vector potential, ϕ^b is the scalar potential, $\omega^{0a}{}_b$ is the time-like part of the spin connection four-vector, and $\widetilde{\omega}^a{}_b$ is the space-like part of the spin connection four-vector. The indices a and b originate in Cartan geometry [2]- [18], [21] and are the indices of the tangent space-time at a point P in the base manifold. These indices indicate polarization states of electromagnetic radiation in ECE theory [2]- [18]. Eq.(7.5) may be written for each index a as:

$$\nabla \cdot \mathbf{E} = \frac{\rho}{\epsilon_0} \tag{7.6}$$

where ρ is the charge density and where ϵ_0 is the S.I. vacuum permittivity. Therefore for each index a, Eq.(7.5) has the same mathematical structure as the standard model Coulomb law [19], [20]. However, the electric field in ECE theory must always be defined by Eq.(7.4), which always involves the spin connection. The electric field is part of spinning space-time.

If attention is restricted to the scalar potential, then for each index a, Eq.(7.4) is:

$$\mathbf{E} = -\boldsymbol{\nabla}\phi + \phi^b \boldsymbol{\omega}_b \tag{7.7}$$

Here ϕ^b is interpreted as a scalar quantity indexed or labeled by b, indicating that the scalar potential applies to this state of polarization of electromagnetic radiation. For a given b index, Eq.(7.7) is:

$$\mathbf{E} = -\boldsymbol{\nabla}\phi + \phi\boldsymbol{\omega} \tag{7.8}$$

Summation over repeated b indices in Eq.(7.7) is implied (Einstein convention) but for the sake of simplicity it has been assumed in Eq.(7.8) that there is only one index and one state of polarization. Therefore we have reduced the complicated Eq.(7.8) to its simplest form (7.4). The result is that the familiar definition of the electric field in the standard model Coulomb law:

$$\mathbf{E} = -\nabla\phi \tag{7.9}$$

is supplemented by a term in the vector part of the spin connection, the vector $\boldsymbol{\omega}$. Eqs.(7.6) and (7.8) give the second order differential equation:

$$\nabla^2\phi - \boldsymbol{\nabla}\cdot(\phi\boldsymbol{\omega}) = -\frac{\rho}{\epsilon_0} \tag{7.10}$$

which compares with the standard model Poisson equation [19], [20]:

$$\nabla^2\phi = -\frac{\rho}{\epsilon_0} \tag{7.11}$$

Eq.(7.10) is an equation of general relativity. Eq.(7.11) is an equation of special relativity. The mathematical properties of Eq.(7.10) include the ability to give resonance, whereas Eq.(7.11) has no resonance solutions. This is a key difference. Resonance is the key to the production of free electrons from ECE space-time, providing a new source of electric power for engineering.

The spin connection vector in Cartesian and spherical polar coordinates is:

$$\begin{aligned}\boldsymbol{\omega} &= \omega_x\mathbf{i} + \omega_y\mathbf{j} + \omega_z\mathbf{k}\\ &= \omega_r\mathbf{e}_r + \omega_\phi\mathbf{e}_\phi + \omega_\theta\mathbf{e}_\theta\end{aligned} \tag{7.12}$$

where ω_r is the radial component of $\boldsymbol{\omega}$. If the latter is assumed to be purely radial, for simplicity of argument, then:

$$\boldsymbol{\omega} = \omega_r\mathbf{e}_r \tag{7.13}$$

and in spherical polar coordinates [23]:

$$\boldsymbol{\omega}\cdot\boldsymbol{\nabla}\phi = \omega_r\frac{\partial\phi}{\partial r} \tag{7.14}$$

$$\phi\boldsymbol{\nabla}\cdot\boldsymbol{\omega} = \frac{\phi}{r^2}\frac{\partial}{\partial r}\left(r^2\omega_r\right) \tag{7.15}$$

$$\nabla^2\phi = \frac{1}{r^2}\frac{\partial}{\partial r}\left(r^2\frac{\partial\phi}{\partial r}\right) = \frac{\partial^2\phi}{\partial r^2} + \frac{2}{r}\frac{\partial\phi}{\partial r} \tag{7.16}$$

The dimensions of $\boldsymbol{\omega}$ are inverse meters 2-18, so the simplest model of the vector spin connection is:

$$\omega_r = \frac{A}{r} \tag{7.17}$$

where A is a dimensionless scaling factor. Eqs.(7.13) to (7.17) give the result:

$$\frac{\partial^2 \phi}{\partial r^2} + (2 - A)\frac{1}{r}\frac{\partial \phi}{\partial r} - \frac{A\phi}{r^2} = -\frac{\rho}{\epsilon_0} \tag{7.18}$$

in spherical polar coordinates. Eq.(7.18) contains second and first order partial derivatives in the scalar potential ϕ. In the special case

$$A = 2 \tag{7.19}$$

Eq.(7.18) becomes:

$$\frac{\partial^2 \phi}{\partial r^2} - \frac{2\phi}{r^2} = -\frac{\rho}{\epsilon_0} \tag{7.20}$$

in which the second term on the left hand side is a REPULSION term. This means that the familiar Coulomb attraction between a proton and an electron in an H atom develops a repulsive component due to the presence of the spin connection vector. Eq.(7.18) has a similar structure to the well known one-dimensional Schrödinger equation for motion in an effective potential with repulsive centrifugal term [20] in the H atom. So the spin connection may be interpreted similarly. If the repulsion term in Eq.(7.20) becomes strong enough, the H atom ionizes, releasing a free electron. Eq.(7.18) is similar to the well-known [24] class of linear inhomogeneous differential equations that give resonance - the damped driven oscillator equations. Eq.(7.20) is a special case - the undamped driven oscillator. In order to induce resonance, the charge density *rho* must be initially oscillatory [24]. In the H atom model we are considering the source of this small original oscillation may be considered to be zitterbewegung (jitterbugging) from quantum electrodynamics [20]. In a molecule it could be a rotational frequency or vibrational bond frequency. At space-time resonance the initially small oscillation is greatly amplified [2]– [18], [24] and kinetic energy is absorbed into the atom or molecule from ECE space-time. If this energy is greater than the ionization potential energy of H (13.6 eV) the electron breaks free of the proton and may be used in a circuit to produce electric power from space-time through the intermediacy of the H atom. This concept may be generalized to any material which contains electrons which are easily released by ionization. The skill in material design revolves around this need. The engineering skill consists in devising a design to induce the resonance and this has been accomplished recently in a repeatable manner [26]. The output power in such experiments [26] may exceed the input power by as much as a factor of one hundred thousand, an amplification that illustrates dramatically the resonance of the spin connection in classical electrodynamics. Care has been taken to ensure that this experiment is repeatable and the apparatus has been observed independently [26] in different laboratories. Every effort has been made to eliminate artifact, and reproducible amplification by five orders of magnitude is unlikely to be artifact. The standard model (MH theory) has no explanation for this phenomenon, even on a qualitative level. Its explanation in general relativity (ECE theory) relies on resonating the spin connection as described already.

In summary of this section therefore the Poisson equation of the standard model (Eq.(7.11)) is modified in the simplest instance to the following ECE equation of general relativity:

$$\frac{\partial^2 \phi}{\partial r^2} = \frac{2\phi}{r^2} - \frac{\rho}{\epsilon_0} \tag{7.21}$$

introducing a repulsive term:

$$\rho_{\text{eff}} = 2\epsilon_0 \frac{\phi}{r^2} \tag{7.22}$$

If the charge density ρ is very small, Eq.(7.21) takes on the approximate mathematical form:

$$\frac{\partial^2 \phi}{\partial r^2} \sim \frac{2\phi}{r^2} \tag{7.23}$$

which has an analytical solution:

$$\phi \sim \frac{\beta}{r} + \alpha r^2 \tag{7.24}$$

where α and β are constants. When r is very small, the potential ϕ becomes very large and a large amount of POSITIVE potential energy may be inputted into the H atom from the spin connection, depending on the value of β. If:

$$\beta \geq \frac{e}{4\pi\epsilon_0} \tag{7.25}$$

then the positive repulsion potential becomes equal to or greater than the negative attraction potential, releasing the electron from the proton. The standard model inverse square Coulomb law is very precise in the vast majority of experiments in macroscopic classical electrodynamics [19] but the recent experiments carried out in ref. [26] indicate that it does not hold in general.

The ECE theory reduces straightforwardly to the standard Coulomb law as follows. In ECE theory the electric field is defined in the simplest instance by:

$$\mathbf{E} = -\boldsymbol{\nabla}\phi + \phi\boldsymbol{\omega} \tag{7.26}$$

and in the standard Coulomb law it is defined by:

$$\mathbf{E} = -\boldsymbol{\nabla}\phi \tag{7.27}$$

Therefore if:

$$\boldsymbol{\nabla}\phi = -\phi\boldsymbol{\omega} \tag{7.28}$$

the mathematical form of the standard Coulomb law is obtained:

$$\mathbf{E} := -2\boldsymbol{\nabla}\phi \tag{7.29}$$

This simply means that the scalar potential is defined by:

$$\Phi := 2\phi \tag{7.30}$$

This makes no difference to the observable force (inverse square law). If:

$$\Phi := \frac{e}{4\phi\epsilon_0 Z} \tag{7.31}$$

and

$$\nabla\Phi = -\Phi\boldsymbol{\omega} \tag{7.32}$$

then:

$$\omega_A = \frac{1}{Z} \tag{7.33}$$

The important conclusion is reached that a spin connection of the type (7.33) is ALWAYS observed in the Coulomb law, which becomes a law of general relativity as required by objectivity in physics. Therefore any experimental departure from the inverse square Coulomb law would indicate that the spin connection is no longer given by Eq.(7.33)In general relativity (ECE theory) the electric field must always be defined according to Eq.(7.26) and the vast majority of experimental data have confirmed the inverse square law of Coulomb for over two hundred years. In general relativity this means that the data show that the spin connection must be of the form (7.33) experimentally. This type of spin connection, conversely, gives the inverse square law of Coulomb. Eq.(7.28) is similar to the operator equivalence of quantum mechanics, and means that:

$$\boldsymbol{\omega} \rightarrow -\boldsymbol{\nabla} \tag{7.34}$$

The operator equivalence is:

$$\mathbf{p} \rightarrow -i\hbar\boldsymbol{\nabla} \tag{7.35}$$

Therefore in general relativity the electric field can be defined equivalently in two ways:

$$\mathbf{E} = -\boldsymbol{\nabla}\Phi = \Phi\boldsymbol{\omega} \tag{7.36}$$

and this is the fundamental definition of the electric field in general relativity. These considerations confirm that ECE theory is correct to very high precision, and give a simple meaning to the spin connection. The result of general relativity, Eq.(7.26), is preferred to the result of special relativity, Eq.(7.27), on several grounds, notably that other aspects of electrodynamics such as the inverse Faraday effect and Eddington effect require a generally covariant unified field theory for their objective interpretation on the classical level.

Dramatically new results such as those by the Mexican group are also accounted for by ECE theory by considerations of resonance as in previous work.

It is significant that spin connections of the type (7.33) also occur as Christoffel connections of the Schwarzschild metric of spherically symmetric space-time. These are well known to indicate a dynamic space-time. Electrodynamics is also now known to be a phenomenon of dynamic space-time, and similarly for the natural, engineering and life sciences. The two simplest Christoffel connections in a spherically symmetric space-time are [2]- [18]:

$$\Gamma^2{}_{12} = \Gamma^3{}_{13} = \frac{1}{r} \tag{7.37}$$

and these are similar to the connection (7.33) of the Coulomb law in ECE theory. For any spherically symmetric space-time the non-vanishing Christoffel connections are:

$$\Gamma^2{}_{12} = \Gamma^3{}_{13} = \frac{1}{r}, \quad \Gamma^0{}_{00} = \frac{1}{c}\frac{\partial \alpha}{\partial t}, \quad \Gamma^0{}_{01} = \frac{\partial \alpha}{\partial X}, \quad \Gamma^1{}_{01} = \frac{1}{c}\frac{\partial B}{\partial t},$$

$$\Gamma^1{}_{11} = \frac{\partial \beta}{\partial X}, \quad \Gamma^0{}_{11} = e^{2(\beta-\alpha)}\frac{1}{c}\frac{\partial \beta}{\partial t}, \quad \Gamma^1{}_{22} = -re^{-2\beta}, \tag{7.38}$$

$$\Gamma^1{}_{33} = \Gamma^2{}_{33} = \Gamma^3{}_{23} = f(\theta),$$

$$ds^2 = -e^{2\alpha}dt^2 + e^{2\beta}dr^2 + r^2 d\Omega^2$$

In the particular case of the Schwarzschild metric [2]- [18]:

$$e^{2\alpha} = \left(1 - \frac{2GM}{c^2 r}\right), \tag{7.39}$$

$$e^{2\beta} = \left(1 - \frac{2GM}{c^2 r}\right)^{-1}, \tag{7.40}$$

and in spherical polar coordinates:

$$\Gamma^0{}_{01} = \partial_1 \alpha, \quad \Gamma^0{}_{11} = e^{2(\beta-\alpha)}\partial_0 \beta,$$
$$\Gamma^1{}_{00} = e^{2(\alpha-\beta)}\partial_1 \alpha, \quad \Gamma^1{}_{11} = \partial_1 \beta. \tag{7.41}$$

However, the ECE Coulomb law is derived from the Cartan torsion, while the Christoffel connections are for the Cartan curvature in the absence of Cartan torsion.

These considerations of the Coulomb law of generally covariant electrostatics can be extended as follows to the generally covariant Ampère Law of magneto-statics. In ECE theory the magnetic field is:

$$\mathbf{B}^a = \boldsymbol{\nabla} \times \mathbf{A}^a - \boldsymbol{\omega}^a{}_b \times \mathbf{A}^b \tag{7.42}$$

and the Ampère Law of magneto-statics takes on a generally covariant form as follows:

$$\boldsymbol{\nabla} \times \mathbf{B}^a = \frac{\mu_0}{c}\tilde{\mathbf{J}}^a \tag{7.43}$$

The magnetic field in general relativity must always be defined by Eq.(7.42) with a non-zero spin connection. The latter is always present in general relativity. In magneto-statics we are dealing with rotational motion, so Eq.(7.42) may be written as:

$$\mathbf{B}^a = \boldsymbol{\nabla} \times \mathbf{A}^a + g\mathbf{A}^b \times \mathbf{A}^c \tag{7.44}$$

where the parameter g is defined as [2]- [18]:

$$g = \frac{\kappa}{A^{(0)}} \tag{7.45}$$

For rotational motion the spin connection is dual to the tetrad if it is assumed that the electromagnetic and gravitational fields are independent. In Eq.(7.45) $A^{(0)}$ is a magnitude and κ has the units of inverse meters. Eq.(7.44) is therefore:

$$\mathbf{B}^a = \boldsymbol{\nabla} \times \mathbf{A}^a + \boldsymbol{\omega}^b \times \mathbf{A}^c \tag{7.46}$$

where

$$\boldsymbol{\omega}^b = \frac{\kappa}{A^{(0)}} \mathbf{A}^b \tag{7.47}$$

If gravitation and electromagnetism are inter-dependent, the general equation (7.42) must be used, and it cannot be assumed that ${\boldsymbol{\omega}^a}_b$ is dual to $\mathbf{A}^c$.

If it is assumed that:

$$\boldsymbol{\nabla} \times \mathbf{A}^a = \boldsymbol{\omega}^b \times \mathbf{A}^c \tag{7.48}$$

then:

$$\mathbf{B}^a = 2\boldsymbol{\nabla} \times \boldsymbol{A}^a \tag{7.49}$$

which is the standard model result for each a. The potential is:

$$\mathbf{A}_{\mathrm{mh}} := 2\mathbf{A}_{\mathrm{ECE}} \tag{7.50}$$

and so:

$$\mathbf{B} = \boldsymbol{\nabla} \times \mathbf{A}_{\mathrm{MH}} \tag{7.51}$$

as is usual in the standard model [2]– [18]. So ECE reduces to the standard model of magneto-statics provided the spin connection obeys Eq.(7.48). This is an important result, because in the vast majority of experiments since the eighteenth century both the Coulomb and Ampère laws hold to very high accuracy. So ECE theory must be able to reduce to these well known results. So both laws are now understood to be very precise laws of general relativity (ECE theory) and not special relativity (Maxwell Heaviside theory). The key advance is that the ECE theory is a generally covariant unified field theory that enables electro-statics and magneto-statics to be unified with all other fields, notably the gravitational field.

The indices a, b and c in Eq.(7.42) originate [2]– [18] in the tangent space of Cartan geometry, and can be defined in the complex circular basis:

$$a, b, c = (1), (2), (3) \tag{7.52}$$

in which the magnetic field is:

$$\begin{aligned} \mathbf{B}^{(1)*} &= \boldsymbol{\nabla} \times \mathbf{A}^{(1)*} - i\boldsymbol{\omega}^{(2)} \times \mathbf{A}^{(3)} \\ \mathbf{B}^{(2)*} &= \boldsymbol{\nabla} \times \mathbf{A}^{(2)*} - i\boldsymbol{\omega}^{(3)} \times \mathbf{A}^{(1)} \\ \mathbf{B}^{(3)*} &= \boldsymbol{\nabla} \times \mathbf{A}^{(3)*} - i\boldsymbol{\omega}^{(1)} \times \mathbf{A}^{(2)} \end{aligned} \tag{7.53}$$

The complex circular basis is defined by the unit vectors:

$$\begin{aligned} \mathbf{e}^{(1)} &= \frac{1}{\sqrt{2}} (\mathbf{i} - i\mathbf{j}) \\ \mathbf{e}^{(2)} &= \frac{1}{\sqrt{2}} (\mathbf{i} + i\mathbf{j}) \\ \mathbf{e}^{(3)} &= \mathbf{k} \end{aligned} \tag{7.54}$$

with $O(3)$ symmetry:

$$\begin{aligned} \mathbf{e}^{(1)} \times \mathbf{e}^{(2)} &= i\mathbf{e}^{(3)*} \\ \mathbf{e}^{(2)} \times \mathbf{e}^{(3)} &= i\mathbf{e}^{(1)*} \\ \mathbf{e}^{(3)} \times \mathbf{e}^{(1)} &= i\mathbf{e}^{(2)*} \end{aligned} \tag{7.55}$$

Here **i**, **j** and **k** are the Cartesian unit vectors. The following are self-consisted vector potential solutions of Eq.(7.53):

$$\mathbf{A}^{(1)} = \frac{A^{(0)}}{\sqrt{2}} \left(\mathbf{i} - i\mathbf{j}\right) e^{-i\kappa Z} \tag{7.56}$$

$$\mathbf{A}^{(2)} = \mathbf{A}^{(1)*} = \frac{A^{(0)}}{\sqrt{2}} \left(\mathbf{i} + i\mathbf{j}\right) e^{i\kappa Z} \tag{7.57}$$

$$\mathbf{A}^{(3)} = A^{(0)}\mathbf{k} \tag{7.58}$$

with vector spin connections:

$$\boldsymbol{\omega}^{(1)} = \frac{\omega^{(0)}}{\sqrt{2}} \left(\mathbf{i} - i\mathbf{j}\right) e^{-i\kappa Z}, \tag{7.59}$$

$$\boldsymbol{\omega}^{(2)} = \frac{\omega^{(0)}}{\sqrt{2}} \left(\mathbf{i} + i\mathbf{j}\right) e^{i\kappa Z}, \tag{7.60}$$

$$\boldsymbol{\omega}^{(3)} = \omega^{(0)}\mathbf{k} \tag{7.61}$$

Using the de Moivre Theorem:

$$\begin{aligned} e^{-i\kappa Z} &= \cos\left(\kappa Z\right) - i\sin\left(\kappa Z\right) \\ e^{i\kappa Z} &= \cos\left(\kappa Z\right) + i\sin\left(\kappa Z\right) \end{aligned} \tag{7.62}$$

Eq.(7.56) has a real and physical component:

$$\text{Real}\mathbf{A}^{(1)} = \frac{A^{(0)}}{\sqrt{2}} \left(\cos\left(\kappa Z\right)\mathbf{i} + \sin\left(\kappa Z\right)\mathbf{j}\right) \tag{7.63}$$

which is a rotating potential with phase angle:

$$\theta = \kappa Z \tag{7.64}$$

It is seen that at $\theta = 0$, $\mathbf{A}^{(0)}$ is in the **i** axis, and if $\theta = \pi/2$, $\mathbf{A}^{(1)}$ is in the **j** axis, and so has rotated by 90°. With these definitions it is seen that:

$$\begin{aligned} \boldsymbol{\nabla} \times \mathbf{A}^{(1)*} &= -i\boldsymbol{\omega}^{(2)} \times \mathbf{A}^{(3)} \\ \boldsymbol{\nabla} \times \mathbf{A}^{(2)*} &= -i\boldsymbol{\omega}^{(3)} \times \mathbf{A}^{(1)} \end{aligned} \tag{7.65}$$

so the (1) and (2) magnetic fields are:

$$\begin{aligned} \mathbf{B}^{(1)} &= 2\boldsymbol{\nabla} \times \mathbf{A}^{(1)}, \\ \mathbf{B}^{(2)} &= 2\boldsymbol{\nabla} \times \mathbf{A}^{(2)}, \end{aligned} \tag{7.66}$$

having the same mathematical form as the standard model. However, general relativity (ECE theory) gives a new result:

$$\mathbf{B}^{(3)*} = i\boldsymbol{\omega}^{(1)} \times \mathbf{A}^{(2)} \tag{7.67}$$

which does not occur in special relativity. Eq.(7.65) may be written as:

$$\mathbf{B} = \boldsymbol{\nabla} \times \mathbf{A}_{\text{MH}} \tag{7.68}$$

From Eq.(7.64) the magnitude of the spin connection is the wave-number:

$$\omega^{(0)} = \kappa \tag{7.69}$$

with units of inverse meters.

It is important to note the existence of the $\mathbf{B}^{(3)}$ field in general relativity, Eq.(7.66). In electrodynamics [2]– [18] this is the ECE spin field of electromagnetic radiation. In magneto-statics, to which the Ampère law applies, a magnetic field may also be defined through the spin connection using Eq.(7.66). Using Eqs.(7.56) to (7.60) the field in Eq.(7.66) is:

$$\mathbf{B}^{(3)} = B^{(0)}\mathbf{k} \tag{7.70}$$

BUT:

$$\nabla \times \mathbf{A}^{(3)} = \mathbf{0} \tag{7.71}$$

In electrodynamics the $\mathbf{B}^{(3)}$ spin field is [2]– [18]:

$$\mathbf{B}^{(3)*} = -i\frac{\kappa}{A^{(0)}}\mathbf{A}^{(1)} \times \mathbf{A}^{(2)} \tag{7.72}$$

where

$$\mathbf{A}^{(1)} = \mathbf{A}^{(2)*} = \frac{A^{(0)}}{\sqrt{2}}\left(\mathbf{i} - i\mathbf{j}\right)e^{i(\omega t - \kappa Z)} \tag{7.73}$$

where Ω is the electromagnetic angular frequency. In magneto-statics (Eq.(7.56) the angular frequency Ω is zero. The magneto-static potential rotates (as we have seen), but the electromagnetic potential rotates and also translates along an axis such as Z.

The electrodynamic spin field $\mathbf{B}^{(3)}$ is observed by its magnetization of matter in the inverse Faraday effect [2]– [18]. This observation shows that classical electrodynamics and non-linear optics are manifestations of general relativity. The spin connection of the inverse Faraday effect is:

$$\boldsymbol{\omega}^{(1)}_{\text{IFE}} = \frac{\kappa}{A^{(0)}}\mathbf{A}^{(1)} \tag{7.74}$$

and without the spin connection there is no inverse Faraday effect. Since all physics must be independent of observer influence (must be objective and covariant under the general coordinate transformation), all physics, including electrodynamics, must be general relativity.

This means that the electromagnetic field under any circumstance must originate in a spinning space-time described by Cartan torsion [2]– [18]. In turn this means that the spin connection is non-zero under any circumstance, as emphasized in this section for the electro-statics and magneto-statics. If the spin connection is non-zero the $\mathbf{B}^{(3)}$ spin field is always non-zero. In Maxwell Heaviside field theory on the other hand the spin connection is zero because the electromagnetic field is philosophically different, it is an entity superimposed on Minkowski space-time, and in this space-time there is no $\mathbf{B}^{(3)}$ field, contrary to observation.

Having shown that the spin connection has a $1/Z$ dependence for the standard model Coulomb law, this type of spin connection may now be used in the resonance equation (7.10) for the potential in general relativity. Therefore in the resonance equation:

$$\nabla^2\phi - \boldsymbol{\omega}\cdot\boldsymbol{\nabla}\phi - (\boldsymbol{\nabla}\cdot\boldsymbol{\omega})\,\phi = -\frac{\rho}{\epsilon_0} \tag{7.75}$$

a vector spin connection of the following type may be used self-consistently

$$\boldsymbol{\omega} = \frac{A}{Z}\mathbf{k} \tag{7.76}$$

where A is a scaling factor. The initial driving charge density may be defined for convenience as:

$$\rho = -\rho_0\cos(\kappa Z) \tag{7.77}$$

In the H atom this cosinusoidal dependence may be assumed to originate in the jitterbugging motion (zitterbewegung) that has its rigorous origins in quantum electrodynamics. In a molecule such as water it may be assumed to originate in a rotational or vibrational frequency of the molecule. Therefore the resonance equation becomes:

$$\nabla^2\phi - \frac{A}{Z}\frac{\partial\phi}{\partial Z} + \frac{A}{Z^2}\phi = \frac{\rho_0}{\epsilon_0}\cos(\kappa Z) \tag{7.78}$$

The simplest example [2]– [18] of a resonance equation is the linear inhomogeneous differential equation:

$$\ddot{x} + 2\beta\dot{x} + \omega_0^2 x = \alpha\cos\omega t \tag{7.79}$$

This is a forced damped oscillator with driving term $\alpha\cos\omega t$. The damping term is $2\beta\dot{x}$ and the Hooke's law term is $\omega_0^2 x$. The frequency resonance from Eq.(7.78) is well known [2]– [18] to occur at:

$$\omega_R = \left(\omega_0^2 - 2\beta^2\right)^{1/2} \tag{7.80}$$

and the kinetic energy resonance occurs at:

$$\omega_E = \omega_0 \tag{7.81}$$

Therefore at some fixed values:

$$\frac{A}{Z} = \frac{A}{Z_0}, \quad \frac{A}{Z^2} = \frac{A}{Z_0^2} \tag{7.82}$$

Eq.(7.77) becomes:

$$\nabla^2\phi - \frac{A}{Z_0}\frac{\partial\phi}{\partial Z} + \frac{A}{Z_0^2}\phi = \frac{\rho_0}{\epsilon_0}\cos(\kappa Z) \tag{7.83}$$

Thus wave-number resonance occurs from Eq.(7.79) at:

$$\kappa_R = \frac{A}{\sqrt{2}}\frac{1}{Z_0} \tag{7.84}$$

and kinetic energy resonance at a wave-number:

$$\kappa_E = \frac{A^{1/2}}{Z_0} \tag{7.85}$$

At resonance the particular (or transient) solution for the potential is [2]- [18]:

$$\phi(\kappa) = \frac{1}{\epsilon_0} \frac{\rho_0 \cos(\kappa Z - \delta)}{\left(\left(\frac{A}{Z_0^2} - \kappa^2\right)^2 + \frac{A^2}{Z_0^2}\kappa^2\right)^{1/2}} \tag{7.86}$$

where

$$\delta = \tan^{-1}\left(\frac{A\kappa/Z_0}{\kappa^2 - A/Z_0^2}\right) \tag{7.87}$$

7.3 Graphical results and discussion

To study the effects of spacetime resonance in Hydrogen on a quantitative level, appropriate numerical code has been developed. Energy levels and radial wave functions are obtained from the solution of the Schrödinger equation. For the Coulomb potential of standard theory a well-known analytical solution exists. Resonance effects, however, require the resonance potential of equation (7.85) to be added to the standard (non-resonant) Coulomb potential. Then an analytical solution of the Schrödinger equation is no longer possible and the equation has to be solved numerically. So the overall numerical approach is to modify the standard Coulomb potential by the analytical form given in equations (7.86), (7.87) and compute the energies and radial wave functions by the numerical code. The latter has been developed as described in detail in [27]. Throughout the presentation of the results atomic units are used, i.e. length is measured in Bohr radius and energy in Rydberg units ($1Ryd = 13.605eV$).

In all calculations equation (7.86) has been multiplied by an exponential function $e(-r/1.25)$ in order to bring the charge oscillations to zero in the limit $r \longrightarrow \infty$. Otherwise there were unphysical oscillations in the potential at large radii which would lead to unbound free-space solutions of the Schrödinger equation.

The first question to answer is why the Coulomb law of standard theory is valid to high precision although the spin connection is present. One answer is given by equations (7.28), (7.29): half of the Coulomb potential in the general relativistic (ECE) description consists of the spin connection. If we allow for a varying strength of the spin connection, (variable A in equation (7.17)) we can study this effect numerically. For this purpose equation (7.18), which describes the radial part of the Coulomb potential, was programmed numerically and solved for several values of A. In the case $\rho = 0$ and $A = 0$ the $1/r$ potential is obtained to high precision. To see how A effects deviations from the $1/r$ form of the Coulomb potential we have chosen the same initial conditions as for the case $A = 0$ and integrated the equation from a radius $r = r_{max}$ down to a radius near to $r = 0$. The results for different A's are presented in Table 7.1 for certain radius values. Significant deviations are only visible for very small radii ($r \approx 0.1$). At $r = 1$ the deviation from the $1/r$ potential is only 1 part in $10,000$ for relevant A values (up to $A = 2$). Even for $A = 10$ changes are remarkable only near to the center. For experiments this means that the spin connection has no measurable effect in off-resonance except very near to the position of a charge.

The resonance curve of the atomic energy levels is shown in Fig. 7.1. The wave number κ has been varied as indicated on the x axis, the other parameters

r	-1/r	VC(r), A=0	A=1	A=2	A=10
0.1	-1.000000E+01	-1.000000E+01	-9.965750E+00	-9.945510E+00	-9.897040E+00
1.0	-1.000000E+00	-1.000000E+00	-9.999656E-01	-9.999449E-01	-9.998938E-01
3.0	-3.333333E-01	-3.333333E-01	-3.333321E-01	-3.333314E-01	-3.333296E-01
10.0	-1.000000E-01	-1.000000E-01	-9.999999E-02	-9.999998E-02	-9.999995E-02

Table 7.1: Values of ECE Coulomb potential at certain radii in dependence of A (atomic units)

were fixed: $A = 1$, $Z_0 = 1.5$, $\rho_0 = 0.1$. For large κ one obtains the off-resonance case with the well-known energy levels of H. For decreasing κ (and on setting resonance effects) the degeneracy of the l quantum number disappears, resulting in a splitting between l levels of the same principal quantum number. The 1s energy is greatly lifted, the maximum occurs at the resonance wave number $\kappa = 0.75$. Interestingly there is a sharp peak to negative energy values directly below the kinetic energy resonance so that the resonance shows up a pole-like behaviour. The middle of both extrema lies at $\kappa = 0.67$ which is the kinetic energy resonance given by equation (7.85) for the parameters mentioned above. The width of the pole is determined by the damping term of the governing differential equation (7.83). Obviously the width depends on the quantum number, it is smaller for the $2p$ state than for the $1s$ state. Besides the 1s state, also the $2s$ and $3s$ state are lifted. This leads to an inversion of s and p energy levels compared to multi-electron atoms where the s states have lower energies than the p states. The same holds for the $3d$ states.

In the limit $\kappa \to 0$ there is again an increase of the energy levels, but this is due to the fact that the resonance potential in this case tends to a constant positive value (see discussion of Fig. 7.7 below). So this is an artifact of the variation parameter of the graph.

The question is what happens to the atomic wave functions (radial functions) in case of resonance. This is explained in Figs. 7.2,7.3,7.4. There some orbitals are shown for three κ values, and additionally the orbitals for the atom without a resonance potential (denoted by 0). The κ values are 0.6 (lower resonance peak, A), 0.75 (upper resonance peak, B) and 2 (off-resonance, C). As can be seen for the three graphed quantum states $(1s, 2s, 2p)$, the charge density is shifted inwards to the core for the lowered energy and shifted outwards for the lifted energy levels. The resonance effect is quite drastic for the $1s$ state because the energy shift is highest in this case. A second local maximum occurs for $r \approx 4$ which is not present in off-resonance. The characteristic of the atomic state is altered completely. Nevertheless it is a valid eigenstate for the angular momentum $l = 0$ because there is no passing through zero. A corresponding result holds for the $2s$ state.

The lower energy resonance of the $2p$ state (Fig. 7.4) deserves particular attention. It shows that, at this point of resonance, states are significantly more localized than in the undisturbed atom. The localization radius here coincides with the value of $Z_0 = 1.5$. This is not so obvious for the other states.

The reason for the energy and wavefunction shifts can be found by looking at the potentials. In Fig. 7.5 the resonance potential according to equation (7.86) is shown. The negative resonance appears due to a negative bump in the potential while at the positive resonance the potential becomes significantly repulsive near to the core. In off-resonance the potential is more pure oscillatory, averaging the impact on the energies and wavefunctions. It can nicely be seen also that the oscillation has slown down for increasing radii by means of the additional exponential damping.

Fig. 7.6 presents the total potential including the standard Coulomb term. A repulsive maximum is seen in the positive resonant case. For the negative resonance the potential is lowered in a certain range. The switching between both states takes place in a very narrow region of κ values as has been seen in Fig. 7.1. The reason is the phase jump of the resonance potential being described by equation (7.87). According to the well-known theory of forced oscillations, the phase changes rapidly by 180 degrees in this region. This can further be seen from Fig. 7.7 where the maximum resonance amplitude has been plotted in dependence of κ. There is a sign change at $\kappa = 0.67$. In addition the integral over the resonance potential

$$\int_0^{r_{\max}} \phi(\kappa, Z) dZ \tag{7.88}$$

is shown. There is an even more significant jump in this curve at the same κ. It is important to note that the integral tends to zero for large wavenumbers. This means that the net contribution to the charge density is zero as is required for charge neutrality. In other words, the exponential damping function has no detrimental effect on charge neutrality.

In the following diagrams several further resonance curves are given. The effect of the strength of spin connection is visible from Fig. 7.8. According to equation (7.85) there is a connection between the resonant κ, the spin connection strength A, and the radius parameter Z_0:

$$\kappa = \frac{A^{1/2}}{Z_0} \tag{7.89}$$

With $\kappa = 0.75$ and $Z_0 = 1.5$ this gives a resonant A value of

$$A_E = \kappa_E^2 Z_0^2 = 1.27 \tag{7.90}$$

The maximum of the $1s$ energy, however, occurs at $A = 1$, which corresponds to $\kappa = 0.67$, the resonance value of Fig. 7.1.

Variation of the fixed radius Z_0 is shown in Fig. 7.9. According to the choice of the other parameters the maximum 1s energy is at $Z_0 = 1.5$ as expected. It should be noted that for $Z_0 \to 0$ the energy splitting for the main quantum numbers disappears. This is plausible from equations (7.86), (7.87) where Φ and δ tend to zero for $Z_0 \to 0$. Consequently, one obtains the original energy levels of the H atom in this case.

The next two figures (Figs. 7.10 and 7.11) reveal the characteristics of the resonance itself for varying parameter configurations. For fixed $A = 1$ the value of Z_0 was varied, while at the same time the value of κ was taken to be at resonance according to equation (7.85). Since the energy values are quite different at wavenumbers near to the resonance (the behaviour is pole-like as explained earlier) we have drawn two curves, one for the lower energies at a value

$$\kappa = \frac{A^{1/2}}{Z_0} - 0.05 \tag{7.91}$$

(see Fig. 7.10) and one for the upper energies at

$$\kappa = \frac{A^{1/2}}{Z_0} + 0.05 \tag{7.92}$$

(Fig. 7.11). From Fig. 7.10 it can be seen that the energy minima come to lie at different Z_0 values in dependence of the orbital. Because the latter differ in their radial charge distribution, they are affected in an individual way. The energy inversion of s, p and d states has already been mentioned.

The corresponding curve for the resonance maxima (Fig. 7.11) looks differently. There is no maximum energy in the range of Z_0 values. For increasing Z_0 the wave functions are shifted more and more outwards, leading to a continuous increase of energies.This may be considered as a transition to the ionization process where the electron is stripped off of the atom.

The effect of the oscillation strength ρ_0 is studied in Figs. 7.12,7.13. Similar as in Figs. 7.10,7.11 we have chosen κ values at the resonance minimum and maximum: $\kappa = 0.61$ and $\kappa = 0.72$. Switching on the oscillation leads to a mostly linear decrease of energy levels for the minimum (Fig. 7.12), while energies increase to a maximum value in the other case (left half of Fig. 7.13). For $\rho_0 > 0.1$ there is no further change since the wave functions have shifted to the outer region where changes in the repulsion potential near to the core have no effect. Again there is an inversion of the angular momentum dependence. For $\kappa_0 \rightarrow 0$ one obtains the original orbital energies of the H atom as expected.

Finally we make some general remarks on the resonance effect in Hydrogen. The numerical results have shown that there is a resonance effect as predicted by the theory and found experimentally in solid materials by the Mexican group. The mechanism of raising the binding energy of the valence electron by resonance has been demonstrated in several resonance diagrams. The ionization effect itself has not been modeled because this requires the inclusion of non-normalizable continuum states. In addition to the raising of energy, there is always a drop of the binding energy at the opposite side of the resonant wavenumber.

If the frequency of excitation charge density ρ has a certain bandwidth, it is plausible that two effects are initiated at the same time: production of conduction electrons by raising energies and production of very deeply bound electrons by lowering energies. Both effects are a consequence of resonant interaction with spacetime. The lowering of electronic states reduces thermal vibration amplitudes which may result in a macroscopic temperature reduction. Thus the experimental finding can be explained that effects of spacetime resonances are often accompanied by a decrease of temperature, an effect, which seems to contradict thermodynamics if one tries to explain it by standard theory.

After the spacetime resonance of the H atom has been studied, the next steps are to investigate these effects in many-electron atoms, molecules and solids. For this, the covariant form of the Coulomb law (equation (7.10)) has to be used in numerical methods which are available today for calculating electronic properties of these materials.

Acknowledgements The British Parliament, Prime Minister and Head of State are thanked fro the award of a Civil List Pension in recognition of distinguished service to science. The staff of AIAS are thanked for many interesting and formative discussions.

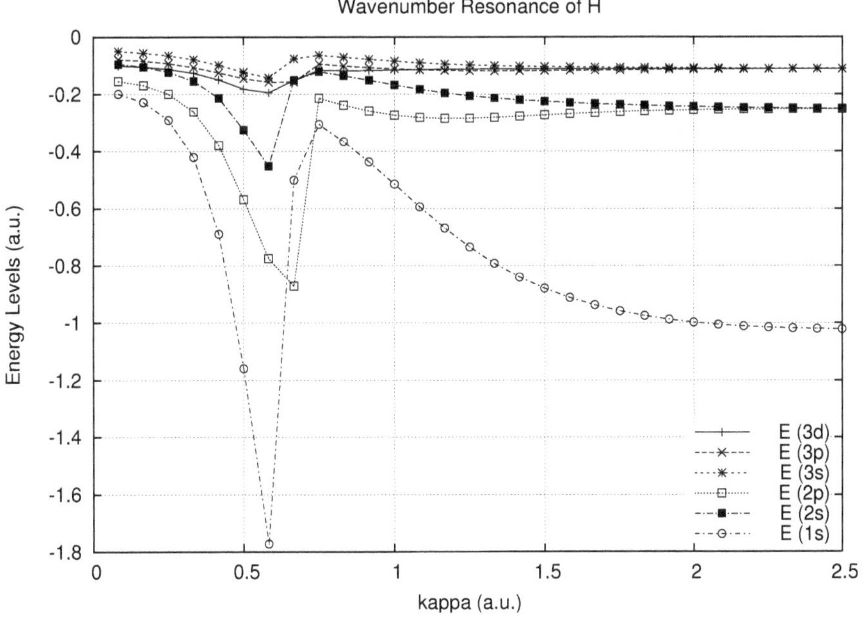

Figure 7.1: Wavenumber resonance of H

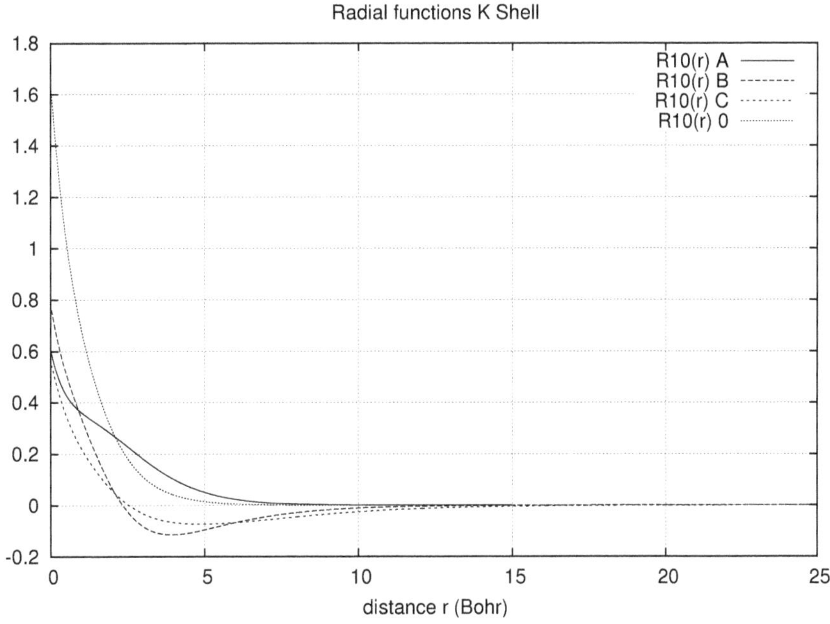

Figure 7.2: Radial 1s wavefunction. A: $\kappa = 0.60$ B: $\kappa = 0.75$ C: $\kappa = 2.0$ O: off-resonance

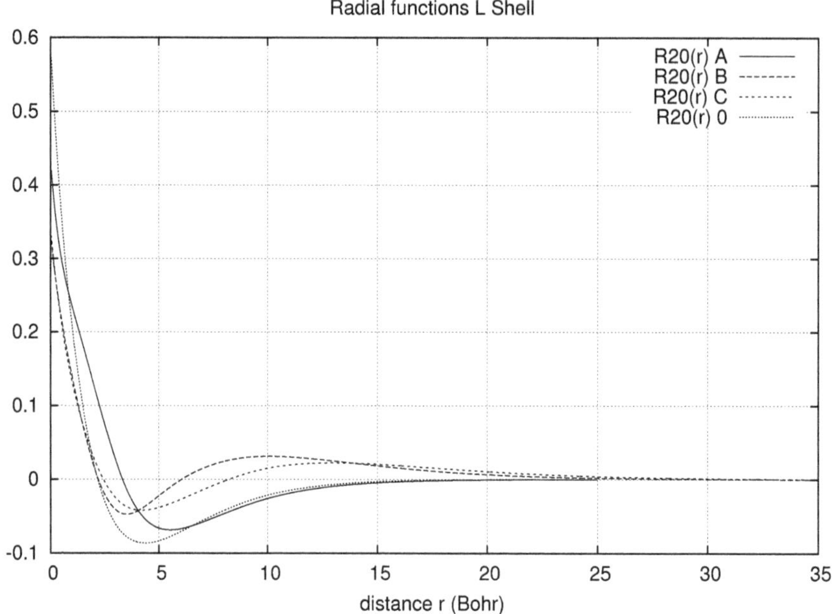

Figure 7.3: Radial 2s wavefunction. A: $\kappa = 0.60$ B: $\kappa = 0.75$ C: $\kappa = 2.0$ O: off-resonance

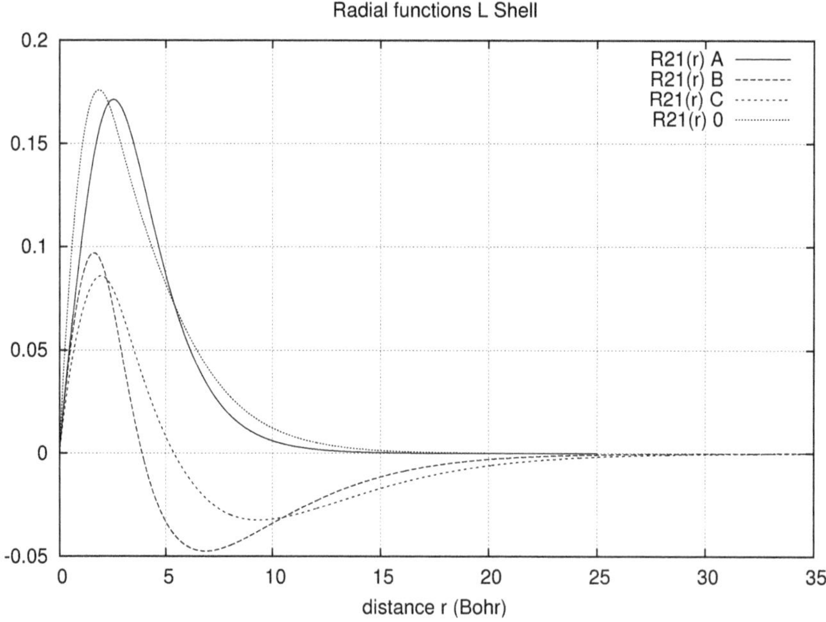

Figure 7.4: Radial 2p wavefunction. A: $\kappa = 0.60$ B: $\kappa = 0.75$ C: $\kappa = 2.0$ O: off-resonance

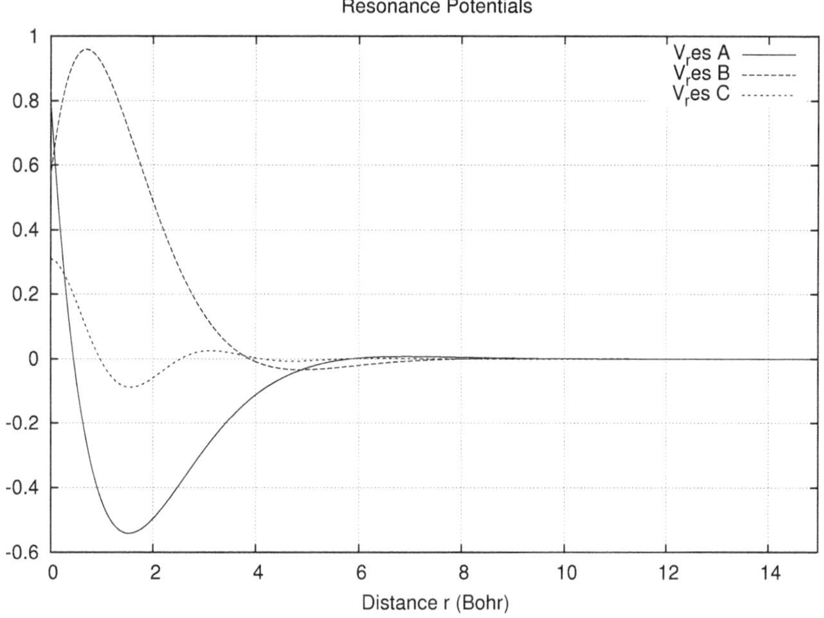

Figure 7.5: Resonant potential. A: $\kappa = 0.60$ B: $\kappa = 0.75$ C: $\kappa = 2.0$

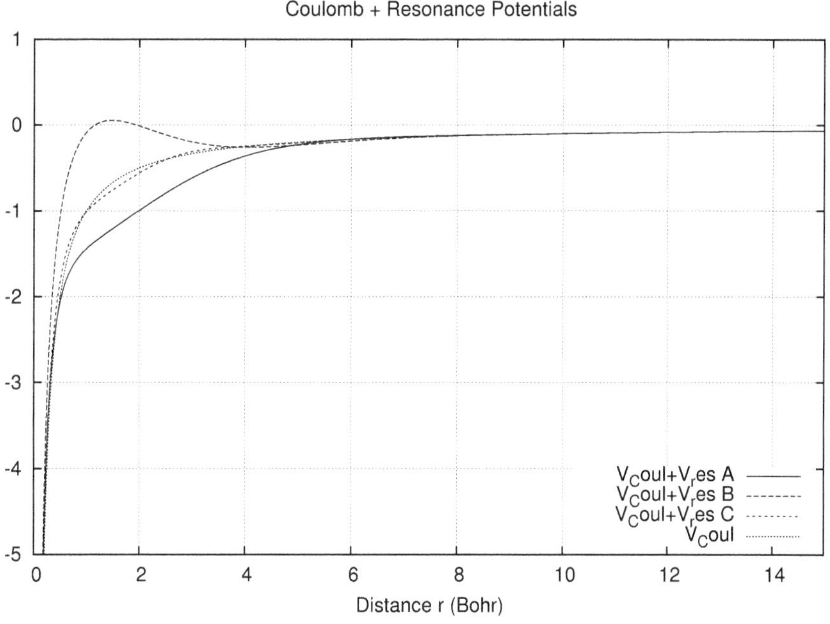

Figure 7.6: Total potential. A: $\kappa = 0.60$ B: $\kappa = 0.75$ C: $\kappa = 2.0$

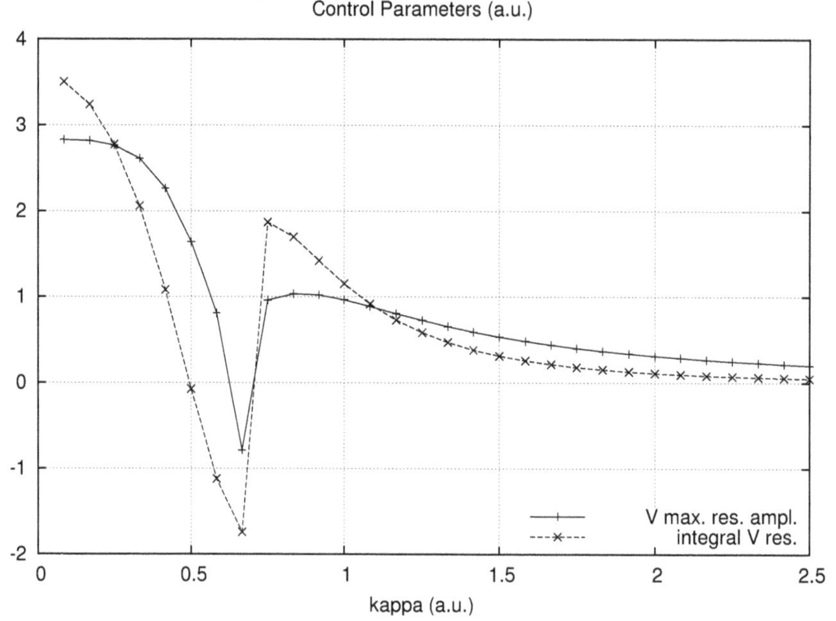

Figure 7.7: Control parameters

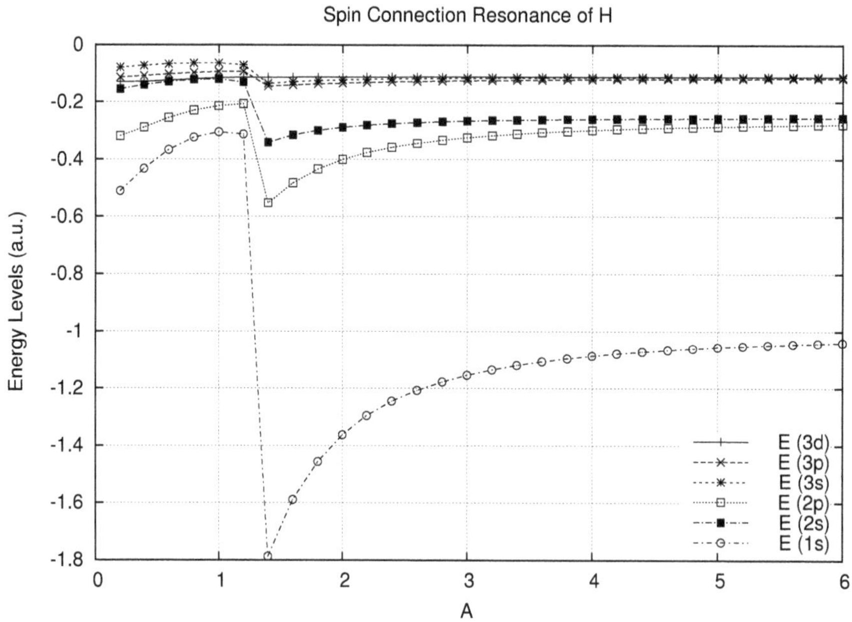

Figure 7.8: Spin connection resonance

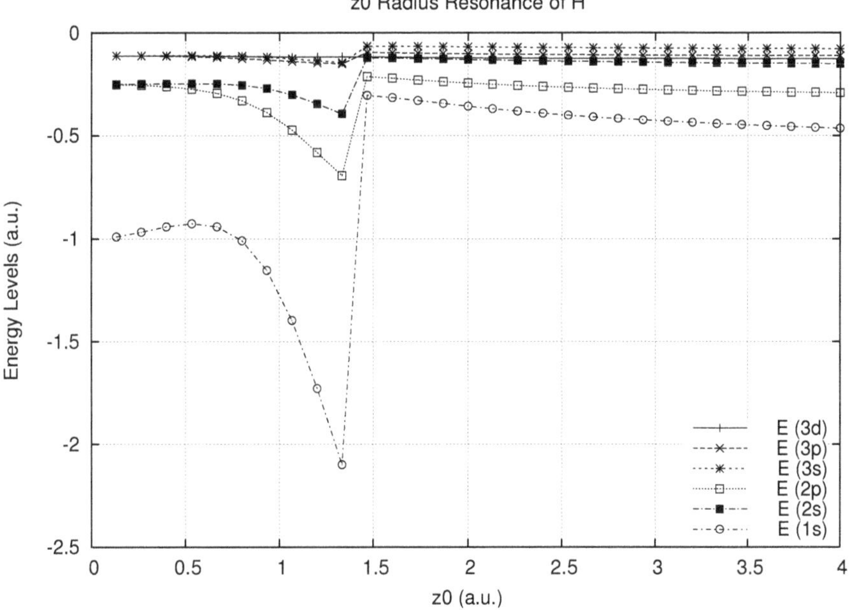

Figure 7.9: z_0 radius resonance

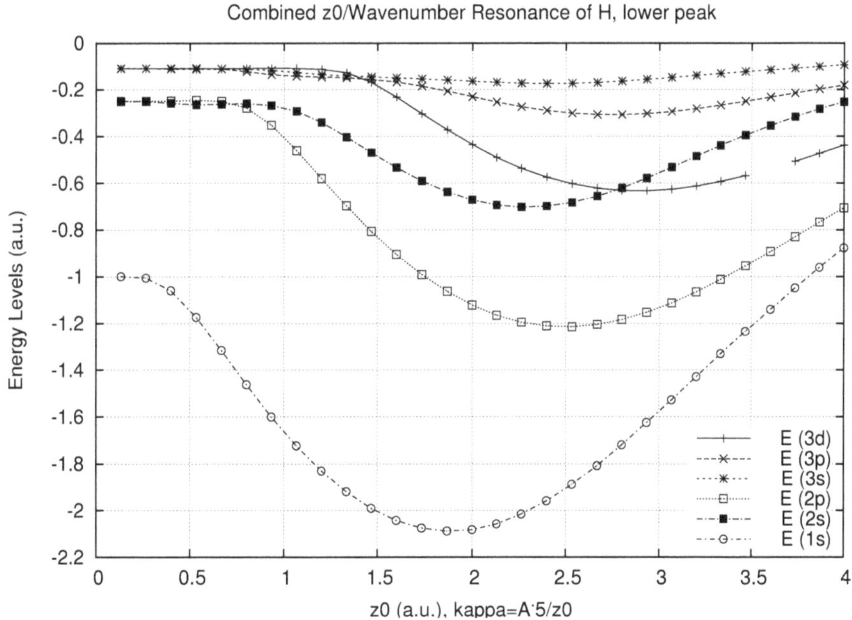

Figure 7.10: Combined z_0/wavenumber resonance, lower energies

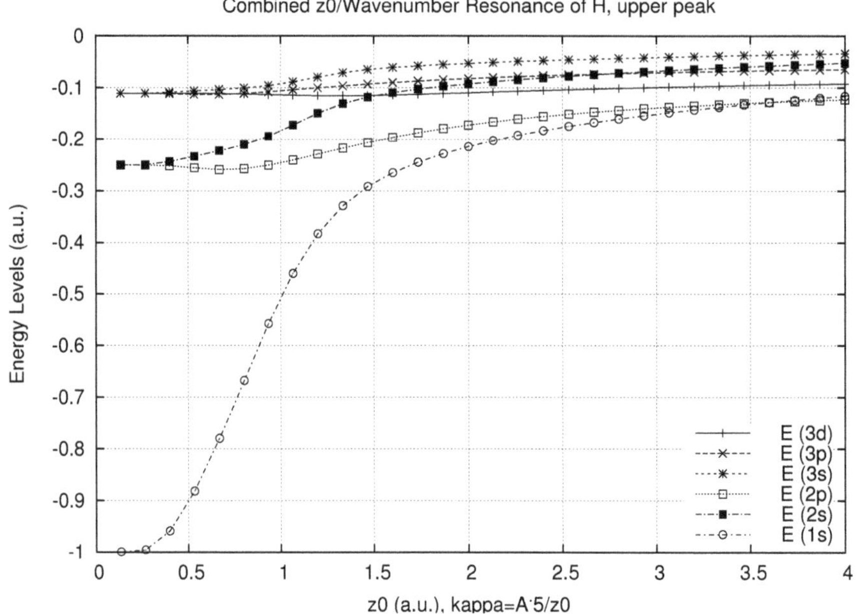

Figure 7.11: Combined z_0/wavenumber resonance, upper energies

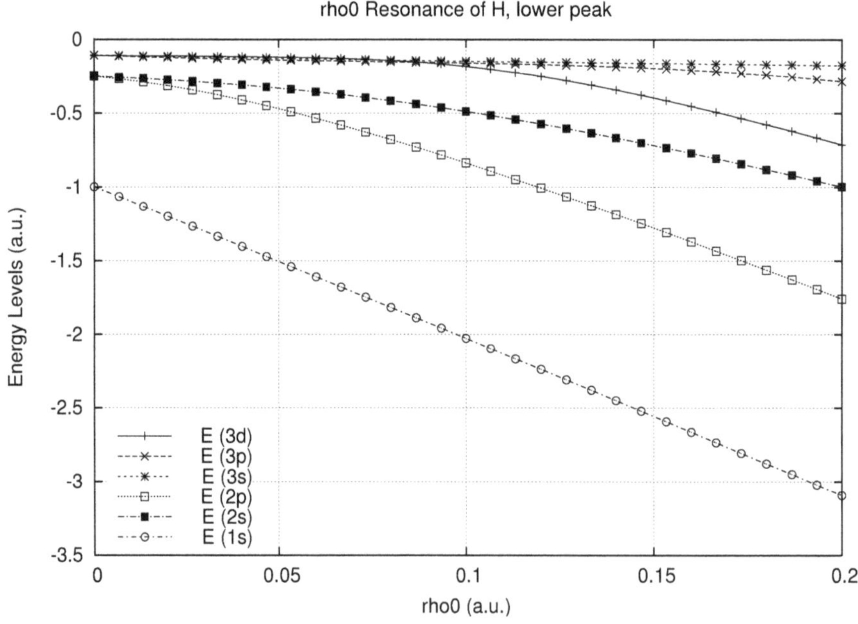

Figure 7.12: Combined ρ_0 resonance, lower energies

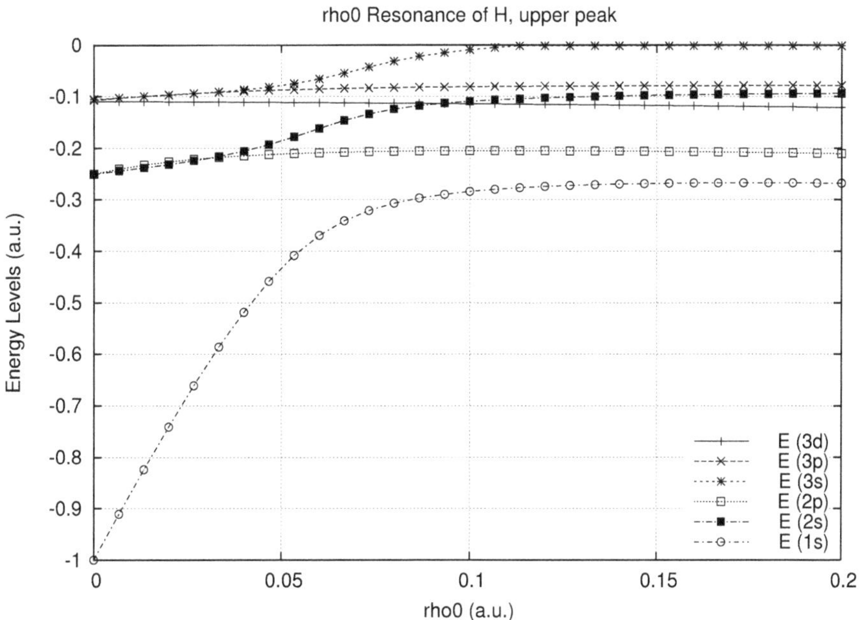

Figure 7.13: ρ_0 resonance, upper energies

Bibliography

[1] NASA Cassini experiments (2002 to present).

[2] M. W. Evans, Generally Covariant Unified Field Theory (Abramis 2005), vol. 1.

[3] M. W. Evans, Generally Covariant Unified Field Theory (Abramis, 2006), vol. 2.

[4] M. W. Evans, Generally Covariant Unified Field Theory (Abramis, 2006), vol. 3.

[5] M. W. Evans, Generally Covariant Unified Field Theory (Abramis, 2007, in prep., preprint papers 55 to 70 available to date on www.aias.us and www.atomicprecision.com). , vol. 4.

[6] L. Felker, The Evans Equations of Unified Field Theory (preprint on www.aias.us and www.atomicprecision.com).

[7] H. Eckardt and L. Felker, papers on www.aias.us and www.atomicprecision.com.

[8] M. W. Evans, Generally Covariant Dynamics(paper 55 of the ECE series, www.aias.us and www.atomicprecision.com)

[9] M. W. Evans, Geodesics and the Aharonov Bohm Effects ibid., paper 56).

[10] M. W. Evans, Canonical and Second Quantization in Generally Covariant Quantum Field Theory (ibid., paper 57).

[11] M. W. Evans, The Effect of Torsion on the Schwarzschild Metric and Light Deflection due to Gravitation (ibid., paper 58).

[12] M. W. Evans, The Resonance Coulomb Law from ECE Theory: Application to the Hydrogen Atom. (ibid., paper 59).

[13] M. W. Evans, Application of Einstein Cartan Evans (ECE) Theory to Atoms and Molecules: Free Electrons at Resonance (ibid., paper 60).

[14] M. W. Evans, papers and letters in Foundations of Physics and Foundations of Physics Letters, 1993 to present.

[15] M. W. Evans, (ed.), Modern Non-Linear Optics, a special topical issue in three parts of I. Prigogine and S. A. Ricc, (series eds.), Advances in Chemical Physics (Wiley-Interscience, New York, 2001, 2nd. ed.), vols. 119(1), 119(2), 119(3) (endorsed by the Royal Swedish Academy).

[16] M. W. Evans and L. B. Crowell, Classical and Quantum Electrodynamics and the $\boldsymbol{B}^{(3)}$ Field (World Scientific, Singapore, 2001).

[17] M. W. Evans and J.-P. Vigier, The Enigmatic Photon (Kluwer, Dordrecht, 1994 to 2002, hardback and softback), in five volumes.

[18] M. W. Evans and S. Kielich (eds.), first edition of ref. (15) (Wiley Interscience, New York, 1992, 1993, 1997) vols. 85(1), 85(2) and 85(3).

[19] L. H. Ryder, Quantum Field Theory (Cambridge Univ. Press, 2nd. Ed., 1996).

[20] J. D. Jackson, Classical Electrodynamics (Wiley, 1999, 3rd. ed.) chapter one.

[21] S. P. Carroll, Space-time and Geometry, an Introduction to General Relativity (Addison-Wesley, New York, 2004).

[22] Feedback sites to www.aias.us, and www.atomicprecision.com recording essentially complete acceptance of ECE theory worldwide as a workable unified field theory

[23] E. G. Milewski (Chief Ed.), The Vector Analysis Problem Solver (Research and Education Association, New York, 1987).

[24] J. B. Marion and S. T. Thornton, Classical Dynamics of Particles and Systems (HB College Publishers, New York, 1988, 3rd ed.) Chapter 3.

[25] M. W. Evans, Physica B, 182, 227, 237 (1992).

[26] Conference on the Mexican group devices, Mexico City Feb. 2006, and in Europe, observers from the U.S., Canada and Europe, also demonstrated at the U.S. Navy.

[27] S. E. Koonin, Computational Physics, Benjamin/Cummings Publishing Company, 1986, pp. 49-62.

Chapter 8

Application of the ECE Lemma to the fermion and electromagnetic fields

(Paper 62)
by
Myron W. Evans
Alpha Institute for Advanced Study (AIAS).
(emyrone@aol.com, www.aias.us, www.atomicprecision.com)

Abstract

The Einstein Cartan Evans (ECE) Lemma is illustrated with reference to the fermion field (for example a Dirac electron) and the electromagnetic field. In both cases the scalar curvature R of the Lemma must be identically non-zero in a generally covariant unified field theory. In the case of the electromagnetic field this result means that the mass of the photon must be identically non-zero. If not, the electric and magnetic components of the electromagnetic field vanish.
Keywords: Einstein Cartan Evans (ECE) Lemma, fermion field, electromagnetic field, generally covariant unified field theory, photon mass.

8.1 Introduction

The Einstein Cartan Evans (ECE) field theory [1]– [18] is a generally covariant unified field theory as required by objectivity in the natural sciences. ECE theory unifies quantum mechanics and general relativity using the ECE Lemma and ECE wave equation. The former is derived straightforwardly (Section 8.2) from the well known [19] tetrad postulate of Cartan's geometry (differential geometry). In this paper the ECE Lemma is illustrated with respect to the fermion and electromagnetic fields (Section 8.3) and found to be self-consistent in its generally covariant description of both fields. In general relativity (ECE field theory) the photon mass must be identically non-zero and this result is derived self-consistently from the ECE Lemma through the finding that its eigenvalues, the scalar curvatures R, must be identically non-zero. The ECE Lemma shows that if the photon mass is identically zero, the electric and magnetic components of the electromagnetic field disappear. In the limit of zero Cartan torsion, ECE theory reduces to Einstein Hilbert (EH) field theory, the original and well known theory of gravitational general relativity published in 1916. The EH theory explains the bending of light by gravity to a precision of one part in 100,000 in the Solar System. This semi-classical explanation self-consistently depends on identically non-zero photon mass, without which the bending of light does not occur semi-classically.

8.2 Derivation of the ECE Lemma from the tetrad postulate

The derivation of the ECE Lemma [1]- [18] from the tetrad postulate is straightforward and is given here for convenience of reference. Begin with the tetrad postulate [19]:

$$D_\mu q^a{}_\nu = 0 \tag{8.1}$$

where D_μ denotes covariant derivative and $q^a{}_\nu$ is the tetrad form. The latter is a mixed index tensor, a vector-valued one-form of differential geometry [1]- [19]. The basic rules of covariant differentiation of a rank two mixed index tensor are given in ref. [19] and applied self-consistently in ECE field theory [1]- [18]. These rules mean that:

$$D_\mu q^a{}_\lambda = \partial_\mu q^a{}_\lambda + \omega^a{}_{\mu b} q^b{}_\lambda - \Gamma^\nu_{\mu\lambda} q^a{}_\nu = 0 \tag{8.2}$$

where $\Gamma^\nu_{\mu\lambda}$ is the general gamma connection of Riemann geometry and where $\omega^a{}_{\mu b}$ is the spin connection of Cartan geometry. Summation over repeated contravariant-covariant indices is implied as usual (the Einstein convention). This summation applies to both Greek and Latin repeated indices wherever they occur in an equation. The ECE Lemma is an identity:

$$D^\mu (D_\mu q^a{}_\nu) := 0 \tag{8.3}$$

formed by covariant differentiation of the tetrad postulate. For a scalar [1]- [19]:

$$D^\mu \phi = \partial^\mu \phi \tag{8.4}$$

All elements in Eq.(8.1) are zero. Applying covariant differentiation to each element:

$$D^\mu 0 = \partial^\mu 0 = 0 \tag{8.5}$$

Therefore Eq.(8.3) becomes:

$$\partial^\mu \left(\partial_\mu q^a{}_\lambda + \omega^a{}_{\mu b} q^b{}_\lambda - \Gamma^\nu_{\mu\lambda} q^a{}_\nu \right) = 0 \tag{8.6}$$

This result has been checked for internal self consistency [1]- [18] by applying the rules [19] for covariant differentiation of a rank three mixed index tensor as follows:

$$\begin{aligned} D_\mu (D^\mu q^a{}_\nu) = \\ \partial_\mu (D^\mu q^a{}_\nu) + \Gamma^\mu_{\mu\lambda} D^\lambda q^a{}_\nu + \omega^a{}_{\mu b} D^\mu q^b{}_\nu - \Gamma^\lambda_{\mu\nu} D^\mu q^a{}_\lambda \\ = \partial_\mu (D^\mu q^a{}_\nu) \end{aligned} \tag{8.7}$$

using the Lemma (8.3) in Eq.(8.7) produces:

$$\partial^{\mu}(D_{\mu}q^{a}{}_{\nu}) = 0 \tag{8.8}$$

which is Eq.(8.6) again, Q.E.D. Various other cross-checks of the Lemma are given in the literature [1]- [18]. Eq.(8.6) can be rewritten as:

$$\Box q^{a}{}_{\lambda} = \partial^{\mu}\left(\Gamma^{\nu}_{\mu\lambda}q^{a}{}_{\nu} - \omega^{a}{}_{\mu b}q^{b}{}_{\lambda}\right) \tag{8.9}$$

where

$$\Box := \partial^{\mu}\partial_{\mu} \tag{8.10}$$

is the d'Alembertian operator. Now define the scalar curvature by:

$$Rq^{a}{}_{\lambda} := \partial^{\mu}\left(\Gamma^{\nu}_{\mu\lambda}q^{a}{}_{\nu} - \omega^{a}{}_{\mu b}q^{b}{}_{\lambda}\right) \tag{8.11}$$

Using the rule [19] for tetrad normalization:

$$q^{a}{}_{\lambda}q^{\lambda}{}_{a} = 4 \tag{8.12}$$

multiply both sides of Eq.(8.11) by $q^{\lambda}{}_{a}$ to obtain:

$$R := \frac{1}{4}q^{\lambda}{}_{a}\partial^{\mu}\left(\Gamma^{\nu}_{\mu\lambda}q^{a}{}_{\nu} - \omega^{a}{}_{\mu b}q^{b}{}_{\lambda}\right) \tag{8.13}$$

With this definition the ECE Lemma is [1]- [18]

$$\Box q^{a}{}_{\lambda} = Rq^{a}{}_{\lambda} \tag{8.14}$$

and is an Eigen-equation or wave equation of generally covariant unified field theory. It unifies quantum mechanics and general relativity. The ECE Lemma is the subsidiary proposition that leads to the ECE wave equation [1]- [18]:

$$(\Box + kT)q^{a}{}_{\lambda} = 0 \tag{8.15}$$

where k is Einstein's constant [20] and T is the index contracted canonical energy-momentum density. In ECE theory the Einstein Ansatz:

$$R = -kT \tag{8.16}$$

is applied to the unified field.

8.3 The fermion and electromagnetic fields

The scalar curvature in Eq.(8.13) may be expressed as:

$$R = q^{\nu}{}_{a}R^{a}{}_{\nu} \tag{8.17}$$

where:

$$R^{a}{}_{\nu} = \partial^{\mu}\left(\Gamma^{a}_{\mu\nu} - \omega^{a}{}_{\mu\nu}\right) \tag{8.18}$$

and

$$\begin{aligned}\Gamma^{a}_{\mu\lambda} &= \Gamma^{\nu}_{\mu\lambda}q^{a}{}_{\nu}\\ \omega^{a}{}_{\mu\lambda} &= \omega^{a}{}_{\mu b}q^{b}{}_{\lambda}\end{aligned} \tag{8.19}$$

The mixed index tensor R^{a}_{ν} is similar to the well known Ricci tensor of gravitational general relativity. Now assume that the tetrad is a traveling wave in the Z axis:

$$q^{a}{}_{\lambda} = q^{a}{}_{\lambda}(0)e^{i(\omega t - \kappa Z)} \tag{8.20}$$

and let

$$\kappa = \frac{\omega}{v}$$

where v is the phase velocity. Then:

$$\begin{aligned}\Box q^a{}_\lambda &= \left(\frac{1}{c^2}\frac{\partial^2}{\partial t^2} - \frac{\partial^2}{\partial Z^2}\right) q^a{}_\lambda(0) e^{i(\omega t - \kappa Z)} \\ &= \left(\kappa^2 - \frac{\omega^2}{c^2}\right) q^a{}_\lambda\end{aligned} \tag{8.21}$$

Therefore:

$$R = \kappa^2 - \frac{\omega^2}{c^2} = \omega^2 \left(\frac{1}{v^2} - \frac{1}{c^2}\right) \tag{8.22}$$

and

$$R^a{}_\nu = \omega^2 \left(\frac{1}{v^2} - \frac{1}{c^2}\right) q^a{}_\nu \tag{8.23}$$

From Eq.(8.18):

$$\begin{aligned}(\Gamma^a - \omega^a)_{30} &= R \int q^a{}_0(0) e^{i(\omega t - \kappa Z)} dZ \\ &= \frac{iR}{\kappa} q^a{}_0(0) e^{i(\omega t - \kappa Z)}\end{aligned} \tag{8.24}$$

whose real part is:

$$Re(\Gamma^a - \omega^a)_{30} = -\frac{R}{\kappa} q^a{}_0(0) sin(\omega t - \kappa Z) \tag{8.25}$$

These results are useful to understand the Dirac equation. The latter is obtained in the following limit of the ECE wave equation [1]– [18]:

$$R = -\left(\frac{mc}{\hbar}\right)^2 \tag{8.26}$$

From Eqs.(8.22) and (8.26):

$$\kappa^2 - \frac{\omega^2}{c^2} = -\frac{m^2 c^2}{\hbar^2} \tag{8.27}$$

i.e.

$$\frac{\omega^2}{c^2} = \kappa^2 + \frac{m^2 c^2}{\hbar^2} \tag{8.28}$$

Now use the well known Planck/Einstein and de Broglie quantum relations:

$$E = \hbar \omega \ , \ p = \hbar \kappa \tag{8.29}$$

and Eq.(8.28) becomes the well known Einstein equation of special relativity:

$$E^2 = p^2 c^2 + m^2 c^4 \tag{8.30}$$

where E is the total relativistic energy, p is the relativistic momentum and:

$$E_0 = mc^2 \tag{8.31}$$

is the rest energy. The well known Compton wavelength is:

$$\lambda = \frac{\hbar}{mc} \tag{8.32}$$

The Dirac equation is therefore [1]– [18]:

$$\left(\Box+\frac{m^2c^2}{\hbar^2}\right)q^a{}_\nu=0 \tag{8.33}$$

where:

$$\begin{aligned} q^a{}_\nu &= q^a{}_\nu(0)e^{i(\omega t-\kappa Z)} \\ &:= q^a{}_\nu(0)e^{ix_\mu p^\mu} \end{aligned} \tag{8.34}$$

By convention [21] the positive energy plane wave spinor is:

$$q^a{}_\nu=q^a{}_\nu(0)e^{-ix_\mu p^\mu} \tag{8.35}$$

and the negative energy plane wave spinor is:

$$q^a{}_\nu=q^a{}_\nu(0)e^{ix_\mu p^\mu} \tag{8.36}$$

So the Ricci type curvature tensor for the Dirac electron is:

$$R^a{}_\nu=\left(\kappa^2-\frac{\omega^2}{c^2}\right)q^a{}_\nu \tag{8.37}$$

and the connection element in Eq.(8.25) is given for the Dirac equation and any fermion by using Eq.(8.22) for R.

The Dirac equation is therefore a manifestation of ECE space-time [1]– [18] in the particular case defined by Eq.(8.22) for R and Eq.(8.37) for R^a_ν. The Einstein equivalence principle [20] states that this particular case is the limit when the fermion field has become independent of the gravitational field. This is the limit of special relativity according to Einstein's original equivalence principle. More generally in ECE theory it is the case where the fermion field has become independent of ALL other fields, not only of the gravitational field. In order to apply the ECE Lemma to the electromagnetic field the ECE Ansatz is used [1]– [18]:

$$A^a_\mu=A^{(0)}q^a{}_\mu \tag{8.38}$$

following a suggestion made by Cartan to Einstein in well known correspondence that the electromagnetic field be the Cartan torsion within a factor $A^{(0)}$. Here $cA^{(0)}$ has the units of volts and is referred to in ECE theory as the primordial voltage. Therefore for the electromagnetic field the ECE Lemma is:

$$\Box A^a_\mu=RA^a_\mu \tag{8.39}$$

In the limit of special relativity Eq.(8.39) becomes the Proca equation [1]– [18]:

$$\left(\Box+\frac{m^2c^2}{\hbar^2}\right)A^a_\mu=0 \tag{8.40}$$

where m is the identically non-zero mass of the photon. If it is asserted for the sake of argument that in Eq.(8.39):

$$R=0 \tag{8.41}$$

it follows that:

$$\partial^\mu\left(\Gamma^a_{\mu\nu}-\omega^a{}_{\mu\nu}\right)=0 \tag{8.42}$$

so:

$$\Gamma^a_{\mu\nu} = \omega^a{}_{\mu\nu} + \zeta^a_{\mu\nu}(0) \tag{8.43}$$

where $\zeta^a_{\mu\nu}(0)$ is a constant of integration independent of x^μ. It follows that in ECE theory:

$$\Gamma^a_{\mu\nu} = \omega^a{}_{\mu\nu} \tag{8.44}$$

because if $\zeta^a_{\mu\nu}(0)$ is independent of x^μ it cannot present curvature or torsion. Any ECE field must be curvature or torsion or a combination thereof. If there is no curvature or torsion there is no field present and the constant of integration may be set to zero. So:

$$\Gamma^a_{\mu\nu} = \omega^a{}_{\mu\nu} \text{ if } R = 0 \tag{8.45}$$

for identically zero scalar curvature. It follows from Eq.(8.42) that:

$$\partial_\nu q^a{}_\mu = 0 \tag{8.46}$$

in this limit. In gravitational theory the tetrad is the fundamental field and Eq.(8.46) indicates that the tetrad is independent of x^ν, meaning again no curvature and no field. So if the scalar curvature is identically zero there is no gravitational field, a self-consistent result. Similarly for identically zero R the electromagnetic field obeys:

$$\partial_\nu A^a_\mu = 0 \tag{8.47}$$

and A^a_μ is independent of x^ν, meaning that there is no Cartan torsion and no electric or magnetic field present. The absence of R means the absence of photon mass, so the absence of photon mass means in turn the absence of electric and magnetic components of the electromagnetic field. Therefore in a generally covariant unified field theory the photon mass must be identically non-zero. This is again a self-consistent result because without mass in general relativity there is no field. In special relativity (Maxwell Heaviside theory) there is no concept of photon mass, and the Proca equation for identically zero photon mass is used for the free electromagnetic field:

$$\Box A^a_\mu = 0 \tag{8.48}$$

This is however inconsistent with general relativity and is therefore not objective physics. If there are interacting fields (as in the bending of light by gravity) the T term in the ECE wave equation must be constructed from contributions from all fields, including interactive terms, as originally inferred [20] by Einstein himself. In Maxwell Heaviside (MH) theory there is no mechanism for this, even on a conceptual level, because in MH theory the electromagnetic field is an entity superimposed on flat or Minkowski space-time.

Acknowledgements The British Parliament, Prime Minister and Head of State are thanked for the award of a Civil List pension in recognition of distinguished contributions to Britain and the Commonwealth in science. The AIAS intellectual environment is thanked for many interesting and formative discussions.

Bibliography

[1] M. W. Evans, Generally Covariant Unified Field Theory (Abramis, 2005), vol. 1.

[2] M. W. Evans, Generally Covariant Unified Field Theory (Abramis, 2006), vol. 2.

[3] M. W. Evans, Generally Covariant Unified Field Theory (Abramis, 2006), vol. 3.

[4] M. W. Evans, Generally Covariant Unified Field Theory (Abramis, 2007, in press, preprint of papers 55 to 70 on www.aias.us and www.atomicprecision.com), vol. 4.

[5] L. Felker, The ECE Equations of Unified Field Theory (preprint on www.aias.us and www.atomicprecision.com). ; H. Eckardt and L. Felker, papers on these websites.

[6] M. W. Evans, Generally Covariant Dynamics (paper 55 of the ECE series on www.aias.us and www.atomicprecision.com).

[7] M. W. Evans, Geodesics and the Aharonov Bohm Effects (ibid., paper 56).

[8] M. W. Evans, Canonical and Second Quantization in Generally Covariant Quantum Field Theory (ibid., paper 57).

[9] M. W. Evans, The Effect of Torsion on the Schwarzschild Metric and Light Deflection due to Gravitation (ibid., paper 58).

[10] M. W. Evans, The Resonance Coulomb Law from ECE Theory: Application to the Hydrogen Atom (ibid., paper 59).

[11] M. W. Evans, Application of Einstein Cartan Evans (ECE) Theory to Atoms and Molecules: Free Electrons at Resonance (ibid., paper 60).

[12] M. W. Evans and H. Eckardt, Space-time Resonances in the Coulomb Law (ibid., paper 61).

[13] M. W. Evans, papers and letters in Foundations of Physics and Foundations of Physics Letters from 1994 to present on $\boldsymbol{B}^{(3)}$ theory, O(3) electrodynamics and ECE theory.

[14] M. W. Evans (ed.), Modern Non-linear Optics, a special topical issue of I Prigogine and S. A. Rice (Series Eds.), Advances in Chemical Physics (Wiley-Interscience, New York, 2001, 2nd. Ed.), vols. 119(1) to 119(3) (endorsed by the Royal Swedish Academy in honor of the late Jean-Pierre Vigier).

[15] M. W. Evans and S. Kielich (eds.), ibid., first edition (Wiley-Interscience, New York, 1992, 1993, 1997), vols. 85(1) to 85(3), awarded a prize for excellence by the Polish Government.

[16] M. W. Evans and L. B. Crowell, Classical and Quantum Electrodynamics and the $\boldsymbol{B}^{(3)}$ Field (World Scientific, Singapore, 2001).

[17] M. W. Evans and J.-P. Vigier, The Enigmatic Photon (Kluwer, Dordrecht, 1994 to 2002, hardback and softback), in five volumes.

[18] M. W. Evans and A. A. Hasanein, The Photomagneton in Quantum Field Theory (World Scientific, Singapore, 1994).

[19] S. P. Carroll, Space-time and Geometry, an Introduction to General Relativity (Addison-Wesley, New York, 2004).

[20] A. Einstein, The Meaning of Relativity (Princeton Univ. Press, 1921 to 1954).

[21] L. H. Ryder, Quantum Field Theory (Cambridge Univ. Press, 2nd ed., 1996).

Chapter 9

The resonant Coulomb Law of Einstein Cartan Evans Field Theory

(Paper 63)
by
Myron W. Evans and Horst Eckardt
Alpha Institute for Advanced Study (AIAS).
(emyrone@aol.com, www.aias.us, www.atomicprecision.com)

Abstract

Einstein Cartan Evans (ECE) field theory is used to show that in general relativity the structure of the laws of electricity, magnetism and electromagnetism is changed fundamentally. This result is demonstrated analytically and numerically with the Coulomb Law. In ECE theory the electromagnetic field is spinning space-time, characterised by the spin connection of Cartan geometry. The spin connection is shown in this paper to change the Poisson equation into a differential equation capable of giving resonance. Off resonance, the standard Poisson equation is observed, and the standard Coulomb Law. At space-time resonance the scalar potential in volts is greatly amplified with fundamental consequences in the natural, engineering and life sciences.

Keywords: Einstein Cartan Evans (ECE) field theory, generally covariant unified field theory, electricity, magnetism, electromagnetism, classical electrodynamics, Coulomb Law, space-time resonance.

9.1 Introduction

The need for objectivity in the natural, engineering and life sciences means that all the fundamental laws of physics must be laws of general relativity, where objectivity is represented by geometry. This includes the laws of classical electrodynamics, the laws of electricity, magnetism and electromagnetism. General relativity means that these laws must be generally covariant, must retain their form under any type of coordinate transformation. They must therefore be laws of a generally covariant unified field theory. Recently [1]- [19] the Einstein Cartan Evans (ECE) theory has been developed along these well known guidelines. Objectivity in ECE field theory is maintained through the use of standard Cartan geometry [20]. The electromagnetic field is represented as the Cartan torsion within a scalar factor $A^{(0)}$. Here $cA^{(0)}$ is a primordial voltage. This procedure follows a well known suggestion by Cartan to Einstein that the electromagnetic field be the Cartan torsion form, a vector valued two-form of differential geometry [1]- [20]. ECE theory applies Cartan's suggestion systematically to all the laws of physics.

The well known Maxwell Heaviside (MH) field theory is used in the standard model [21] to represent the laws of electricity, magnetism and electromagnetism. The MH theory is a nineteenth century theory of special relativity and is neither generally covariant nor unified with other fundamental fields such as gravitation. It is a Lorentz covariant theory in which the electromagnetic field is considered to be an entity separate from the frame of reference in a Minkowski space-time. The latter is often referred to as flat space-time, because it has neither curvature nor torsion. It is also a static space-time. The laws of gravitation on the other hand are described in the standard model by the Einstein Hilbert (EH) field theory of general relativity [1]- [20] and the gravitational field is the frame itself, not something separate from the frame as in MH theory. EH space-time has curvature but no torsion, and is a dynamic space-time. Objectivity in EH field theory is based on Riemann geometry with a Christoffel connection. This assumption implies a zero torsion tensor [20] and means that gravitation cannot be unified with electromagnetism in EH theory. As we have argued, electrodynamics cannot be unified with gravitation in MH theory. In ECE theory [1]- [19] a unified description of all fields has been developed straightforwardly using ECE space-time in which curvature and torsion are simultaneously non-zero. The generally covariant unified field of ECE theory is the frame itself, as required by general relativity and objectivity. Gravitation is described by curvature, the electromagnetic, weak and strong fields by torsion using the appropriate representation spaces (respectively $O(3)$, $SU(2)$ and $SU(3)$). The two Cartan structure equations and the two Bianchi identities of differential geometry control all the laws of physics [1]- [19], including those of quantum mechanics through the tetrad postulate [20]. The latter is the fundamental requirement that the complete vector field in n dimensions be independent of the components and basis elements chosen to represent it. Thus ECE theory has unified quantum mechanics with general relativity and provides a generally covariant unified field theory. ECE theory has therefore been accepted as mainstream physics [22].

In Section 9.2, the Coulomb law of electricity is developed with the spin connection incorporated as required by general relativity, by the fact that the complete electromagnetic field is spinning space-time. A spinning of space-time means a spinning of the frame of reference itself. This means that the spin connection must always be non-zero.

The Coulomb Law is derived in ECE theory [1]- [19] from the first Cartan structure equation and the first Bianchi identity. Use of vector notation and some simplifying assumptions [1]- [19] lead to the initial equations of Section 9.2. These give a resonance equation whose properties are developed analytically to give the required resonance solution. The latter is fundamentally important in the natural, engineering and life sciences because the Coulomb law is the basis of all quantum chemistry, and therefore the basic law of computational quantum chemistry.

In Section 9.3, numerical solutions of the resonance Coulomb law are developed to illustrate a novel space-time resonance spectrum. Off resonance the standard Coulomb Law is recovered. The standard Coulomb law is well known [21] to be among the most precise laws of physics, so it must be recovered from ECE theory in a given limit. This is achieved by identifying the radial spin connection as the one that gives the standard Coulomb law off resonance. In the off resonant condition the spin connection effectively doubles the value of the electric field, so its presence cannot be detected experimentally. At space-time resonance however the scalar potential in volts of the Coulomb Law is greatly amplified, leading to a surge in voltage that cannot be explained by MH theory. This phenomenon has been reported experimentally by several independent groups [23]. If this resonant amplification of the scalar potential occurs inside an atom or molecule, electrons may be released by ionization. In this Section the process is illustrated by the radial wave-functions of the hydrogen atom in anticipation of the systematic development of density functional code incorporating the resonant Coulomb law of ECE theory. A short review of density functional methods is also given in this section. Finally a discussion is given of how to induce space-time resonance in circuits and materials.

9.2 The Resonant Coulomb Law

In the simplest instance [1]– [19] the Coulomb law in ECE theory is given by:

$$\boldsymbol{\nabla} \cdot \mathbf{E} = \frac{\rho}{\epsilon_0} \tag{9.1}$$

where

$$\mathbf{E} = -\left(\boldsymbol{\nabla} + \boldsymbol{\omega}\right)\phi. \tag{9.2}$$

Here ϕ is the scalar potential in volts, $\boldsymbol{\omega}$ is the vector spin connection in inverse meters, $\mathbf{E}$ is the electric field strength in volts m^{-1}, ρ is the charge density in Cm^{-3}, and ϵ_0 is the S. I. vacuum permittivity:

$$\epsilon_0 = 8.854 \times 10^{-12} J^{-1} C^2 m^{-1}. \tag{9.3}$$

Thus:

$$\boldsymbol{\nabla} \cdot \left(\left(\boldsymbol{\nabla} + \boldsymbol{\omega}\right)\phi\right) = -\frac{\rho}{\epsilon_0} \tag{9.4}$$

i.e.

$$\nabla^2 \phi + \boldsymbol{\nabla} \cdot \left(\phi \boldsymbol{\omega}\right) = -\frac{\rho}{\epsilon_0} \tag{9.5}$$

If there is no spin connection, Eq.(9.5) is the Poisson equation [21] of the standard model. Otherwise:

$$\nabla^2 \phi + \boldsymbol{\omega} \cdot \boldsymbol{\nabla}\phi + \left(\boldsymbol{\nabla} \cdot \omega\right)\phi = -\frac{\rho}{\epsilon_0} \tag{9.6}$$

which is an equation capable of giving resonant solutions [1]– [19], [24] from the spin connection vector. The Poisson equation does not give resonant solutions.

Eq.(9.6) is first developed in one (Z) dimension of the Cartesian coordinate system, and then for the radial component of the spherical polar coordinate system [25].

In one Z dimension Eq.(9.6) becomes:

$$\frac{\partial^2 \phi}{\partial Z^2} + \omega_Z \frac{\partial \phi}{\partial Z} + \left(\frac{\partial \omega_Z}{\partial Z} \right) \phi = -\frac{\rho}{\epsilon_0} \tag{9.7}$$

The spin connection in Eq.(9.7) must be:

$$\omega_Z = \frac{2}{Z} \tag{9.8}$$

in order to recover the standard Coulomb law off resonance. This is because:

$$\phi = \frac{-e}{4\pi\epsilon_0 Z}, \quad \frac{\partial \phi}{\partial Z} = \frac{e}{4\pi\epsilon_0 Z^2} = -\frac{\omega_z}{2}\phi \tag{9.9}$$

in the off resonant condition, giving Eq.(9.8). In the off resonant condition the role of the spin connection is to change the sign of the electric field according to Eq.(9.9). The way in which the field $\mathbf{E}$ and potential ϕ are related is changed, but this has no experimental effect since $\mathbf{E}$ is effectively replaced by $-\mathbf{E}$. With the spin connection (9.8) Eq.(9.7) becomes:

$$\frac{\partial^2 \phi}{\partial Z^2} + \frac{2}{Z}\frac{\partial \phi}{\partial Z} - \frac{2}{Z^2}\phi = -\frac{\rho}{\epsilon_0} \tag{9.10}$$

Now assume that the charge density is initially oscillatory:

$$\rho = \rho(0)\cos(\kappa Z) \tag{9.11}$$

where κ is a wave-number. Thus:

$$\frac{\partial^2 \phi}{\partial Z^2} + \frac{2}{Z}\frac{\partial \phi}{\partial Z} - \frac{2}{Z^2}\phi = -\rho(0)\cos(\kappa Z) \tag{9.12}$$

Since ϕ depends only on Z the partial derivatives can be replaced by total derivatives to give an ordinary differential equation [24], [26]:

$$\frac{\partial^2 \phi}{\partial Z^2} + \frac{2}{Z}\frac{\partial \phi}{\partial Z} - \frac{2}{Z^2}\phi = -\rho(0)\cos(\kappa Z) \tag{9.13}$$

using the well known Euler method [24], [26] this equation can be reduced to an undamped oscillator equation that has resonant solutions. Define a change of variable [24], [26] by:

$$\kappa Z = e^{i\kappa x} \tag{9.14}$$

Thus:

$$\frac{dx}{dZ} = -\frac{i}{\kappa Z} \tag{9.15}$$

Now use:

$$\frac{d\phi}{dZ} = \frac{d\phi}{dx}\frac{dx}{dZ} = -\frac{i}{\kappa Z}\frac{d\phi}{dx} \tag{9.16}$$

and construct the second derivative:

$$\frac{d^2\phi}{dZ^2} = \frac{i}{\kappa Z^2}\frac{d\phi}{dx} - \frac{i}{\kappa Z}\frac{d}{dZ}\left(\frac{d\phi}{dx}\right) \tag{9.17}$$

Isotropy means that:

$$\frac{d}{dZ}\left(\frac{d\phi}{dx}\right)=\frac{d^2\phi}{dZdx}=\frac{d^2\phi}{dxdZ}=\frac{d}{dx}\left(\frac{d\phi}{dZ}\right) \tag{9.18}$$

so

$$\frac{d^2\phi}{dZ^2}=\frac{i}{\kappa Z^2}\frac{d\phi}{dx}-\frac{1}{\kappa^2 Z^2}\frac{d^2\phi}{dx^2}. \tag{9.19}$$

Thus:

$$Z\frac{d\phi}{dZ}=-\frac{i}{\kappa}\frac{d\phi}{dx}, \tag{9.20}$$

$$Z^2\frac{d^2\phi}{dZ^2}=\frac{i}{\kappa}\frac{d\phi}{dx}-\frac{1}{\kappa^2}\frac{d^2\phi}{dx^2}. \tag{9.21}$$

Now substitute Eqs.(9.20) and (9.21) in Eq.(9.13) to give:

$$\frac{d^2\phi}{dx^2}+2\kappa^2\phi=\frac{\rho(0)}{\epsilon_0}\text{Real}\left(e^{2i\kappa x}\cos\left(e^{i\kappa x}\right)\right) \tag{9.22}$$

which the undamped oscillator equation [24], [26]:

$$\begin{aligned}\frac{d^2\phi}{dx^2}+2\kappa^2\phi=\frac{\rho(0)}{\epsilon_0}&\left(\cos(2\kappa x)\cos(\cos(\kappa x))\cosh(\sin(\kappa x))\right.\\&\left.+\sin(2\kappa x)\sin(\cos(\kappa x))\sinh(\sin(\kappa x))\right).\end{aligned} \tag{9.23}$$

Assume that the particular integral of this equation is:

$$\phi_p(x)=\frac{A_\rho(0)}{\epsilon_0}\left(\frac{\cos(\kappa' x)}{2\kappa^2-\kappa'^2}\right) \tag{9.24}$$

where κ' is defined by the identity:

$$A\cos(\kappa' x):=\text{Real }\left(e^{2i\kappa x}\cos\left(e^{i\kappa x}\right)\right) \tag{9.25}$$

where A is a function of κ and x in general.

From Eqs.(9.22) to (9.25) it is found that:

$$\frac{d^2 f}{dx^2}+2\kappa^2 f=\cos\kappa' x \tag{9.26}$$

where:

$$f=\frac{\cos\kappa' x}{2\kappa^2-\kappa'^2} \tag{9.27}$$

Eq.(9.27) is, self-consistently, the solution of Eq.(9.26), Q.E.D.

Therefore Eq.(9.24) is a valid particular integral of Eq.(9.23) if κ' is defined by:

$$\kappa'=\frac{1}{x}\cos^{-1}\left(\frac{1}{A}\text{Real }\left(e^{2i\kappa x}\cos\left(e^{i\kappa x}\right)\right)\right) \tag{9.28}$$

At resonance:

$$2\kappa^2=\kappa'^2 \tag{9.29}$$

and ϕ_p theoretically infinite. For a given A, resonance in ϕ occurs when x is defined by:

$$
\begin{aligned}
A\cos\left(\sqrt{2}\kappa x\right) &= \cos(2\kappa x)\cos(\cos(\kappa x))\cosh(\sin(\kappa x)) \\
&\quad + \sin(2\kappa x)\sin(\cos(\kappa x))\sinh(\sin(\kappa x))
\end{aligned}
\tag{9.30}
$$

It is seen that A can be greater than unity because $\cosh y$ and $\sinh y$ can be greater than unity.

Secondly consider the radial component r in three dimensions of the spherical polar coordinate system. In this system [24]:

$$
\left.
\begin{aligned}
\nabla^2\phi &= \frac{\partial^2\phi}{\partial r^2} + \frac{2}{r}\frac{\partial\phi}{\partial r}, \\
\boldsymbol{\omega}\cdot\boldsymbol{\nabla}\phi = \omega_r\frac{\partial\phi}{\partial r}, &\quad (\boldsymbol{\nabla}\cdot\boldsymbol{\omega})\,\phi = \frac{\phi}{r^2}\frac{\partial}{\partial r}\left(r^2\omega_r\right)
\end{aligned}
\right\}
\tag{9.31}
$$

so Eq.(9.6) becomes:

$$
\frac{\partial^2\phi}{\partial r^2} + \left(\frac{2}{r} + \omega_r\right)\frac{d\phi}{dr} + \frac{\phi}{r^2}\left(2r\omega_r + r^2\frac{d\omega_r}{dr}\right) = -\frac{\rho}{\epsilon_0}
\tag{9.32}
$$

Now choose a radial spin connection:

$$
\omega_r = -\frac{1}{r}
\tag{9.33}
$$

to obtain:

$$
\frac{\partial^2\phi}{dr^2} + \frac{1}{r}\frac{\partial\phi}{\partial r} - \frac{1}{r^2}\phi = -\frac{\rho}{\epsilon_0}
\tag{9.34}
$$

This equation has the same mathematical structure as Eq.(9.13), and since ϕ is a function only of r can be written as an ordinary differential equation:

$$
\frac{\partial^2\phi}{dr^2} + \frac{1}{r}\frac{\partial\phi}{\partial r} - \frac{1}{r^2}\phi = -\frac{\rho}{\epsilon_0}
\tag{9.35}
$$

Assume now that the initial charge is oscillatory as follows:

$$
\rho = \rho(0)\cos(\kappa_r r)
\tag{9.36}
$$

and use the change of variable:

$$
\kappa_r r = e^{i\kappa_r R}
\tag{9.37}
$$

to obtain the undamped oscillator equation:

$$
\frac{d^2\phi}{dR^2} + \kappa_r^2\phi = \frac{\rho(0)}{\epsilon_0}\text{Real}\left(e^{2i\kappa_r R}\cos\left(e^{i\kappa_r R}\right)\right)
\tag{9.38}
$$

where

$$
R = \frac{1}{\kappa_r}\cos^{-1}(\kappa_r r)
\tag{9.39}
$$

Now define:

$$
A\cos(\kappa' R) := \text{Real}\left(e^{2i\kappa_r R}\cos e^{i\kappa_r R}\right)
\tag{9.40}
$$

to obtain the particular integral:

$$\phi_\rho(R) = \frac{A_\rho}{\epsilon_0}\frac{\cos(\kappa' R)}{\kappa_r^2 - \kappa'^2} \tag{9.41}$$

Resonance occurs at:

$$\kappa_r = \kappa' \tag{9.42}$$

where:

$$\begin{aligned} A\cos(\kappa_r R) = &\cos(2\kappa_r R)\cos(\cos(\kappa_r R))\cosh(\sin(\kappa_r R)) \\ &+ \sin(2\kappa_r R)\sin(\cos(\kappa_r R))\sinh(\sin(\kappa_r R)) \end{aligned} \tag{9.43}$$

The particular integral of Eq.(9.38) may be obtained by first assuming that the solution has the form:

$$\phi = \frac{A_\rho(0)}{\epsilon_0}\text{Real }\left(e^{2i\kappa_r R}\cos\left(e^{i\kappa_r R}\right)\right) \tag{9.44}$$

where A is to be determined. Substituting Eq.(9.44) in Eq.(9.38) gives:

$$\begin{aligned} A &= \frac{\text{Real }\left(e^{2i\kappa_r R}\cos\left(e^{i\kappa_r R}\right)\right)}{\kappa^2\left(-3e^{2i\kappa_r R}\cos\left(e^{i\kappa_r R}\right) + 5e^{3i\kappa_r R}\sin\left(e^{i\kappa_r R}\right) + e^{4i\kappa_r R}\cos\left(e^{i\kappa_r R}\right)\right)} \\ &= \frac{r^2\cos(\kappa_r r)}{\kappa_r^4 r^4\cos(\kappa_r r) + 5\kappa_r^3 r^3\sin(\kappa_r r) - 3\kappa_r^2 r^2\cos(\kappa_r r)} \end{aligned} \tag{9.45}$$

Therefore the particular integral is:

$$\phi = \frac{\rho(0)}{\epsilon_0}\left(\frac{\kappa_r^2 r^4\cos^2(\kappa_r r)}{\kappa_r^4 r^4\cos(\kappa_r r) + 5\kappa_r^3 r^3\sin(\kappa_r r) - 3\kappa_r^2 r^2\cos(\kappa_r r)}\right) \tag{9.46}$$

which has the correct S.I. units of volts $= JC^{-1}$. Resonance occurs in the scalar potential in volts of Eq.(9.46) when:

$$\kappa_r^4 r^4\cos(\kappa_r r) + 5\kappa_r^3 r^3\sin(\kappa_r r) = 3\kappa_r^2 r^2\cos(\kappa_r r) \tag{9.47}$$

If

$$x := \kappa_r r \tag{9.48}$$

the structure of Eq.(9.47) is as follows:

$$\left.\begin{aligned} \left(x^2 - 3\right)\cos x + 5x\sin x = 0, \\ \text{i.e.}\quad \tan x = \frac{3 - x^2}{5x} \end{aligned}\right\} \tag{9.49}$$

and in general will show peaks as a function of x. The analytical solution (9.46) also shows many sharp peaks (Section 9.3), all of which denote surges in voltage (scalar potential). These peaks in voltage can be used in the equivalent circuits of Eqs.(9.35) or (9.38) to produce new power.

$$\phi \longrightarrow \infty \quad \text{or maximized} \tag{9.50}$$

These equations are analyzed numerically in Section 9.3. To obtain this result it has been assumed that the initial driving charge density oscillates according to a cosinal function on the right hand side of Eq.(9.36). A more complicated initial driving function may be used according to circuit design or similar. The important result is that resonance occurs in the voltage, and this surge in voltage is caused by the spin connection of space-time. The voltage obtained in this way may be used for new energy.

In the limit:

$$r \to \infty, \tag{9.51}$$

$$\kappa = \text{constant}, \tag{9.52}$$

Eq.(9.35) reduces to the Poisson equation:

$$\frac{d^2\phi}{dr^2} = -\frac{\rho}{\epsilon_0} \tag{9.53}$$

used in the standard model. Eq.(9.46) may be rewritten as:

$$\phi = \frac{\rho(0)}{\epsilon_0} \frac{\cos^2(\kappa r)}{\left(\kappa^2 \cos(\kappa r) + \frac{5\kappa}{r}\sin(\kappa r) - \frac{3}{r^2}\cos(\kappa r)\right)} \tag{9.54}$$

and in infinite r limit this equation becomes:

$$\phi \ \overrightarrow{r \to \infty} \ \frac{\rho(0)}{\epsilon_0}\left(\frac{\cos(\kappa r)}{\kappa^2}\right) \tag{9.55}$$

so that:

$$\frac{d^2\phi}{dr^2} = \frac{-\rho(0)}{\epsilon_0}\cos(\kappa r) = \frac{-\rho}{\epsilon_0} \tag{9.56}$$

Q.E.D. In this limit is known that the scalar potential is [21]:

$$\phi = \frac{1}{4\pi\epsilon_0}\int \frac{\rho(\mathbf{r}')}{|\mathbf{r}-\mathbf{r}'|}d^3r' \tag{9.57}$$

so:

$$\left(\frac{\cos(\kappa r)}{\kappa^2}\right) \ \overrightarrow{r \to \infty} \ \frac{1}{4\pi\rho(0)}\int \frac{\rho(\mathbf{r}')}{|\mathbf{r}-\mathbf{r}'|}d^3r' \tag{9.58}$$

This is a mathematical check on the self-consistency of the analytical solution (9.46) of the resonance equation (9.35). Physically however the spin connection cannot vanish unless r becomes the radius of the universe. This is because the electromagnetic field is always spinning space-time in ECE theory. Similarly, the gravitational field is always curving space-time. MH theory (standard model) has no conception of the spin connection.

For multi electron systems and in three space dimensions, consider the equation:

$$\mathbf{E} = -\boldsymbol{\nabla}\phi = -\boldsymbol{\omega}\phi \tag{9.59}$$

The electric field from this equation is [21]:

$$\mathbf{E}(\mathbf{r}) = -\frac{1}{4\pi\epsilon_0}\boldsymbol{\nabla}\int \frac{\rho(\mathbf{r}')}{|\mathbf{r}-\mathbf{r}'|}d^3r' \tag{9.60}$$

and the scalar potential is:

$$\phi = \frac{1}{4\pi\epsilon_0}\int \frac{\rho(\mathbf{r}')}{|\mathbf{r}-\mathbf{r}'|}d^3r' \tag{9.61}$$

Thus:

$$\int \nabla \phi d^3 r' = \int \phi \boldsymbol{\omega} d^3 r' \tag{9.62}$$

where:

$$-\nabla \left(\frac{1}{|\mathbf{r} - \mathbf{r}'|} \right) = \frac{\mathbf{r} - \mathbf{r}'}{|\mathbf{r} - \mathbf{r}'|^3} \tag{9.63}$$

so:

$$\frac{\mathbf{r} - \mathbf{r}'}{|\mathbf{r} - \mathbf{r}'|^3} = -\frac{\boldsymbol{\omega}}{|\mathbf{r} - \mathbf{r}'|} \tag{9.64}$$

and the three dimensional spin connection for an n electron system is:

$$\boldsymbol{\omega} = -\frac{(\mathbf{r} - \mathbf{r}')}{|\mathbf{r} - \mathbf{r}'|^2} \tag{9.65}$$

The resonance equation for n electrons is therefore:

$$\nabla^2 \phi + \boldsymbol{\omega} \cdot \nabla \phi + (\nabla \cdot \boldsymbol{\omega}) \phi = -\frac{\rho}{\epsilon_0} \tag{9.66}$$

where the spin connection is given by Eq.(9.65) and where the charge density is defined in terms of the three dimensional Dirac delta function [21] as follows:

$$\rho(\mathbf{r}) = \sum_{i=1}^{n} q_i \delta(\mathbf{r} - \mathbf{r}_i) \tag{9.67}$$

The electric field is [21]:

$$\begin{aligned} \mathbf{E}(\mathbf{r}) &= \frac{1}{4\pi\epsilon_0} \sum_{i=1}^{n} q_i \frac{(\mathbf{r} - \mathbf{r}')}{|\mathbf{r} - \mathbf{r}'|^3} \\ &= \frac{1}{4\pi\epsilon_0} \int \rho(\mathbf{r}') \frac{(\mathbf{r} - \mathbf{r}')}{|\mathbf{r} - \mathbf{r}'|^3} d^3 r' \end{aligned} \tag{9.68}$$

If Δq is the charge in a small volume

$$d^3 r = \Delta x \Delta y \Delta z \tag{9.69}$$

then:

$$\Delta q = \rho(\mathbf{r}') \Delta x \Delta y \Delta z \tag{9.70}$$

so the three dimensional resonance equation for n electrons is:

$$\nabla^2 \phi + \frac{(\mathbf{r} - \mathbf{r}')}{|\mathbf{r} - \mathbf{r}'|^2} \cdot \nabla \phi + \left(\nabla \cdot \frac{(\mathbf{r} - \mathbf{r}')}{|\mathbf{r} - \mathbf{r}'|^2} \right) \phi = -\frac{1}{\epsilon_0} \left(\sum_{i=1}^{n} q_i \delta(\mathbf{r} - \mathbf{r}_i) \right) \tag{9.71}$$

where the spin connection is:

$$\omega_i = -\frac{(\mathbf{r} - \mathbf{r}')}{|\mathbf{r} - \mathbf{r}'|^2} \tag{9.72}$$

Therefore each electron and proton in an atom or molecule has its spin connection.

In order to clarify the meaning of these equations prior to coding in density function packages (Section 9.3), some detail is given as follows. To prove the vector equation [21]:

$$-\nabla\left(\frac{1}{|\mathbf{r}-\mathbf{r}'|}\right)=\frac{(\mathbf{r}-\mathbf{r}')}{|\mathbf{r}-\mathbf{r}'|^3} \tag{9.73}$$

write it as:

$$-\nabla\left(|\mathbf{r}-\mathbf{a}|\right)^{-1}=\frac{\mathbf{r}-\mathbf{a}}{|\mathbf{r}-\mathbf{a}|^3} \tag{9.74}$$

where

$$|\mathbf{r}-\mathbf{a}|=\left((x-a_x)^2+(y-a_y)^2+(z-a_z)^2\right)^{1/2} \tag{9.75}$$

and

$$-\nabla\left(\frac{1}{|\mathbf{r}-\mathbf{a}|}\right)=-\frac{\partial}{\partial x}\left(\frac{1}{|\mathbf{r}-\mathbf{a}|}\right)\mathbf{i}-\frac{\partial}{\partial y}\left(\frac{1}{|\mathbf{r}-\mathbf{a}|}\right)\mathbf{j}-\frac{\partial}{\partial z}\left(\frac{1}{|\mathbf{r}-\mathbf{a}|}\right)\mathbf{k} \tag{9.76}$$

Consider terms such as:

$$\begin{aligned}&\frac{\partial}{\partial x}\left((x-a_x)^2+(y-a_y)^2+(z-a_z)^2\right)^{1/2}\\&=-\frac{(x-a_x)}{\left((x-a_x)^2+(y-a_y)^2+(z-a_z)^2\right)^{3/2}}\end{aligned} \tag{9.77}$$

to find Eqs.(9.73) and (9.74), Q.E.D. In the resonance equation (9.71) there occurs the term:

$$f(\mathbf{r})=\nabla\cdot\frac{(\mathbf{r}-\mathbf{r}_i)}{|\mathbf{r}-\mathbf{r}_i|^2} \tag{9.78}$$

and this may be developed as follows. Write:

$$f(\mathbf{r})=\nabla\cdot\frac{(\mathbf{r}-\mathbf{a})}{|\mathbf{r}-\mathbf{a}|^2} \tag{9.79}$$

where:

$$|\mathbf{r}-\mathbf{a}|^2=(x-a_x)^2+(y-a_y)^2+(z-a_z)^2 \tag{9.80}$$

and

$$\mathbf{r}-\mathbf{a}=(x-a_x)\,\mathbf{i}+(y-a_y)\,\mathbf{j}+(z-a_z)\,\mathbf{k} \tag{9.81}$$

so:

$$\begin{aligned}f(\mathbf{r})&=\frac{\partial}{\partial x}\frac{(\mathbf{x}-\mathbf{a}_x)}{|\mathbf{r}-\mathbf{a}|^2}+\frac{\partial}{\partial y}\frac{(\mathbf{y}-\mathbf{a}_y)}{|\mathbf{r}-\mathbf{a}|^2}+\frac{\partial}{\partial z}\frac{(\mathbf{z}-\mathbf{a}_z)}{|\mathbf{r}-\mathbf{a}|^2}\\&=\frac{\partial}{\partial x}\frac{(x-a_x)}{(x-a_x)^2+(y-a_y)^2+(z-a_z)^2}+\cdots\end{aligned} \tag{9.82}$$

Using the rules of differentiation:

$$\begin{aligned}f(\mathbf{r})&=\left((x-a_x)^2+(y-a_y)^2+(z-a_z)^2\right)^{-2}\\&\quad-\frac{2(x-a_x)^2}{\left((x-a_x)^2+(y-a_y)^2+(z-a_z)^2\right)^2}+\cdots\\&=\frac{1}{|\mathbf{r}-\mathbf{a}|^2}-\frac{2}{|\mathbf{r}-\mathbf{a}|^2}\\&=\frac{-1}{|\mathbf{r}-\mathbf{a}|^2}\end{aligned} \tag{9.83}$$

Therefore:

$$\nabla \cdot \frac{(\mathbf{r}-\mathbf{r}_i)}{|\mathbf{r}-\mathbf{r}_i|^2} = -\frac{1}{|\mathbf{r}-\mathbf{r}_i|^2} \tag{9.84}$$

so the resonance equation is:

$$\nabla^2\phi + \frac{(\mathbf{r}-\mathbf{r}_i)}{|\mathbf{r}-\mathbf{r}_i|^2} \cdot \nabla\phi - \frac{1}{|\mathbf{r}-\mathbf{r}_i|^2}\phi = -\frac{1}{\epsilon_0}\left(\sum_{i=1}^{n} q_i\delta(\mathbf{r}-\mathbf{r}_i)\right) \tag{9.85}$$

This can be written as:

$$|\mathbf{r}-\mathbf{r}_i|^2 \nabla^2\phi + (\mathbf{r}-\mathbf{r}_i) \cdot \nabla\phi - \phi = \frac{-1}{\epsilon_0}|\mathbf{r}-\mathbf{r}_i|^2 \sum_{i=1}^{n} q_i\delta(\mathbf{r}-\mathbf{r}_i) \tag{9.86}$$

and incorporated in density functional code for the Coulomb potential. In Eq.(9.86) the Dirac delta function is defined as usual [21] by:

$$\delta(\mathbf{r}-\mathbf{r}_1) = \delta(x_1 - X_1)\,\delta(y_1 - Y_1)\,\delta(z_1 - Z_1) \tag{9.87}$$

and the charge density is defined by:

$$\rho(\mathbf{r}) = \sum_{i=1}^{n} e_i\delta(\mathbf{r}-\mathbf{r}_i) \tag{9.88}$$

Using:

$$\mathbf{r}-\mathbf{r}_i = (x-x_i)\,\mathbf{i} + (y-y_i)\,\mathbf{j} + (z-z_i)\,\mathbf{k} \tag{9.89}$$

and:

$$|\mathbf{r}-\mathbf{r}_i|^2 = (x-x_i)^2 + (y-y_i)^2 + (z-z_i)^2 \tag{9.90}$$

the spin connection in Eq.(9.72) an be developed as:

$$\boldsymbol{\omega}_i = -\frac{((x-x_i)\,\mathbf{i} + (y-y_i)\,\mathbf{j} + (z-z_i)\,\mathbf{k})}{(x-x_i)^2 + (y-y_i)^2 + (z-z_i)^2} \tag{9.91}$$

With these definitions the resonance equation in three dimensions and for n electrons and protons in an atom or molecule is therefore:

$$\nabla^2\phi + \frac{(\mathbf{r}-\mathbf{r}_i)}{|\mathbf{r}-\mathbf{r}_i|^2} \cdot \nabla\phi - \frac{1}{|\mathbf{r}-\mathbf{r}_i|^2}\phi = -\frac{1}{\epsilon_0}\sum_{i=1}^{n} q_i\delta(\mathbf{r}-\mathbf{r}_i) \tag{9.92}$$

The potential in volts from this equation can be used to build up the Hartree term in density functional code. This term describes electron electron repulsion through the Coulomb interaction and is (Section 9.3):

$$V = \frac{1}{4\pi\epsilon_0}\int \frac{e^2 n_s(\mathbf{r})}{|\mathbf{r}-\mathbf{r}'|} d^3r' \tag{9.93}$$

where $n_s(\mathbf{r})$ is a number density. Therefore the scalar potential in volts is:

$$\phi_H = \frac{e}{4\phi\epsilon_0}\int \frac{n_s(\mathbf{r})}{|r-r'|} d^2r' \tag{9.94}$$

When there are peaks in ϕ_H, the effect is to greatly amplify the number density and to greatly increase the electron-electron repulsion, resulting in ionization of the atom or molecule into free electrons which can be used for new energy.

The most direct method for acquiring new energy from the spin connection is to construct the circuits equivalent to Eqs.(9.35) or (9.38). The type of circuit needed for Eq.(9.38), whose analytical solution is Eq.(9.46), is that of the undamped oscillator. The simplest undamped oscillator equation [24] is:

$$m\ddot{x} + kx = F \tag{9.95}$$

where m is mass, x is displacement, k is Hooke's constant and F is driving force. Its equivalent circuit equation is [24]:

$$L\ddot{q} + \frac{q}{C} = E \tag{9.96}$$

where q is charge, L is inductance, C is capacitance and E is electromotive force. Eq (9.96) describes an electromotive force in series with a capacitor and induction coil. A material may be incorporated inside the induction coil. Therefore the circuit equivalent to the undamped oscillator (9.38) is the same design, but the electromotive force is synthesized to be the same as the right hand side of Eq.(9.38), i.e. made up of circular and hyperbolic functions. The exact solution (9.46) shows many sharp peaks of voltage (Section 9.3), so the equivalent circuit also shows many sharp peaks of voltage for a small initial electromotive force. These peaks of voltage can be used for new power and fed to the grid from a power plant.

9.3 Numerical results and discussion

In the following we present numerical results for the resonance equations and the application of the resonant Coulomb law to the Hydrogen atom. Then we discuss the general applicability to solid state physics and electrical engineering.

9.3.1 Graphs of resonance effects

As described in Section 9.2, the solution of the ECE Coulomb law (see Eq.(9.12) for Cartesian or Eq.(9.34) for spherical coordinates) shows up resonances. This can be seen in general from the particular integral (9.46). This particular solution is plotted in Fig. 9.1 for four κ values. The higher κ, the more resonances are seen in a certain range of the radial coordinate r. The resonances appear in form of poles of the particular integral, i.e. where the denominator approaches to zero. In general these are the roots of the denominator equation (9.49). These are computed numerically and listed in Table 9.1

Since the variable x in Eq.(9.49) is the product κr, the coordinate r corresponds directly to x in case $\kappa = 1$. This can be seen from Fig. 9.1: For the second κ value the zeros come to lie at the x values of Table 9.1.

The most interesting curve is the resonance behaviour of true solutions of the resonant Coulomb law (9.34). The equation has been transformed by the Euler method to the form of Eq.(9.38) without approximations. This form is easier to solve numerically and it can directly be seen that it is an equation of

n	zero(rad)	zero(degrees)
1	0.660310	37.832983
2	2.807011	160.829901
3	5.502118	315.248163
4	8.409624	481.835945
5	11.416981	654.144835
6	14.474251	829.313494
7	17.558761	1006.042917

Table 9.1: First seven zero values of Eq.(9.49)

a forced oscillation. The "driving term" is the right-hand side of Eq.(9.38). For obtaining the resonance curve the value of κ_r at the left hand side has been kept fixed and the κ value of the driving force has been varied as suggested by Eq.(9.40). The results are shown in Fig. 9.2. The maximum amplitude difference after 15 oscillations has been plotted against the wave number κ of the driving force. In case of the pure cosine term the resonance occurs for $\kappa = \kappa_r$ as expected (we used $\kappa_r = 1$). In principle this is an undamped resonance with infinite resonance amplitude, but since we have restricted the calculation to 15 wavelengths the curve remains bounded. The same holds for the exact driving force. There are two sharp resonances at $\kappa = 0.5$ and $\kappa = 0.25$ and a smaller third at $\kappa = 0.166$. The curve scales with κ_r which was set to unity here. This resonance behaviour is a consequence of the spin connection which is only present in ECE theory, not in standard Maxwell-Heaviside theory.

The resonant increase of the solution Φ of Eq.(9.38) can be studied from Fig. 9.3 where the curve $\Phi(R)$ is presented for four different κ values. The first two κ's were chosen to be the resonance values of Fig. 9.2. Both lead to increasing amplitudes with R. The other two κ values are off resonance, the maximum amplitude does not change with R. The driving force (right-hand side of Eq.(9.38)) is presented in Fig. 9.4. There are two maxima and one minimum per period. Decreasing κ only means a broadening of the form. This driving force is required for experimental construction of equivalent circuits of spin connection resonance, therefore we have calculated the Fourier spectrum (Fig. 9.5). As already seen in Fig. 9.4, changing the κ value does not change the wave form, i.e. the Fourier spectrum remains unaltered for abscissa values of $\kappa = 2\pi/\lambda$, only the right end varies with λ. The three significant peaks of Fig. 9.2 are directly represented in the spectrum. In particular one can see that the first and second resonance peak differ in phase (sign of Fourier coefficient). The phase difference between both amplitudes is 180 degrees as can directly be observed from Fig. 9.3.

So far we have inspected the numerical solution $\Phi(R)$ in the transformed coordinate R as given by Eq.(9.39). The question is how the original function $\Phi(r)$ behaves. We have to perform the back-transformation $R \to r$ which is given by

$$r = \frac{1}{\kappa_r} \cos(\kappa_r R) \tag{9.97}$$

Since the value range of the cosine function is restricted to $[-1, 1]$ we have to provide a rule how to obtain r values for $\kappa_r r$ greater than unity. The proposed solution is to shift the values according to the number of full periods $0 \le \kappa_r R <$

π contained in the argument. Additionally the result is transformed in a way to give a continuous function as is depicted in Fig. 9.6. The back-transformed potential $\Phi(r)$ can be seen in Fig. 9.7 (it corresponds to the left-most part of Fig. 9.3). $\Phi(r)$ has to be interpolated if one wants to have equidistant abscissa values as for example was required for the calculation of the H spectrum. Obviously the cosine function leads to a deformation of Φ where $r(R)$ approaches the horizontal asymptote.

9.3.2 Results for the Hydrogen atom

The energy eigenvalues and radial wavefunctions of atomic Hydrogen have been calculated numerically in presence of a small oscillatory charge density serving as "driving force". The ECE Coulomb potential of the driving force has been added to the proton core potential of the H atom. This is the same proceeding as had already been applied earlier [11]. In that case an approximate method was used to calculate the ECE Coulomb potential with a constant radius parameter. Here we use the numerical solution of Eq.(9.38) as discussed in the preceding section. The value of κ_r was set to 0.5 to obtain a r range in the backtransformation which was broad enough compared to the atomic radius. We used a maximum value of $r = 3.5$. As can be seen from Fig. 9.3, the stationary oscillatory behaviour develops after a certain initial R range which is dominated by initial value effects [24]. To avoid these initial value dependences we have taken the solution $\Phi(R)$ for $R \geq 26.5$ and shifted the range to the coordinate origin of r. For this operation one has to observe that the boundaries of the transformation range shown in Fig. 9.6 are not exceeded in order not to obtain unphysical jumps in Φ. In Fig. 9.8 the total atomic potential is shown for three κ values, two in resonance and one in off-resonance. Because we have chosen $\kappa_r = 0.5$ the resonances come to lie at $\kappa = 0.175$ and $\kappa = 0.225$. In the first resonance the potential is shifted from smaller to greater radii; it takes strongly repulsive values in the second resonance. The third κ value is in off-resonance. The potential leads to the resonance graph presented in Fig. 9.9. There is an oscillating shift of atomic energies with a remarkable lifting of all values at the main resonance peak of $\kappa = 0.225$. This is a more complicated resonance structure compared with our earlier calculation [11] where the analytical form of the particular integral of the resonance equation for a fixed radius was used.

The radial dependence of some orbitals is depicted in Figs. 9.10–9.12. The orbitals are calculated for κ values already discussed with Fig. 9.8. The potential shift in the first case (A) leads to a shift of the $1s$ and $2p$ orbitals to the core because mainly the decrease in potential has an effect for these orbitals. In case of the repulsive resonance (B) all orbitals are significantly pushed outward. In particular the $1s$ orbital is strongly delocalized and is in a transition to an unbound state. This supports the former assumptions on the ionization mechanism. Nevertheless the particular angular momentum character of the orbitals (sign and number of zeros) remains valid.

In Fig. 9.13 some control parameters are shown as was also done in [11]. The resonance potential takes its extrema near to the resonant κ values. The potential integrates out to zero for large r which indicates that charge neutrality is conserved and the calculations are made on a reasonable basis. The overall results are similar to those obtained in [11].

9.3.3 Application to solid state physics and engineering

The resonance effect in Hydrogen was investigated because it serves as a model system for the huge class of atomic, molecular and solid state physics. Resonances in solids are particularly important because the electronic structure of solids is the basis of modern electronic chip design which can be extended to resonant devices on a microstructure scale. The most used method for computation of electronic properties of solids is Density Functional Theory. Therefore we give a short introduction into the theory as contributed by Lothar Fritsche [27]. The method, which in practical applications presupposes the validity of the Born-Oppenheimer approximation (fixed atomic nuclei), is as follows:

1. The interacting N-electron system (ground-state wavefunction $\widehat{\Psi}_0(\boldsymbol{x}_1, \boldsymbol{x}_2, \cdots, \boldsymbol{x}_N)$) is reduced to a non-interacting N-particle system (wavefunction $\Phi_0(\boldsymbol{x}_1, \boldsymbol{x}_2, \cdots, \boldsymbol{x}_N)$ = Slater determinant) by adiabatically switching off the electron-electron interaction while keeping the oneparticle density $\rho(\boldsymbol{r})$ fixed. The latter means

$$\begin{aligned}\rho(\boldsymbol{r}) =& N\sum_\sigma \int \mid \widehat{\Psi}_0((\boldsymbol{r},\sigma), \boldsymbol{x}_2, \cdots, \boldsymbol{x}_N \mid^2 d^4x_2 \cdots d^4x_N = \\ & N\sum_\sigma \int \mid \phi_0((\boldsymbol{r},\sigma), \boldsymbol{x}_2, \cdots, \boldsymbol{x}_N \mid^2 d^4x_2 \cdots d^4x_N\end{aligned} \tag{9.98}$$

Note that the coordinate $\boldsymbol{x}$ stands collectively for $(\boldsymbol{r};\sigma)$ where $\sigma = \pm 1$ denotes the spincoordinate.

If the nuclear Coulomb potential resulting from all atoms of the system is denoted by $V_{ext}(\boldsymbol{r})$ the original Hamiltonian has the form (a.u.: energy: Hartree, length: a_B)

$$\widehat{H}_{interact} = \sum_{k=1}^{N}\left[-\frac{1}{2}\nabla_k^2 + V_{ext}(\boldsymbol{r}_k)\right] + \frac{1}{2}\sum_{k=1,l\neq k}^{N}\frac{1}{\mid \boldsymbol{r}_k - \boldsymbol{r}_l \mid} \tag{9.99}$$

In the intermediate state when the electron-electron interaction is not yet completely switched off, the Hamiltonian attains the form

$$\sum_{k=1}^{N}\left[-\frac{1}{2}\nabla_k^2 + V_{ext}(\boldsymbol{r}_k) + \widehat{V}_{ext}(\lambda, \boldsymbol{r}_k)\right] + \frac{\lambda}{2}\sum_{k,l\neq k}\frac{1}{\mid \boldsymbol{r}_k - \boldsymbol{r}_l \mid} \tag{9.100}$$

where λ controls the strength of the electron-electron interaction, i. e. $0 \leq \lambda \leq 1$. Furthermore, $\widehat{V}_{ext}(\lambda, \boldsymbol{r})$ denotes the additional potential that has to be switched on at coupling strength λ to ensure the conservation of the original interacting density $\rho(\boldsymbol{r})$. If this potential exists, it is unique, as has been shown by Hohenberg and Kohn (1964). However, the actual existence can only be shown for densities that are defined on a lattice of points in real-space [28]. The Hamiltonian of the non-interacting substitute system derives from Eq.(9.100) for $\lambda = 0$:

$$\widehat{H}_{substitute} - \sum_{k=1}^{N}\left[-\frac{1}{2}\nabla_k^2 + V_{ext}(\boldsymbol{r}_k) + \widehat{V}_{ext}(\boldsymbol{r}_k)\right] \tag{9.101}$$

where $\widehat{V}_{ext}(\boldsymbol{r}) \equiv \widehat{V}_{ext}(\lambda = 0, \boldsymbol{r})$.

2. The Schrödinger equation associated with the substitute Hamiltonian (9.101) can be decomposed into N one-particle equations

$$\left[-\frac{1}{2}\nabla^2 + V_{ext}(\boldsymbol{r}) + \widehat{V}_{ext}(\boldsymbol{r})\right]\psi_i(\boldsymbol{r}) = \epsilon_i\psi_i(\boldsymbol{r}) \tag{9.102}$$

which represent the so-called "Kohn-Sham equations". It can be shown that the non-interacting wavefunction $\Phi_0(x_1, x_2, \cdots, x_N)$ which can be cast as a Slater determinant has to be formed from the N energetically lowest lying solutions $\psi_i(\boldsymbol{r})$. Hence one has from Eq.(9.98)

$$\rho(\boldsymbol{r}) = \sum_{i=1}^{N} \mid \psi_i(\boldsymbol{r}) \mid^2 \tag{9.103}$$

3. By using the Hellmann-Feynman theorem it is easy to show that the total energy E_0 of the actual **interacting** system can be cast as

$$E_0 = \underbrace{\langle\phi_0 \mid \sum_{k=1}^{N}\left[-\frac{1}{2}\nabla_k^2\right] \mid \phi_0\rangle}_{\equiv T_0} + \int \rho(\boldsymbol{r})V_{ext}(\boldsymbol{r})d^3r + \langle\overline{V_{e-e}}\rangle \tag{9.104}$$

where

$$T_0 = \sum_{i=1}^{N}\int \psi_i^*(\boldsymbol{r})\left[-\frac{1}{2}\nabla^2\right]\psi_i(\boldsymbol{r})d^3r\ ,\ \rho(\boldsymbol{r}) = \sum_{i=1}^{N} \mid \psi_i(\boldsymbol{r}) \mid^2 \tag{9.105}$$

and

$$\langle\overline{V_{e-e}}\rangle = \iint \frac{\overline{\rho_2(\boldsymbol{r}', \boldsymbol{r})}}{\mid \boldsymbol{r}' - \boldsymbol{r} \mid} d^3r'd^3r \tag{9.106}$$

with $\overline{\rho}_2$ denoting the electronic pair density averaged over the coupling strength:

$$\overline{\rho_2(\boldsymbol{r}', \boldsymbol{r})} = \int_0^1 \rho_2(\lambda, \boldsymbol{r}', \boldsymbol{r})d\lambda \tag{9.107}$$

where

$$\begin{aligned}&\rho_2(\lambda, \boldsymbol{r}', \boldsymbol{r}) = \\ &N(N-1)\sum_{\sigma',\sigma}\int \mid \widehat{\Psi}_0(\lambda, \boldsymbol{x}', \boldsymbol{x}, \boldsymbol{x}_3, \cdots, \boldsymbol{x}_N) \mid^2 d^4x_3 \cdots d^4x_N\end{aligned} \tag{9.108}$$

and $\widehat{\Psi}_0(\lambda, \boldsymbol{x}', \boldsymbol{x}, \boldsymbol{x}^3, \cdots, \boldsymbol{x}^N)$ solves the Schrödinger equation associated with the Hamiltonian (9.99) and ground-state energy E_0.

4. Eqs.(9.102) and (9.104) define the framework of density functional theory. At this stage it is still equivalent to a rigorous (non-relativistic) N-electron theory since no approximations have been made so far. But in practice it is useless in this form, because the density conserving extra potential $\widehat{V}_{ext}(\boldsymbol{r})$ and $\overline{\rho}_2$ are unknown. However, exploiting universal properties of

the wavefunction $\widehat{\Psi}_0(\boldsymbol{x}, \boldsymbol{x}_2, \cdots, \boldsymbol{x}_N)$, in particular its antisymmetry giving rise to "Fermi-holes" in the pair-density $\rho_2(\boldsymbol{r}', \boldsymbol{r})$, one arrives at a surprisingly simple approximation to $\langle\overline{V_{e-e}}\rangle$:

$$\langle\overline{V_{e-e}}\rangle = \frac{1}{2}\iint \frac{\rho(\boldsymbol{r}')\rho(\boldsymbol{r})}{\mid \boldsymbol{r}' - \boldsymbol{r} \mid} d^3r' d^3r - \int \frac{3}{4}\left(\frac{3}{\pi}\right)^{1/3} [\rho(\boldsymbol{r})]^{4/3}\, d^3r \qquad (9.109)$$

For details see [29]. More advanced expressions account for contributions of the gradient of $\rho(\boldsymbol{r})$ ("Generalized Gradient Approximation", "GGA") see [30].

If one inserts the expression (9.109) in Eq.(9.104) and uses Eq.(9.103), E_0 becomes a functional of the orbital set $\{\psi_i(\boldsymbol{r})\}$. Requiring E_0 to be stationary against variations of these orbitals one arrives at

$$\begin{aligned} \delta E_0 = \sum_{i=1}^{N_\sigma} \int \delta\psi_i^*(\boldsymbol{r}) \Big[-\frac{1}{2}\nabla^2 + \\ V_{ext}(\boldsymbol{r}) + V_H(\boldsymbol{r}) + V_{XC}(\boldsymbol{r}, \sigma) \Big] \psi_i(\boldsymbol{r}) d^3r + \text{c.c.} = 0 \end{aligned} \qquad (9.110)$$

Here $V_H(\boldsymbol{r})$ and $V_{xc}(\boldsymbol{r})$ denote, respectively, the so-called Hartree- and exchange-correlation potential which derive from the respective two contributions on the right-hand side of Eq.(9.109). They have the form

$$\begin{aligned} V_H(\boldsymbol{r}) &= \int \frac{\rho(\boldsymbol{r}')}{\mid \boldsymbol{r}' - \boldsymbol{r} \mid} d^3r' \\ &\text{and} \\ V_{XC}(\boldsymbol{r}) &= -\frac{3}{\pi}^{1/3} [\rho(\boldsymbol{r})]^{1/3} \end{aligned} \qquad (9.111)$$

Eq.(9.110) is obviously fulfilled if

$$\widehat{V}_{ext}(\boldsymbol{r}) = V_H(\boldsymbol{r}) + V_{XC}(\boldsymbol{r}) \qquad (9.112)$$

The DFT-scheme is now fully defined and easily accessible to applications. In case of ECE resonances the potentials $V_H(\boldsymbol{r})$ and $V_{XC}(\boldsymbol{r})$ have to incorporate the spin connection effects via the modified charge density $\rho(\boldsymbol{r})$. This in turn is determined from Eqs.(9.103) and (9.102) where the spin connection terms of Eq.(9.92) have to be respected.

It should be observed, however, that the rigorous version requires the pair density $\rho_2(\lambda, \boldsymbol{r}', \boldsymbol{r})$, averaged over the coupling strength to be known. This, in turn, implies the availability of $\widehat{\Psi}_0(\lambda, \boldsymbol{x}', \boldsymbol{x}, \boldsymbol{x}_3, \cdots, \boldsymbol{x}_N)$ for all values of λ, which means one has to solve the N-electron Schrödinger equation associated with the Hamiltonian (9.100). Hence, if the DFT-scheme is defined by the property to solve the N-electron problem without determining the N-electron wavefunction, it constitutes always an approximate theory.

Finally we give some hints how the resonant Coulomb law can be used in electrical engineering. At the end of Section 9.2 it was already stated that the circuit equivalent of Eq.(9.35) or (9.38) is an undamped electrical oscillator. This is a connection in series of an induction coil and a capacitor (see Fig. 9.14).

The differential equation for the oscillating charge q in this circuit is (according to Eq.(9.96)

$$\ddot{q} + \omega_0^2 q = U_{emf} \tag{9.113}$$

with

$$\omega_0 = 2\pi\nu_0 = \frac{1}{\sqrt{LC}} \tag{9.114}$$

being the resonance frequency (ν_0) and resonance angular frequency (ω_0). U_{emf} is the driving voltage representing the electromotive force. According to conventional Maxwell Heaviside theory, resonance occurs if U_{emf} is a harmonic A.C. voltage of the form

$$U_{emf} = U_0 \cos(\omega_0 t) \tag{9.115}$$

with an amplitude U_0. In order to activate a spin connection resonance we have to apply the driving force given at the right hand side of Eq.(9.38) which has been graphed in Fig. 9.4. In the equivalent circuit the length coordinate has to be replaced by time and the spatial resonance frequency κ_r (wave number) by the time frequency ω_0. In the same way the varying wave number of the driving force κ is to be changed to a frequency ω. From Fig. 9.5 we know that there are only three significant frequency contributions in the Fourier spectrum of the driving term. We can compose it by adding three single voltages for a given frequency ω:

$$U_{emf} = U_0(0.997\cos(2\omega t) - 0.503\cos(4\omega t) + 0.040\cos(6\omega t)) \tag{9.116}$$

This Fourier synthesis is graphed in Fig. 9.17. There is no visible difference to the exact form using all Fourier coefficients. Spin connection resonance should occur when ω is chosen in the ratio to ω_0 w.r.t. the resonance frequencies found from Fig. 9.2:

$$\begin{aligned} \omega_1 &= \frac{1}{2}\omega_0 \\ \omega_2 &= \frac{1}{4}\omega_0 \\ \omega_3 &= 0.166\omega_0 \end{aligned} \tag{9.117}$$

In particular the first and strongest resonance is obtained if

$$U_{emf}^{(1)} = U_0\left(0.997\cos(\omega_0 t) - 0.503\cos(2\omega_0 t) + 0.040\cos(3\omega_0 t)\right) \tag{9.118}$$

The voltage enhanced by resonance should occur then at the components of the equivalent circuit and can be tapped at the positions denoted by U_{res1} and U_{res2} in Fig. 9.14. The question remains how the mechanism for spin connection resonance in the circuit works. Some of the circuit material has to be brought to Coulomb resonance so that the material is ionized and electrons are emitted making up the additional current and voltage. How this is achieved has to be found experimentally. The material of the capacitor plates (or foils) could be this as well as the dielectric in-between. Alternatively a suitable material may be incorporated in the induction coil. Then the free electrons would appear in this material and could be extracted by a small voltage (see Fig. 9.15). In addition a feedback loop could be established by connecting the ends of the coil core with U_{emf} (dashed lines in Fig. 9.15). Since the voltage at the coil has a phase shift compared to the exciter emf, some phase shifting elements have to be added (denoted symbolically by P in Fig. 9.15). After having been started by a small initial voltage, such a machine would run by its own. From Eq.(9.118) it is seen that U_{emf} consists of harmonics of the conventional resonance frequency ω_0. This means that the driving force is "compatible" with the circuit and could be enhanced by resonance without impairing its function.

As a last example we present an advanced circuit design suggested by Douglas Mann [31], see Fig. 9.16. The capacitor has been replaced by a secondary coil made by a bifilar wire. This special construction suppresses its own magnetic field nearly completely. Interaction with the inner induction coil is by electric fields of the wires, building a capacity. Thus both elements of the equivalent circuit are present and the resonant medium consists of the wires exclusively. The emf can be applied magnetically by a further induction coil with few windings which surrounds the two other coils. It is expected that devices similar like this will be able to deliver energy from spacetime in the near future.

Acknowledgements The British Head of State, Parliament, Prime Minister and Royal Society are thanked for the award of a Civil List pension in recognition of distinguished contributions to Britain and the Commonwealth in science. The staff and environment of AIAS are thanked for many interesting discussions.

Appendix 1: Analytical solutions of the undamped oscillator

Consider the basic undamped oscillator equation (see text):

$$\frac{d\phi^2}{dR^2} + \kappa^2\phi = \frac{\rho(0)}{\epsilon_0} f(\kappa R) \tag{A-1}$$

where:

$$f(\kappa R) = e^{2i\kappa R}\cos\left(e^{i\kappa R}\right) \tag{A-2}$$

If $f(\kappa R)$ satisfies the Dirichlet condition, i.e. is single valued and continuous in an interval such as $\pi < f(\kappa R) \leq \pi$ it can be expanded in a Fourier series:

$$f(\kappa R) = \frac{a_0}{2} + \sum_{d=1}^{\infty}(a_\alpha \cos(\alpha\kappa R) + b_\alpha \sin(\alpha\kappa R)) \tag{A-3}$$

where:

$$\left.\begin{aligned} a_0 &= \frac{1}{\pi}\int_{-\pi}^{\pi} f(\kappa R)\, d(\kappa R) \\ a_\alpha &= \frac{1}{\pi}\int_{-\pi}^{\pi} f(\kappa R)\cos(\alpha\kappa R)\, d(\kappa R) \\ b_\alpha &= \frac{1}{\pi}\int_{-\pi}^{\pi} f(\kappa R)\sin(\alpha\kappa R)\, d(\kappa R)\,. \end{aligned}\right\} \tag{A-4}$$

These integrals can be computed straightforwardly to any required precision in any interval, the latter is not necessarily constrained to $\pi < f(\kappa R) \leq \pi$, the latter is used for illustration. Therefore Eq.(A-1) becomes:

$$\begin{aligned} \frac{d^2\phi}{dR^2} + \kappa^2\phi = \frac{\rho(0)}{\epsilon_0}\Big(&\frac{a_0}{2} + a_1\cos(\kappa R) + a_2\cos(2\kappa R) \\ &+ \cdots + b_1\sin(\kappa R) + b_2\sin(2\kappa R) \\ &+ \cdots\Big) \end{aligned} \tag{A-5}$$

Assume a solution of the type:

$$\begin{aligned} \phi = \frac{\rho(0)}{\epsilon_0}\Big(&A_0\frac{a_0}{2} + A_1 a_1\cos(\kappa R) + A_2 a_2\cos(2\kappa R) + \\ &\cdots + B_1 b_1\sin(\kappa R) + B_2 b_2\sin(2\kappa R) + \cdots\Big) \end{aligned} \tag{A-6}$$

Substituting Eq.(A-6) in Eq.(A-5) and comparing terms by term:

$$\left.\begin{aligned} \kappa^2 A_0\frac{a_0}{2} &= \frac{a_0}{2} \\ A_1\kappa^2(1 - a_1)\cos(\kappa R) &= \cos(\kappa R) \\ A_1 2\kappa^2(1 - 4a_1)\cos(\kappa R) &= \cos(\kappa R) \\ &\vdots \\ B_n\kappa^2\left(1 - n^2 b_n\right)\sin(n\kappa R) &= \sin(n\kappa R)\,. \end{aligned}\right\} \tag{A-7}$$

Thus:

$$\begin{aligned}\phi = \frac{\rho(0)}{\epsilon_0\kappa^2}\Bigg(&\frac{a_0}{2} + \frac{\cos(\kappa R)}{(1-a_1)} + \frac{\cos(2\kappa R)}{(1-4a_2)} + \cdots \\ &+ \frac{\sin(\kappa R)}{(1-b_1)} + \frac{\sin(2\kappa R)}{(1-4b_2)} + \cdots\Bigg)\end{aligned} \tag{A-8}$$

Infinite resonances occur at:

$$\left.\begin{aligned} a_n &= 1/n^2, \quad n = 1, \cdots, m, \\ b_n &= 1/n^2, \quad n = 1, \cdots, m \end{aligned}\right\} \tag{A-9}$$

In general these resonances occur at:

$$\text{Real}\left(\frac{1}{\pi}\int_{-\pi}^{\pi} e^{2i\kappa R}\cos\left(e^{i\kappa R}\right)\cos(n\kappa R)\, d(\kappa R)\right) = \frac{1}{n^2} \tag{A-10}$$

and:

$$\text{Real}\left(\frac{1}{\pi}\int_{-\pi}^{\pi} e^{2i\kappa R}\cos\left(e^{i\kappa R}\right)\sin(n\kappa R)\, d(\kappa R)\right) = \frac{1}{n^2} \tag{A-11}$$

This analysis can be repeated straightforwardly for any driving term:

$$f(\kappa R) = e^{2i\kappa R} f_1\left(e^{i\kappa R}\right) \tag{A-12}$$

A constrained particular integral of Eq.(A-1) can be obtained for any driving function $f(\kappa R)$. In this case the undamped oscillator is:

$$\frac{d^2\phi}{dR^2} + \kappa^2\phi = \frac{\rho(0)}{\epsilon_0} e^{2i\kappa R} f\left(e^{i\kappa R}\right) \tag{A-13}$$

Assume a solution of the type:

$$\phi = \frac{A\rho(0)}{\epsilon_0} e^{2i\kappa R} f\left(e^{i\kappa R}\right) \tag{A-14}$$

subject to the constraint:

$$\frac{dA}{dR} = 0 \tag{A-15}$$

Then:

$$\frac{d\phi}{dR} = i\kappa\frac{A\rho(0)}{\epsilon_0}\left(2e^{2i\kappa R} f + e^{3i\kappa R} f'\right) \tag{A-16}$$

and:

$$\frac{d^2\phi}{dR^2} = -\kappa^2\frac{A\rho(0)}{\epsilon_0}\left(4e^{2i\kappa R} f + 5e^{3i\kappa R} f' e^{4i\kappa R} f''\right) \tag{A-17}$$

So the particular integral is:

$$\phi = -\frac{\phi(0)}{\epsilon_0}\left(\frac{r^2 f^2}{3f + 5\kappa r f' + \kappa^2 r^2 f''}\right) \tag{A-18}$$

subject to the constraint:

$$\frac{d\phi}{dr} = 0 \tag{A-19}$$

A solution of Eq.(A-19) is the general resonance condition:

$$3f + 5\kappa r f' + \kappa^2 r^2 f'' = 0 \tag{A-20}$$

To explain the notation in Eq.(A-20) consider for example a cosine driving term:

$$f = \cos x, \quad x = \kappa r \tag{A-21}$$

Then the notation means:

$$f' = -\sin x, \quad f'' = -\cos x \tag{A-22}$$

The resonance condition (A-20) then becomes:

$$\tan x = \frac{3 - x^2}{5x} \tag{A-23}$$

to which there is an infinite number of solutions. For a driving term:

$$f = e^{-x}, \quad f' = -e^{-x}, \quad f'' = e^{-x} \tag{A-24}$$

the resonance condition is:

$$x^2 - 5x + 3 = 0 \tag{A-25}$$

and there are two solutions at

$$x = 4.3028, \quad 0.6972 \tag{A-26}$$

For a driving term:

$$\left.\begin{aligned} f &= e^{-x}\cos x \\ f' &= e^{-x}(\cos x - \sin x) \\ f'' &= -2e^{-x}\sin x \end{aligned}\right\} \tag{A-27}$$

the resonance equation is:

$$\tan x = \frac{3 + 5x}{x(5 - 6x)} \tag{A-28}$$

and there are again an infinite number of solutions.

Appendix 2 : Simultaneous equations for ϕ and $\boldsymbol{\omega}$

The general resonance equation for the Coulomb Law is:

$$\nabla^2\phi + \boldsymbol{\nabla}\phi \cdot \boldsymbol{\omega} + (\boldsymbol{\nabla} \cdot \boldsymbol{\omega})\,\phi = -\frac{\rho}{\epsilon_0} \tag{B-1}$$

The limit of the standard model is reached when:

$$\boldsymbol{\nabla}\phi = \boldsymbol{\omega}\phi \tag{B-2}$$

i.e.

$$\nabla^2\phi = \boldsymbol{\nabla}\phi \cdot \boldsymbol{\omega} + (\boldsymbol{\nabla} \cdot \boldsymbol{\omega})\,\phi \tag{B-3}$$

and

$$\phi = \frac{-e}{4\pi\epsilon_0 r} \tag{B-4}$$

Therefore:

$$\boldsymbol{\omega} = -\frac{1}{r}\mathbf{e}_r \tag{B-5}$$

where $\mathbf{e}_r$ is the radial unit vector of the spherical polar coordinate system. So:

$$\omega_r = -\frac{1}{r} \tag{B-6}$$

Eq.(B-3) is a limiting case or boundary value of the general resonance equation (B-1). There is not sufficient information in Eq.(B-1) alone to completely determine ϕ and $\boldsymbol{\omega}$ under all conditions, because there are two variables and only one equation. In the text of the paper it has been assumed in order to proceed that Eq.(B-6) holds under all conditions, so Eq.(B-1) becomes (in spherical polar coordinates):

$$\frac{\partial^2}{\partial r^2} + \frac{1}{r}\frac{\partial\phi}{\partial r} - \frac{1}{r^2}\phi = -\frac{\rho\,(0)}{\epsilon_0}\cos\left(\kappa r\right) \tag{B-7}$$

Eq.(B-7) has been developed in the text into the undamped oscillator:

$$\frac{d^2\phi}{dR^2} + \kappa^2 R = \frac{\rho\,(0)}{\epsilon_0}e^{2i\kappa R}\cos\left(e^{i\kappa R}\right) \tag{B-8}$$

and Eq.(B-8) has been solved numerically and analytically to show the presence in general of an infinite number of resonant voltage peaks of theoretically infinite amplitude at which the voltage becomes infinite.

More information can be obtained about ϕ and $\boldsymbol{\omega}$ by using the ECE Faraday Law of induction:

$$\boldsymbol{\nabla} \times \mathbf{E}^a + \frac{\partial \mathbf{B}^a}{\partial t} = \mu_0 \mathbf{j}^a \tag{B-9}$$

where $\mathbf{j}^a$ is the homogeneous current density of ECE theory. When there is no magnetic field present (as in the Coulomb Law) Eq.(B-9) becomes the electrostatic law:

$$\boldsymbol{\nabla} \times \mathbf{E}^a = \mu_0 \mathbf{j}^a \tag{B-10}$$

Since Eq.(B-10) holds for all a it can be written simply as:

$$\boldsymbol{\nabla} \times \mathbf{E} = \mu_0 \mathbf{j} \tag{B-11}$$

where:

$$\mathbf{E} = -\left(\boldsymbol{\nabla} + \boldsymbol{\omega}\right)\phi \tag{B-12}$$

From Eqs.(B-11) and (B-12):

$$\boldsymbol{\nabla} \times \left(\boldsymbol{\nabla}\phi + \boldsymbol{\omega}\phi\right) = -\mu_0 \mathbf{j} \tag{B-13}$$

Using the vector relations:

$$\boldsymbol{\nabla} \times \boldsymbol{\nabla}\phi = \mathbf{0} \tag{B-14}$$

$$\boldsymbol{\nabla} \times (\phi\boldsymbol{\omega}) = \phi\boldsymbol{\nabla} \times \boldsymbol{\omega} + \boldsymbol{\nabla}\phi \times \boldsymbol{\omega} \tag{B-15}$$

Eq.(B-13) becomes:

$$\phi\boldsymbol{\nabla} \times \boldsymbol{\omega} + \boldsymbol{\nabla}\phi \times \boldsymbol{\omega} = -\mu_0 \mathbf{j} \tag{B-16}$$

Thus Eqs.(B-1 and B-16) are the two simultaneous equations needed to solve for ϕ and $\boldsymbol{\omega}$ under all conditions in general.

If it is assumed that gravitation has no effect on electromagnetism the homogeneous current disappears:

$$\mathbf{j} = \mathbf{0} \tag{B-17}$$

so that Eq.(B-16) simplifies to:

$$\phi\boldsymbol{\nabla} \times \boldsymbol{\omega} + \boldsymbol{\nabla}\phi \times \boldsymbol{\omega} = \mathbf{0} \tag{B-18}$$

Eq.(B-1) and (B-18) must be solved simultaneously by computer in general to find the class of solutions for the spin connection that gives resonance. In the far off resonance condition they reduce to the Poisson equation:

$$\nabla^2\phi = -\frac{\rho}{\epsilon_0} \tag{B-19}$$

and thus to the Coulomb potential (B-4) and spin connection (B-5). The latter is a valid solution of Eqs. (B-1) and (B-19) because from Eq.(B-5):

$$\boldsymbol{\nabla} \times \boldsymbol{\omega} = \mathbf{0} \tag{B-20}$$

so Eq.(B-18) reduces to:

$$\boldsymbol{\nabla}\phi \times \boldsymbol{\omega} = \mathbf{0} \tag{B-21}$$

If:

$$\boldsymbol{\nabla}\phi = \boldsymbol{\omega}\phi \tag{B-22}$$

Eq.(B-21) is true identically. Also, if consideration is restricted to the radial component:

$$\boldsymbol{\nabla} = \frac{\partial}{\partial r}\mathbf{e}_r \tag{B-23}$$

in the spherical polar coordinate system, then Eq.(B-21) is valid for any radial spin connection of the type:

$$\boldsymbol{\omega} = \omega_r \mathbf{e}_r \tag{B-24}$$

because

$$\mathbf{e}_r \times \mathbf{e}_r = \mathbf{0} \tag{B-25}$$

So Eq.(B-18) is true for any radially directed spin connection. The one that gives the standard model as a limit is Eq.(B-5), Q.E.D.

Bibliography

[1] M. W. Evans, Generally Covariant Unified Field Theory: The Geometrization of Physics (Abramis Academic, 2005, 2006) vols. 1 and 2.

[2] ibid. vols 3 and 4 (Abramis Academic, 2006 and 2007, in press, preprints on www.aias.us and www.atomicprecision.com).

[3] L. Felker, The ECE Equations of Unified Field Theory (preprints on www.aias.us and www.atomicprecision.com).

[4] L. Felker and H. Eckardt, papers on www.aias.us and www.atomicprecision.com.

[5] M. W. Evans, Generally Covariant Dynamics (paper 55 (volume 4 chapter 1) of the ECE series, preprints on www.aias.us and www.atomicprecision.com).

[6] M. W. Evans, Geodesics and the Aharonov Bohm Effects (paper 56).

[7] M. W. Evans, Canonical and Second Quantization in Generally Covariant Quantum field Theory (paper 57).

[8] M. W. Evans, The Effect of Torsion on the Schwarzschild Metric and Light Deflection due to Gravitation (paper 58).

[9] M. W. Evans, The Resonance Coulomb Law form ECE Theory: Application to the Hydrogen Atom (paper 59).

[10] M. W. Evans, Application of Einstein Cartan Evans (ECE) Theory to Atoms and Molecules: Free Electrons at Resonance (paper 60).

[11] M. W. Evans and H. Eckardt, Space-time Resonances in the Coulomb Law (paper 61).

[12] M. W. Evans, Application of the ECE Lemma to the Fermion and Electromagnetic Fields (paper 62).

[13] M. W. Evans, Physica B, **182**, 227, 237

[14] M. W. Evans, papers and letters in Foundations of Physics and Foundations of Physics Letters, 1993 to present.

[15] M. W. Evans, (ed.), Modern Non-Linear Optics, a special topical issue in three parts of I. Prigogine and S. A. Rice, (series eds.), Advances in Chemical Physics (Wiley-Interscience, New York, 2001, 2nd. ed.), vols. 119(1), 119(2), 119(3) (endorsed by the Royal Swedish Academy).

[16] M. W. Evans and S. Kielich (eds.), ibid. first edition of ref. (15) (Wiley Interscience, New York, 1992, 1993, 1997) vols. 85(1) to 85(3). Prize for excellence from the Polish Government.

[17] M. W. Evans and L. B. Crowell, Classical and Quantum Electrodynamics and the $\boldsymbol{B}^{(3)}$ Field (World Scientific, Singapore, 2001).

[18] M. W. Evans and J.-P. Vigier, The Enigmatic Photon (Kluwer, Dordrecht, 1994 to 2002, hardback and softback), in five volumes.

[19] M. W. Evans and A. A. Hasanein, The Photomagneton in Quantum Field Theory (World Scientific, Singapore, 1994).

[20] S. P. Carroll, Space-time and Geometry, an Introduction to General Relativity (Addison-Wesley, New York, 2004).

[21] J. D. Jackson, Classical Electrodynamics (Wiley, 1999, 3rd. ed.)

[22] Feedback sites to *www.aias.us*. showing showing worldwide interest.

[23] AIAS group discussion over the last three years.

[24] J. B. Marion and S. T. Thornton, Classical Dynamics of Particles and Systems (HB College Publishers, New York, 1988, 3rd ed.) Chapter 3.

[25] E. G. Milewski (Chief Ed.), The Vector Analysis Problem Solver (Research and Education Association, New York, 1987).

[26] G. Stephenson, Mathematical Methods for Science Students (Longmans, London, 1968).

[27] L. Fritsche, private communication (2006)

[28] J. T. Chayes, L. Chayes, and M. B. Ruskai, J. Stat. Phys. 38, 497 (1985)

[29] L. Fritsche and J. Koller, J. Solid State Chem. 176, 652 (2003).

[30] J. P. Perdew, K. Burke, and M. Ernzerhof, Phys. Rev. Lett. 77, 3865(1996) and Phys. Rev. Lett. 78, 1396 (1997).

[31] D. Mann, private communication (2006)

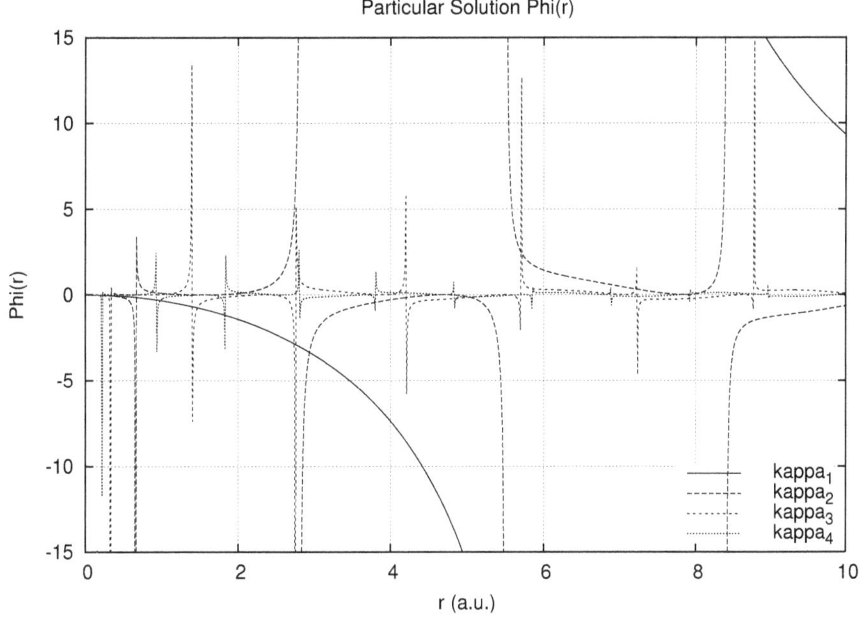

Figure 9.1: Graph of particular solution (9.46) for four κ different values (0.1,1,2,3)

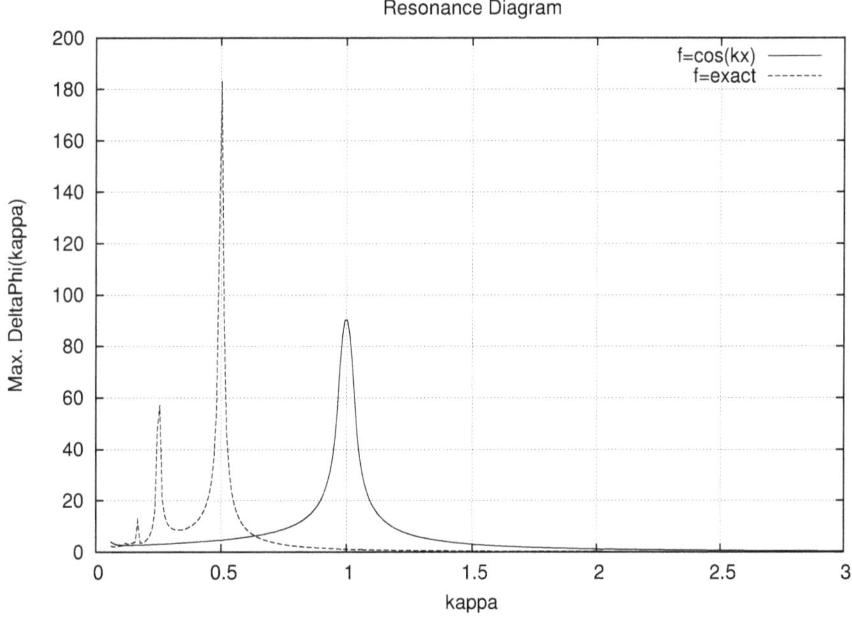

Figure 9.2: Resonance diagram, max. amplitude difference after 15 wavelengths $\lambda = 2\pi/\kappa$

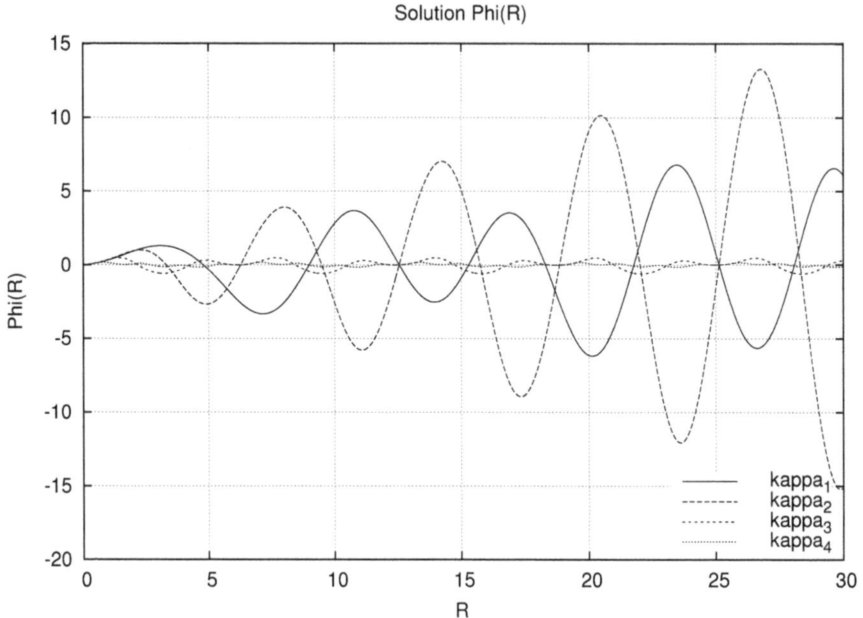

Figure 9.3: Numerical solution of Eq.(9.38) for four different κ values (0.25, 0.5, 1, 2)

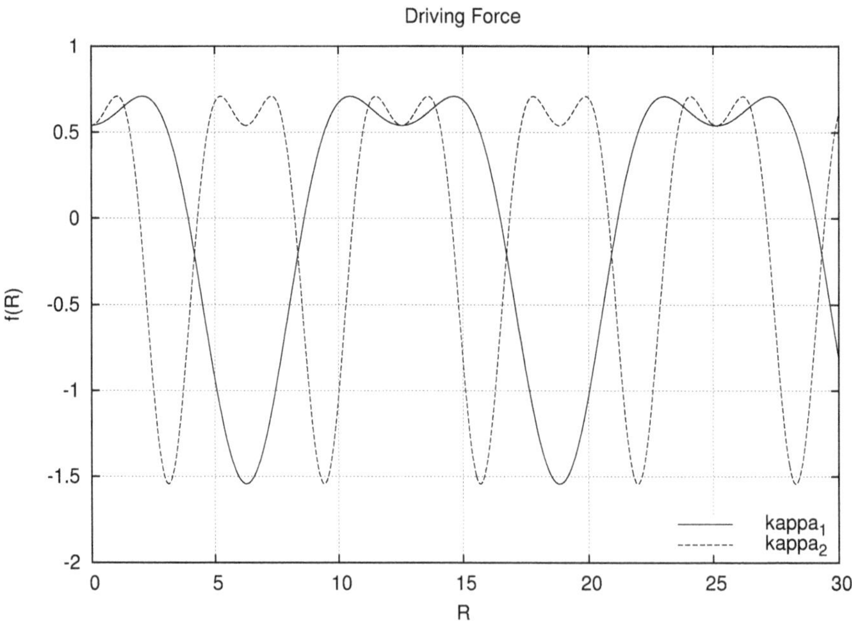

Figure 9.4: Driving force of Eq.(9.38) for $\kappa = 0.25$ and $\kappa = 0.5$

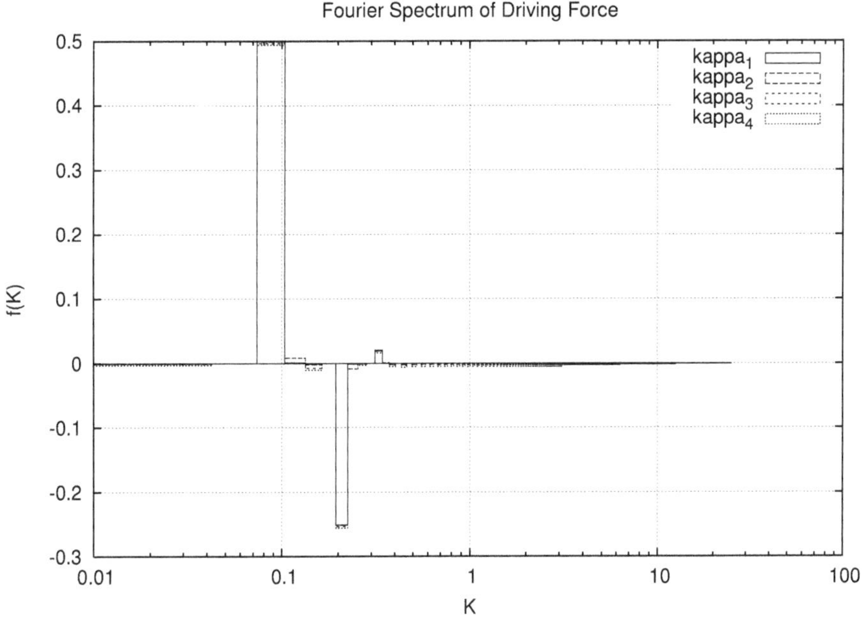

Figure 9.5: Fourier spectrum of driving force (Fig. 9.4) for four different κ values (0.25, 0.5, 1, 2)

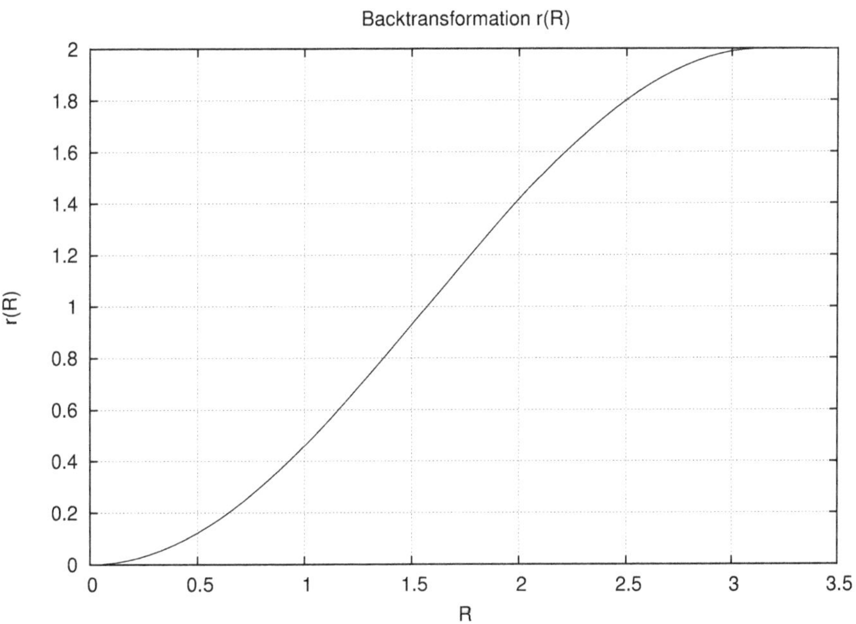

Figure 9.6: Backtransformation $R \rightarrow r$ for $\kappa_r = 1$

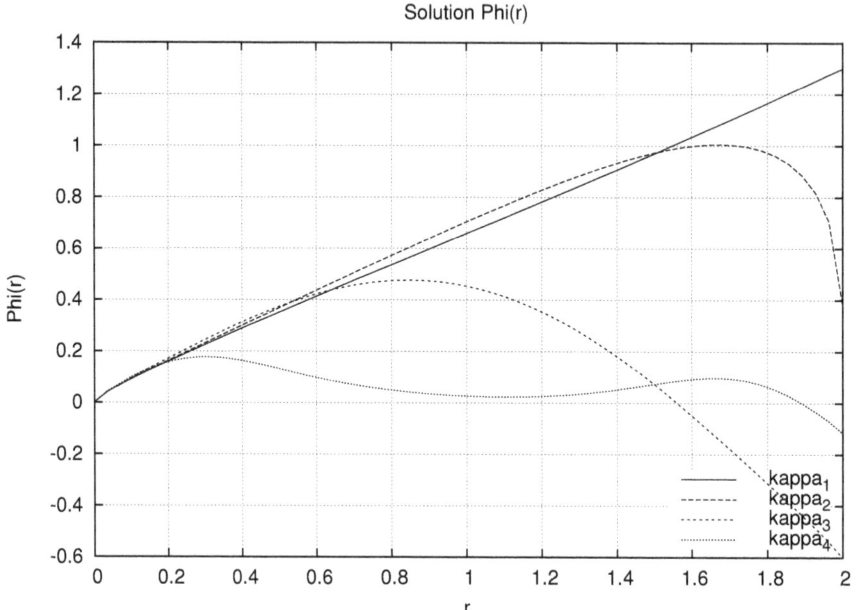

Figure 9.7: Back-transformed $\Phi(r)$ according to Fig. 9.6

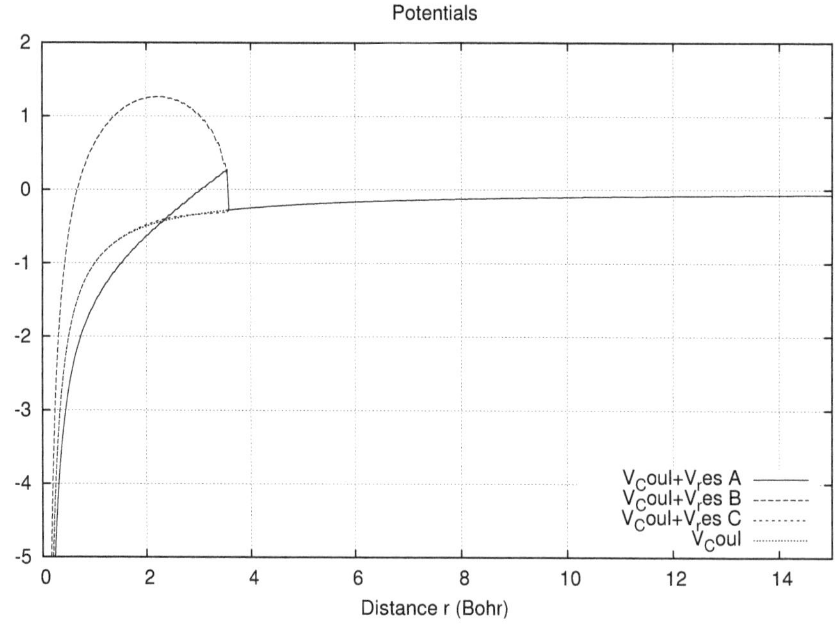

Figure 9.8: Coulomb and resonance potential of H. A: $\kappa = 0.175$, B: $\kappa = 0.225$, C: $\kappa = 1.0$

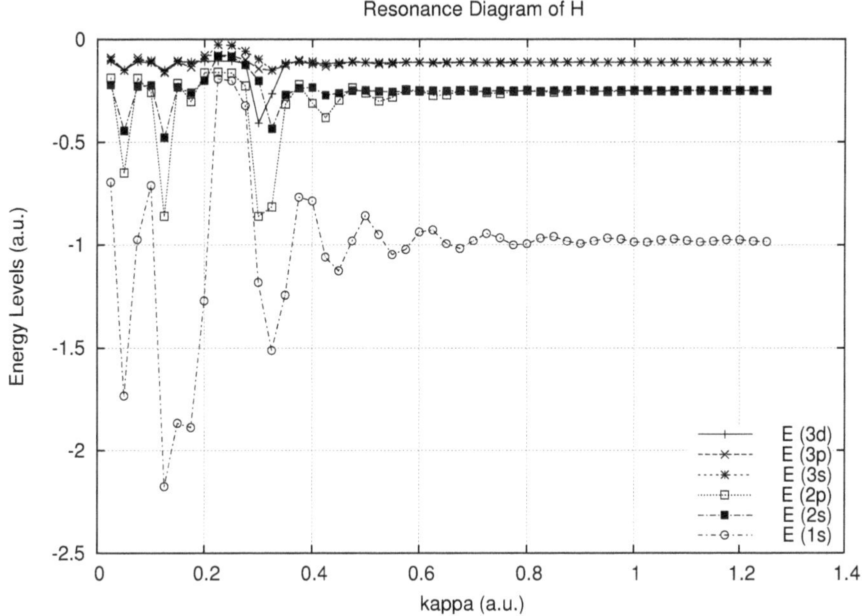

Figure 9.9: Resonance diagram of atomic Hydrogen ($\kappa_r = 0.5$)

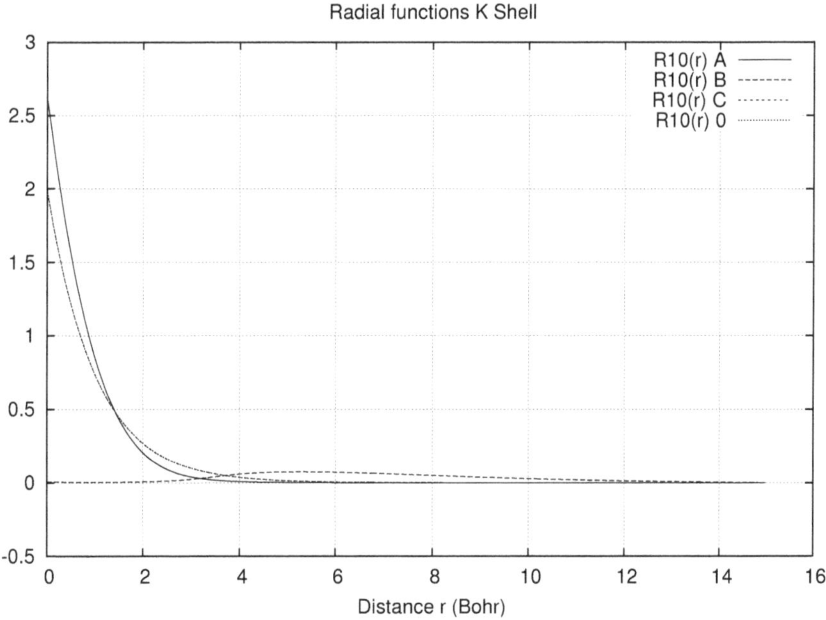

Figure 9.10: Radial 1s wavefunction of H. A: $\kappa = 0.175, B : \kappa = 0.225, C : \kappa = 1.0$

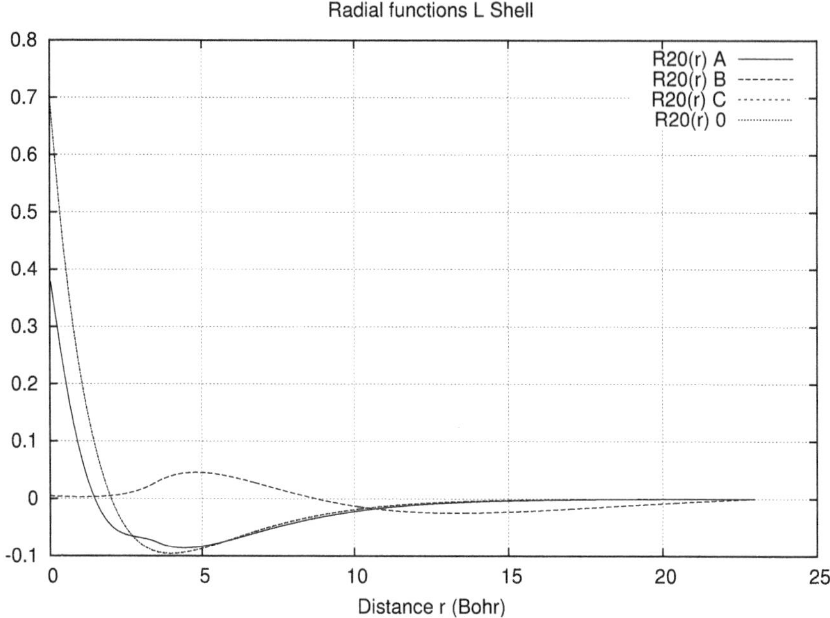

Figure 9.11: Radial 2s wavefunction of H. A: $\kappa = 0.175, B : \kappa = 0.225, C : \kappa = 1.0$

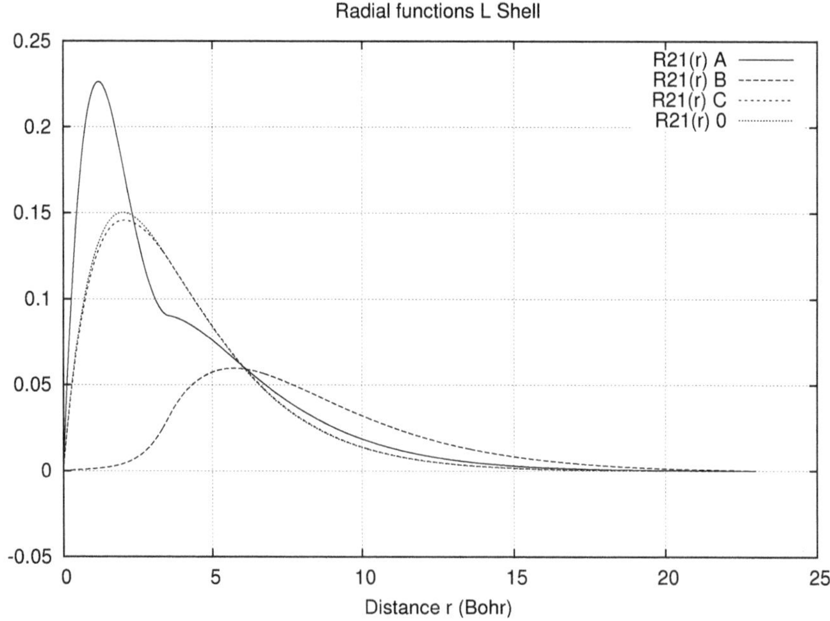

Figure 9.12: Radial 2p wavefunction of H. A: $\kappa = 0.175, B : \kappa = 0.225, C : \kappa = 1.0$

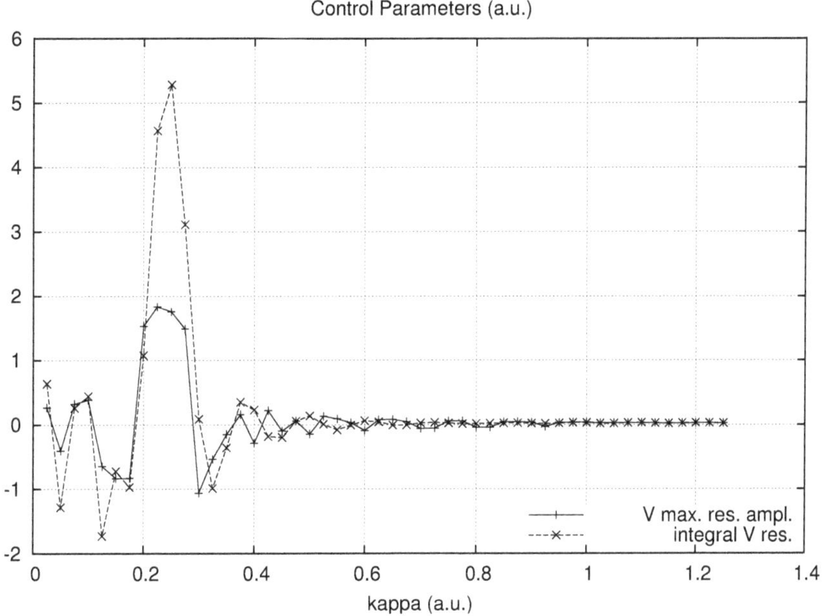

Figure 9.13: Control parameters

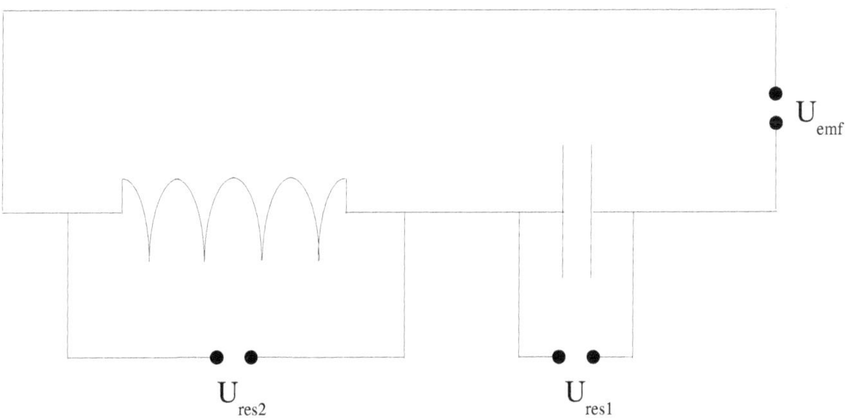

Figure 9.14: Equivalent circuit for the resonant Coulomb law

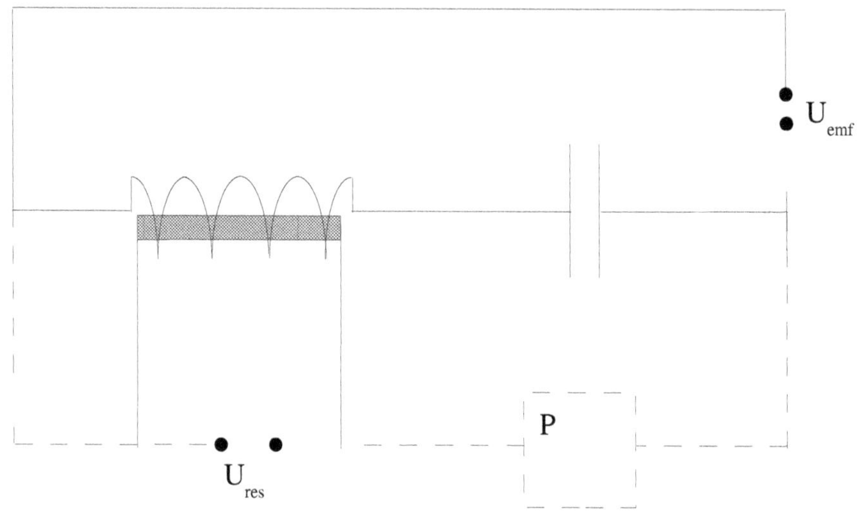

Figure 9.15: Alternative tapping of energy from equivalent circuit

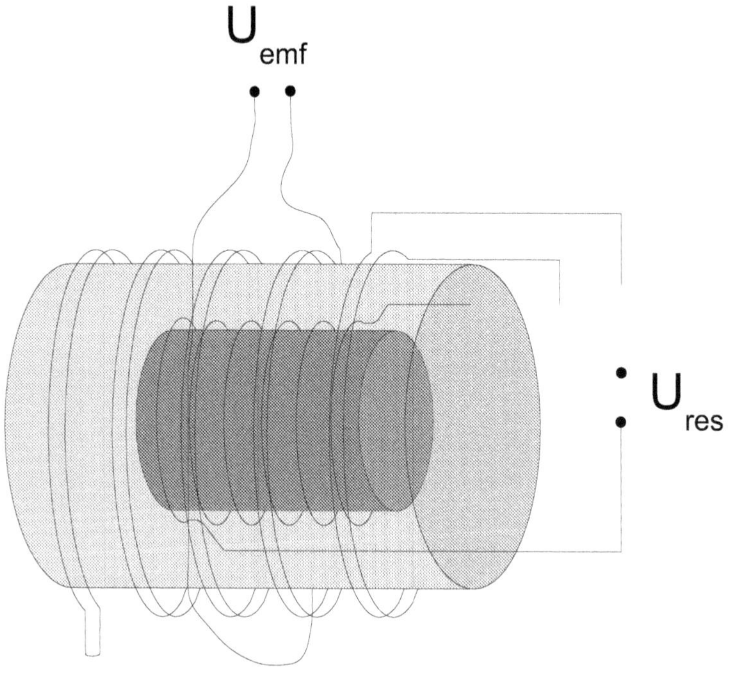

Figure 9.16: Alternative tapping of energy from equivalent circuit

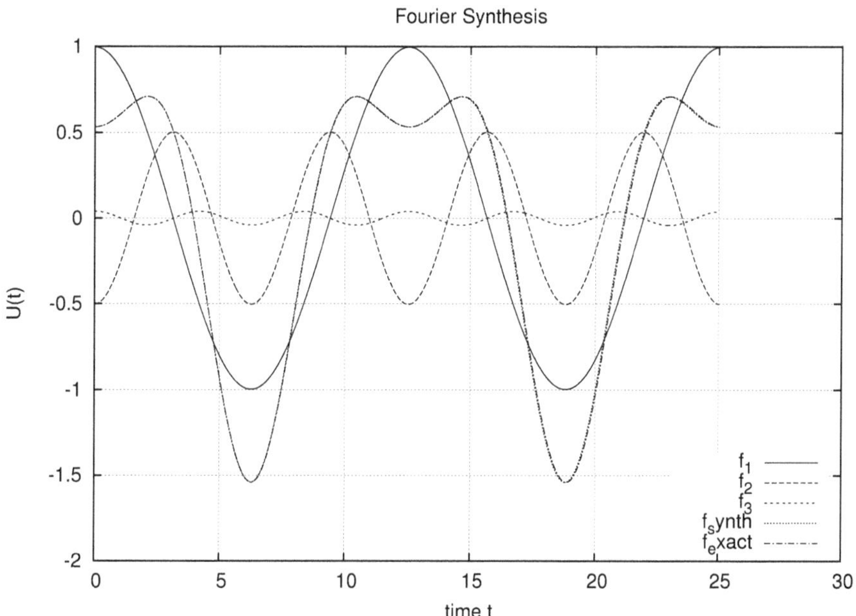

Figure 9.17: Fourier synthesis of the driving force U_{emf}

Chapter 10

Spin connection resonance in gravitational general relativity

(Paper 64)
by
Myron W. Evans
Alpha Institute for Advanced Study (AIAS).
(emyrone@aol.com, www.aias.us, www.atomicprecision.com)

Abstract

The equations of gravitational general relativity are developed with Cartan geometry using the second Cartan structure equation and the second Bianchi identity. These two equations combined result in a second order differential equation with resonant solutions. At resonance the force due to gravity is greatly amplified. When expressed in vector notation, one of the equations obtained from the Cartan geometry reduces to the Newton inverse square law. It is shown that the latter is always valid in the off resonance condition, but at resonance, the force due to gravity is greatly amplified even in the Newtonian limit. This is a direct consequence of Cartan geometry. The latter reduces to Riemann geometry when the Cartan torsion vanishes and when the spin connection becomes equivalent to the Christoffel connection.

Keywords: Spin connection resonance (SCR), Einstein Cartan Evans (ECE) field theory, gravitational general relativity, Newtonian dynamics.

10.1 Introduction

It is well known that gravitational general relativity is based on Riemann geometry with a Christoffel connection. This type of geometry is a special case of Cartan geometry [1] when the torsion form is zero. Therefore gravitational general relativity can be expressed in terms of this limit of Cartan geometry. In this paper this procedure is shown to produce a second order differential equation with resonant solutions [2]- [20]. Off resonance the mathematical form of the Newton inverse square law is obtained from a well defined approximation to the complete theory, but the relation between the gravitational potential field and the gravitational force field is shown to contain the spin connection in general. At resonance the force field can be greatly amplified, or conversely decreased. This is shown in Section 10.2 and given the appellation "spin connection resonance" (SCR). A short discussion is given of possible technological implications in the area of counter gravitation.

10.2 Gravitation and Cartan geometry

The existence of SCR in the theory of gravitation is shown by consideration of the second Cartan structure equation:

$$R^a{}_b = D \wedge \omega^a{}_b \tag{10.1}$$

and the second Bianchi identity of Cartan geometry:

$$D \wedge R^a{}_b := 0 \tag{10.2}$$

Therefore:

$$D \wedge (D \wedge \omega^a{}_b) = 0 \tag{10.3}$$

and this is a second order differential equation with resonant solutions [2]- [20]. In these equations $\omega^a{}_b$ is the spin connection form [1]- [20] and $R^a{}_b$ is the curvature or Riemann form. The symbol $D\wedge$ is the covariant exterior derivative of Cartan geometry and $\wedge$ represents the wedge product of Cartan geometry. If the torsion form T^a of Cartan geometry is zero:

$$T^a = 0 \tag{10.4}$$

Eqs.(10.1) to (10.3) reduce to Riemann geometry, and are fully equivalent to Riemann geometry [1]- [20]. It is well known that Riemann geometry was used by Einstein and independently by Hilbert to obtain the field equation of gravitational general relativity. From Eq.(10.3) it is shown in ths paper that SCR exists in gravitational general theory within Einstein Hilbert (EH) field theory.

If Eqs.(10.1) and (10.2) are developed into tensor and vector equations they lose the basic simplicity of structure of Eqs.(10.1) and (10.2) and the existence of SCR is obscured. So the existence of this important resonance phenomenon has been missed for ninety years. Written out in full, (10.1) and (10.2) are [1]- [20]:

$$R^a{}_b = d \wedge \omega^a{}_b + \omega^a{}_c \wedge \omega^c{}_b \tag{10.5}$$

and

$$d \wedge R^a{}_b + \omega^a{}_c \wedge R^c{}_b - R^a{}_c \wedge \omega^c{}_b := 0 \tag{10.6}$$

where $d\wedge$ is the exterior derivative of Cartan geometry. Eq.(10.4) becomes:

$$T^a = d \wedge q^a + \omega^a{}_b \wedge q^b = 0 \tag{10.7}$$

so in Riemann geometry the following relation exists between the tetrad form q^a and the spin connection form:

$$d \wedge q^a = q^b \wedge \omega^a{}_b \tag{10.8}$$

EH theory relies on a second order description through the symmetric metric:

$$g_{\mu\nu} = q^a{}_\mu q^b{}_\nu \eta_{ab} \tag{10.9}$$

where η_{ab} is the Minkowski metric [1]- [20]. In Cartan geometry the description of gravitation is first order through the tetrad, and is simpler and more elegant. However the physical results of the two representations are the same. Eq.(10.6) can be rewritten as:

$$d \wedge R^a{}_b = j^a{}_b \tag{10.10}$$

$$d \wedge \widetilde{R}^a{}_b = \widetilde{j}^a{}_b \tag{10.11}$$

where:

$$j^a{}_b = R^a{}_c \wedge \omega^c{}_b - \omega^a{}_c \wedge R^c{}_b \tag{10.12}$$

$$\widetilde{j}^a{}_b = \widetilde{R}^a{}_c \wedge \omega^c{}_b - \omega^a{}_c \wedge \widetilde{R}^c{}_b \tag{10.13}$$

The tilde denotes the Hodge dual [1]- [20] of the tensor valued two-form:

$$R^a{}_{b\mu\nu} = -R^a{}_{b\mu\nu} \tag{10.14}$$

in four-dimensions (time and three space dimensions). Eqs.(10.10) and (10.11) are directly analogous to the electrodynamic sector field equations of Einstein Cartan Evans (ECE) field theory [2]- [20]:

$$d \wedge F^a = \mu_0 j^a \tag{10.15}$$

$$d \wedge \widetilde{F}^a = \mu_0 \widetilde{j}^a \tag{10.16}$$

where F^a is the electromagnetic field form, μ_0 is the permeability in vacuo and j^a is the homogeneous current of ECE field theory. The ECE Ansatz defines:

$$F^a = A^{(0)} T^a \tag{10.17}$$

$$A^a = A^{(0)} q^a \tag{10.18}$$

where $cA^{(0)}$ is the primordial or universal voltage of ECE theory and where A^a is the differential form that defines the electromagnetic potential. If there is interaction between electromagnetism and gravitation Eqs.(10.10) to (10.18) are inter-related by the first Bianchi identity of Cartan geometry [1]- [20]:

$$d \wedge T^a + \omega^a{}_b \wedge T^b := R^a{}_b \wedge q^b \tag{10.19}$$

i.e.

$$d \wedge T^a = R^a{}_b \wedge q^b - \omega^a{}_b \wedge T^b \tag{10.20}$$

or

$$d \wedge F^a = A^{(0)} \left(R^a{}_b \wedge q^b - \omega^a{}_b \wedge T^b \right) \tag{10.21}$$

It is seen that:

$$j^a = \frac{A^{(0)}}{\mu_0} \left(R^a{}_b \wedge q^b - \omega^a{}_b \wedge T^b \right) \tag{10.22}$$

and

$$\widetilde{j}^a = \frac{A^{(0)}}{\mu_0}\left(\widetilde{R}^a{}_b \wedge q^b - \omega^a{}_b \wedge \widetilde{T}^b\right) \tag{10.23}$$

Eq.(10.19) links the curvature form $R^a{}_b$ with the torsion form T^a. Eq.(10.17) defines the electromagnetic field F^a as the Cartan torsion T^a within a factor $A^{(0)}$ and Eq.(10.18) defines the electromagnetic potential A^a as the Cartan tetrad q^a within the same factor. In EH theory there is no torsion and the spin connection reduces to the Christoffel connection. This means that Eq.(10.19) reduces to:

$$R^a{}_b \wedge q^b = 0 \tag{10.24}$$

which is the Ricci cyclic equation used in EH theory. In tensor notation Eq.(10.24) is the well known cyclic sum:

$$R_{\sigma\mu\nu\rho} + R_{\sigma\rho\mu\nu} + R_{\sigma\nu\rho\mu} = 0 \tag{10.25}$$

where

$$R_{\sigma\mu\nu\rho} = g_{\sigma\kappa} R^\kappa{}_{\mu\nu\rho} \tag{10.26}$$

etc.

Here $R^\kappa{}_{\mu\nu\rho}$ is the Riemann tensor related to the Riemann form by:

$$R^\kappa{}_{\mu\nu\rho} = q^\kappa{}_a q^b{}_\mu R^a{}_{b\nu\rho} \tag{10.27}$$

Ricci inferred tensors in the late nineteenth century and Einstein used tensor notation from about 1906 to 1916 to develop the EH field equation from the second Bianchi identity (10.2) in tensor notation [1]- [20]. Cartan geometry was not available to Einstein until well after 1916, when Einstein and Cartan developed their well known correspondence. It was during this correspondence that Cartan suggested that the electromagnetic field be related to the torsion. The ECE Ansatz (10.17) was developed independently in early 2003 [2]- [20] without knowledge of Cartan's suggestion. The Ansatz is the simplest way in which F^a and T^a can be related by direct proportionality. The EH field equation is inferred from a comparison of this tensorial Bianchi identity of Riemann geometry with the Noether Theorem:

$$D_\mu G^{\mu\nu} = k D_\mu T^{\mu\nu} \quad := 0 \tag{10.28}$$

The EH field equation is a possible solution of Eq.(10.28) and is:

$$G^{\mu\nu} = \kappa T^{\mu\nu} \tag{10.29}$$

Here $G^{\mu\nu}$ is the Einstein tensor, k the Einstein constant and $T^{\mu\nu}$ the canonical energy-momentum density in tensorial form. Eq.(10.29) is well known, but much less transparent than the equivalent Cartan equation:

$$D \wedge \omega^a{}_b = kD \wedge T^a{}_b \quad := 0 \tag{10.30}$$

The use of tensor notation obscured the existence of SCR for ninety years. From Eqs. (10.5) and (10.10) the basic SCR equation of EH theory is:

$$d \wedge (d \wedge \omega^a{}_b + \omega^a{}_c \wedge \omega^c{}_b) = j^a{}_b \tag{10.31}$$

and its Hodge dual. It is shown in this section that Eq.(10.31) produces an infinite number of resonance peaks of infinite amplitude in the gravitational potential [2]– [20]. To show this numerically, Eq.(10.31) is developed in vector notation. Attention is restricted to the equivalent of the Coulomb Law in ECE gravitational theory. This equivalent is:

$$\nabla \cdot \boldsymbol{R}(\text{orbital}) = J^0 \tag{10.32}$$

where the orbital curvature vector [2]– [20] is defined from the Schwarzschild metric as:

$$\boldsymbol{R}(\text{orbital}) = R^0{}_1{}^{01}\boldsymbol{i} + R^0{}_2{}^{02}\boldsymbol{j} + R^0{}_3{}^{03}\boldsymbol{k} \tag{10.33}$$

The analogy of Eq.(10.32) in electrodynamics is the Coulomb law [2]– [20]:

$$\nabla \cdot \boldsymbol{E} = \frac{\rho}{\epsilon_0} \tag{10.34}$$

where $\boldsymbol{E}$ is the electric field strength in volts per meter, ρ is the charge density in coulombs per cubic meter and ϵ_0 is the vacuum permittivity. In ECE theory:

$$\boldsymbol{E} = -(\nabla + \boldsymbol{\omega})\phi \tag{10.35}$$

where $\boldsymbol{\omega}$ is the spin connection vector. Far off resonance [2]– [20]:

$$\nabla\phi = \boldsymbol{\omega}\phi \tag{10.36}$$

There are direct gravitational analogues of Eqs. (10.34) to (10.36). These are found from the vector equivalent of the second Cartan structure equation, Eq.(10.1), which is:

$$\boldsymbol{R}^a{}_b = -\frac{1}{c}\frac{\partial \boldsymbol{\omega}^a{}_b}{\partial t} - \nabla\omega^{0a}{}_b - \omega^{0a}{}_c\boldsymbol{\omega}^c{}_b + \boldsymbol{\omega}^a{}_c\omega^{0c}{}_b \tag{10.37}$$

in vector notation. The spin connection form is the four vector:

$$\omega^a{}_{\mu b} = (\omega^a{}_{0b}, \boldsymbol{\omega}^a{}_b) \tag{10.38}$$

with time-like component $\omega^a{}_{0b}$ and space-like component $\boldsymbol{\omega}^a{}_b$. For each a and b the time-like component is a scalar and the space-like component is a vector in three dimensions. The electromagnetic analogue of Eq.(10.37) is [2]– [20]:

$$\boldsymbol{E}^a = -\frac{\partial \boldsymbol{A}^a}{\partial t} - c\nabla A^{0a} - c\omega^{0a}{}_b\boldsymbol{A}^b + cA^{0b}\boldsymbol{\omega}^a{}_b \tag{10.39}$$

where $\boldsymbol{A}^a$ is the vector potential and

$$\phi^a = cA^{0a} \tag{10.40}$$

is the scalar potential. It is seen that $\boldsymbol{\omega}^a{}_b$ in gravitational theory plays the role of $\boldsymbol{A}^a$ in electromagnetic theory. The role of ϕ^a in electromagnetic theory is played by $\omega^{0a}{}_c$ in gravitational theory. The role of $\boldsymbol{E}^a$ in electromagnetic theory is played by $\boldsymbol{R}^a{}_b$ in gravitational theory. The vector $\boldsymbol{R}$(orbital) is a particular case of $\boldsymbol{R}^a{}_b$ defined by Eq.(10.33). This particular case fixes the indices a and b.

Attention is now confined to the equivalent in gravitational theory of the electrostatic limit in electromagnetic theory, a limit in which $\boldsymbol{E}^a$ is expressed in terms of the scalar potential only:

$$\boldsymbol{E}^a = -\boldsymbol{\nabla}\phi^a + \boldsymbol{\omega}^a{}_b\phi^b \tag{10.41}$$

The equivalent of Eq.(10.41) in gravitational theory is:

$$\boldsymbol{R}^a{}_b = -\boldsymbol{\nabla}\omega^{0a}{}_b + \boldsymbol{\omega}^a{}_c\omega^{0c}{}_b \tag{10.42}$$

The gravitational scalar potential is therefore identified as the time-like part of the spin connection

$$\Phi^a{}_b = \omega^{0a}{}_b \tag{10.43}$$

Thus:

$$\boldsymbol{R}^a{}_b := -\boldsymbol{\nabla}\Phi^a_b + \boldsymbol{\omega}^a{}_c\Phi^c{}_b \tag{10.44}$$

It is convenient to use a negative sign for the vector part of the spin connection, so:

$$\boldsymbol{R}^a{}_b := -(\boldsymbol{\nabla}\Phi^a{}_b + \boldsymbol{\omega}^a{}_c\Phi^c{}_b) \tag{10.45}$$

This is the direct gravitational analogue of the electromagnetic equation (10.35). The off resonance condition [2]– [20] is now defined by:

$$\boldsymbol{\nabla}\Phi^a{}_b = \boldsymbol{\omega}^a{}_c\Phi^c{}_b \tag{10.46}$$

This is the analogue of Eq.(10.36) in electrodynamics. Summation over repeated indices are implied, so:

$$\boldsymbol{\omega}^a{}_c\Phi^c{}_b = \boldsymbol{\omega}^a{}_0\Phi^0{}_b + \cdots + \boldsymbol{\omega}^a{}_3\Phi^3{}_b \tag{10.47}$$

Since a and b occur in the same way in all terms, Eq.(10.47) can be written as:

$$(\boldsymbol{\omega}_c\Phi^c)^a_b = (\boldsymbol{\omega}_0\Phi^0 + \cdots + \boldsymbol{\omega}_3\Phi^3)^a_b \tag{10.48}$$

The indices a and b are found by the use of the Schwarzschild metric in defining $\boldsymbol{R}$(orbital) in Eq.(10.33). For each a and b we define:

$$\boldsymbol{\omega}\Phi := \omega_0\Phi^0 + \cdots + \boldsymbol{\omega}_3\Phi^3 \tag{10.49}$$

With these definitions:

$$\boldsymbol{R}(\text{orbital}) = -(\boldsymbol{\nabla} + \boldsymbol{\omega})\Phi = R^0{}_1{}^{01}\boldsymbol{i} + R^0{}_2{}^{02}\boldsymbol{j} + R^0{}_3{}^{03}\boldsymbol{k} \tag{10.50}$$

From Eqs. (10.32) and (10.50):

$$\nabla^2\Phi - \boldsymbol{\nabla}\cdot(\boldsymbol{\omega}\Phi) = -J^0 \tag{10.51}$$

which is the gravitational analogue of the electromagnetic:

$$\nabla^2\phi - \boldsymbol{\nabla}\cdot(\boldsymbol{\omega}\phi) = -\frac{\rho}{\epsilon_0} \tag{10.52}$$

The Newton inverse square law is obtained from units analysis [2]– [20]:

$$\boldsymbol{F} = m^2 G\boldsymbol{R}(\text{orbital}) \tag{10.53}$$

where $\boldsymbol{F}$ is force, m is mass, and G is the Newton gravitational constant. Here $\boldsymbol{R}$(orbital) has the units of inverse square meters. The Newtonian limit is defined as the off resonance condition:

$$\boldsymbol{\nabla}\Phi = \boldsymbol{\omega}\Phi \tag{10.54}$$

with gravitational potential:

$$\Phi = -\frac{1}{r} \tag{10.55}$$

along the radial direction. Therefore the spin connection vector is found to be:

$$\boldsymbol{\omega} = -\frac{1}{r}\boldsymbol{e}_r \tag{10.56}$$

where $\boldsymbol{e}_r$ is the radial unit vector in spherical polar coordinates. Therefore:

$$\boldsymbol{\nabla}\Phi = \boldsymbol{\omega}\Phi = \frac{1}{r^2}\boldsymbol{e}_r \tag{10.57}$$

and:

$$\boldsymbol{R}(\text{orbital}) = -\frac{2}{r^2}\boldsymbol{e}_r \tag{10.58}$$

The Newton inverse square law emerges from Eqs. (10.53) and (10.58) by identifying:

$$m^2 = m_1 m_2 \tag{10.59}$$

where m_1 and m_2 are two gravitating masses, and by using half the value of $\boldsymbol{R}$(orbital) in Eq.(10.58). In the non-relativistic (classical) theory of Newtonian dynamics the spin connection is missing, and the classical relation of force and gravitational potential is:

$$\boldsymbol{F} = -m_1 m_2 G \boldsymbol{\nabla}\Phi \tag{10.60}$$

In gravitational general relativity, gravitation is due to curvature, and the spin connection is always non-zero. This means that the force in general relativity is twice the force in the classical theory for a given potential Φ. The extra contribution is due to the spin connection as follows:

$$\boldsymbol{F}(\text{classical}) = -m_1 m_2 G \boldsymbol{\nabla}\Phi \tag{10.61}$$

$$\boldsymbol{F}(\text{relativistic}) = -m_1 m_2 G (\boldsymbol{\nabla} + \boldsymbol{\omega})\Phi \tag{10.62}$$

Off resonance therefore the presence of the spin connection cannot be detected because it simply doubles the definition of the potential. The force is the quantity which is detected experimentally and as long as the force is expressed in terms of the potential as in Eq.(10.57) the inverse square law is not changed. Similar considerations apply [2]– [20] to the Coulomb law of ECE theory. The factor two also occurs in the original EH theory [1], but is arrived at via a different route in a much more complicated way. In the bending of light by gravity the relativistic result is now known to be twice the classical result to a precision of one part in 100,000. Again, the tensorial theory of this well known result is much more complicated than the simple route used here and elsewhere [2]– [20] in this series of papers. SCR emerges finally from the radial form of Eq.(10.51), which is:

$$\frac{\partial^2 \Phi}{\partial r^2} - \frac{1}{r}\frac{\partial \Phi}{\partial r} + \frac{1}{r^2}\Phi = -J^0 \tag{10.63}$$

where it is assumed that there is an oscillatory driving term:

$$J^0 = J^0(0)cos(\kappa r) \tag{10.64}$$

The electromagnetic analogue of Eq.(10.63) is:

$$\frac{\partial^2\phi}{\partial r^2} - \frac{1}{r}\frac{\partial\phi}{\partial r} + \frac{1}{r^2}\phi = -\frac{\rho(0)}{\epsilon_0}cos(\kappa r) \tag{10.65}$$

which has been solved recently using analytical and numerical methods [2]– [20]. These solutions for ϕ and Φ show the presence of an infinite number of resonance peaks, each of which become infinite in amplitude at resonance.

Acknowledgements The British Government is thanked for the award of a Civil List pension for distinguished services to science, and the staff of AIAS are thanked for many interesting discussions.

Bibliography

[1] S. P. Carroll, Space-time and Geometry, an Introduction to General Relativity (Addison-Wesley, New York, 2004).

[2] M., W. Evans, Generally Covariant Unified Field Theory: the Geometrization of Physics (Abramis Academic, 2005 and 2006), vols 1 and 2.

[3] ibid., vols 3 and 4 (in press, preprints on www.aias.us and www.atomicprecision.com).

[4] L. Felker, The ECE Equations of Unified Field Theory (preprints on www.aias.us and www.atomicprecision.com).

[5] L. Felker and H. Eckardt, papers on www.aias.us and www.atomicprecision.com).

[6] M. W. Evans, Generally Covariant Dynamics(paper 55 (volume 4 chapter 1) of the ECE series on www.aias.us and www.atomicprecision.com).

[7] M. W. Evans, Geodesics and the Aharonov Bohm Effects (ibid., paper 56).

[8] M. W. Evans, Canonical and Second Quantization in Generally Covariant Quantum Field Theory (ibid., paper 57).

[9] M. W. Evans, The Effect of Torsion on the Schwarzschild Metric and Light Deflection due to Gravitation (ibid., paper 58).

[10] M. W. Evans, The Resonance Coulomb Law from ECE Theory: Application to the Hydrogen Atom (ibid., paper 59).

[11] M. W. Evans, Application of Einstein Cartan Evans (ECE) Theory to Atoms and Molecules: Free Electrons at Resonance (ibid., paper 60).

[12] M. W. Evans and H. Eckardt, Space-time Resonances in the Coulomb Law (ibid., paper 61).

[13] M. W. Evans, Application of the ECE Lemma to the Fermion and Electromagnetic Fields (ibid, paper 62).

[14] M. W. Evans and H. Eckardt, The Resonant Coulomb Law of Einstein Cartan Evans Field Theory (ibid., paper 63).

[15] M. W. Evans, papers and letters in Foundations of Physics and Foundations of Physics Letters, 1994 to present.

[16] M. W. Evans (ed.), Modern Non-linear Optics, a special topical issue in three parts of I. Prigogien and S. A. Rice (sereis eds.), Advances in Chemical Physics (Wiley Interscience, New York, 2001, 2nd ed.,) vols. 119(1) to 119(3).

[17] M. W. Evans and S. Kielich (eds.), first edition of ref. (16) (Wiley Interscience, New York, 1992, 1993, 1997), vols. 85(1) to 85(3).

[18] M. W. Evans and L. B. Crowell, Classical and Quantum Electrodynamics and the $\boldsymbol{B}^{(3)}$ Field (World Scientific, Singapore, 2001).

[19] M. W. Evans and J.-P. Vigier, The Enigmatic Photon (Kluwer, Dordrecht, 1994 to 2002, hardback and softback), in five volumes.

[20] M. W. Evans and A. A. Hasanein, The Photomagneton in Quantum Field Theory (World Scientifc, Singapore, 1994).

Chapter 11

Spin connection resonance (SCR) in magneto-statics

(Paper 65)
by
Myron W. Evans
Alpha Institute for Advanced Study (AIAS).
(emyrone@aol.com, www.aias.us, www.atomicprecision.com)

Abstract

The existence of spin connection resonance (SCR) is demonstrated in magneto-statics by developing the Gauss Law of magnetism and the Ampère Law in the context of generally covariant unified field theory, i.e. Einstein Cartan Evans (ECE) field theory. One example is given where an analytical solution is possible. At SCR in magneto-statics the magnetic field is greatly amplified for a given driving current density. Tuning to SCR means tuning a wave-number or frequency, and constructing an equivalent circuit based on the relevant resonance equations.

Keywords: Spin connection resonance (SCR), Magneto-statics, Einstein Cartan Evans (ECE) field theory, generally covariant unified field theory.

11.1 Introduction

Recently, classical electrodynamics has been made generally covariant [1]- [18] following the rules of Cartan geometry. This methodology parallels the development by Einstein and others of gravitational theory using the rules of Riemann geometry. The type of Riemann geometry used by Einstein is a limiting case of Cartan geometry when the Cartan torsion [19] vanishes. Geometry is used as the foundation stone of objectivity in natural philosophy. Without objectivity science means all things to all people and there would be no science. The well known Maxwell Heaviside (MH) field theory of classical electrodynamics is a theory of special relativity, and not a theory of general relativity as required by objectivity. The MH theory is Lorentz covariant, but not generally covariant. This means that the field equations of MH theory are covariant under the Lorentz transformation of special relativity [19] but are not covariant under any type of coordinate transformation as required by general relativity and objectivity. In consequence MH theory loses a great deal of information. The Einstein Cartan Evans (ECE) field theory [1]- [18] is a generally covariant unified field theory as required by objectivity in science. This means that all the foundational fields of physics are developed in a rigorously objective manner, and in consequence become aspects of a unified field, i.e. aspects of Cartan geometry. Thus, electromagnetism may be influenced by gravitation on the classical level, as observed in light bending by gravitation. Classical electromagnetism is recognized to be the Cartan torsion form [1]- [19] within a C negative scalar factor denoted $A^{(0)}$. In S.I. units the quantity $cA^{(0)}$ has the units of volts, and in ECE theory is referred to as the primordial voltage of the universe. Voltage originates in geometry, the geometry of four dimensional space-time (known as ECE space-time to distinguish it from the torsion-less space-time used by Einstein in the theory of gravitation, and from the flat or Minkowski space-time of MH theory). In order to develop a generally covariant theory of classical electro-magnetism, it is recognized in ECE theory that the electromagnetic field is due to the spinning of space-time. Similarly it was recognized by Einstein that the gravitational field is due to the curving of space-time. Spinning space-time is characterized by the torsion form, and curving space-time by the Cartan curvature form, or Riemann form [1]- [19]. Experimental evidence for the spinning of space-time is available from a number of sources summarized in refs. ([1]- [18]), notably the magnetization of matter by an electromagnetic field, the inverse Faraday effect. The latter has been shown to be due to the ECE spin field $\boldsymbol{B}^{(3)}$ [1]- [18] generated by the spin connection which defines the Cartan torsion and curvature. In MH theory there is no spin connection, no spin field, and no inverse Faraday effect, contrary to data.

The phenomenon of spin connection resonance (SCR) is due to the fundamental structure of the equations of generally covariant or objective electrodynamics as developed in ECE theory [1]- [18]. The resonance equations emerge directly from Cartan geometry. The ECE field theory of classical electrodynamics is Cartan geometry itself within the factor $A^{(0)}$. In consequence SCR exists in the laws of electrodynamics. It also exists in the laws of gravitation. It has been developed for the Coulomb Law to show the existence of an undamped driven oscillator equation [1]- [18]. At resonance the voltage becomes theoretically infinite. If the experimental means were found to generate SCR, a new source of electric power would become available.

In Section 11.2, SCR is demonstrated for magneto-statics by developing the Gauss Law of magnetism and the Ampère Law into simultaneous, generally covariant, field equations using ECE theory. The spin connection is first simplified by using the approximation of an electromagnetic field uninfluenced by the gravitational field. In this approximation there are equations [1]- [18] which inter-relate tetrad and spin connection elements. Before proceeding to magneto-statics the role of the spin connection is illustrated in electro-statics, and it is shown that in ECE theory, the static electric field becomes characterized by a vector boson similar to that already known from gauge theory in the weak field. After simplifying the spin connection SCR in magneto-statics is illustrated using an example to which there is an analytical solution. It is shown from this solution that the magnetic field becomes theoretically infinite at resonance. In practice the magnetic field would be enhanced at resonance because of inevitable circuit inefficiencies. Finally in Section 11.3 this methodology is extended to electrodynamics using the Faraday Law of induction as an example.

11.2 Magneto-Statics

The spin connection is the most important quantity in generally covariant electrodynamics and for practical applications it is important to simplify the theory and to define the spin connection precisely. In the Coulomb law for example [1]–[18] the basic equations of ECE theory are:

$$A^a{}_\mu = A^{(0)} q^a{}_\mu \tag{11.1}$$

and

$$\boldsymbol{E}^a = -\boldsymbol{\nabla}\phi^a - \boldsymbol{\omega}^a{}_b\phi^b. \tag{11.2}$$

Here $A^a{}_\mu$ is the potential one-form and $q^a{}_\mu$ is the tetrad one-form, the fundamental field. In the absence of a vector potential (space-like part of $A^a{}_\mu$) the generally covariant electric field is the vector:

$$\boldsymbol{E}^a = \boldsymbol{E}^a \; (a = 1, 2, 3) \tag{11.3}$$

where ϕ^a denotes the scalar potential and $\boldsymbol{\omega}^a{}_b$ denotes the spin connection vector, the space-like part of the complete spin connection [1]– [18]. In electrostatics only the scalar potential is considered, so Eq.(11.1) becomes:

$$\phi^a{}_0 = \phi^{(0)} q^a{}_0. \tag{11.4}$$

Therefore only the time-like part of the tetrad,$q^a{}_0$, is considered in electrostatics. When the electromagnetic and gravitational fields are independent in ECE theory there exists the following relation [1]– [18] between the tetrad and spin connection:

$$\omega^a{}_{\mu b} = -\frac{\kappa}{2}\epsilon^a{}_{bc} q^c{}_\mu \tag{11.5}$$

where κ has the units of inverse meters. The anti-symmetric tensor $\epsilon^a{}_{bc}$ is defined by:

$$\epsilon^a{}_{bc} = \eta^{ad}\epsilon_{dbc} \tag{11.6}$$

where η^{ad} is the Minkowski metric. Eq.(11.5) can be developed into the following six relations between elements [1]– [18]:

$$\omega^0{}_1 = -\frac{\kappa}{2}\left(q^2 + q^3\right) = -\omega^1{}_0, \tag{11.7}$$

$$\omega^0{}_2 = -\frac{\kappa}{2}\left(q^3 - q^1\right) = -\omega^2{}_0, \tag{11.8}$$

$$\omega^0{}_3 = -\frac{\kappa}{2}\left(-q^1 - q^2\right) = -\omega^3{}_0, \tag{11.9}$$

$$\omega^1{}_2 = \frac{\kappa}{2}\left(q^0 + q^3\right) = -\omega^2{}_1 \tag{11.10}$$

$$\omega^1{}_3 = \frac{\kappa}{2}\left(-q^2 + q^0\right) = -\omega^3{}_1, \tag{11.11}$$

$$\omega^2{}_3 = \frac{\kappa}{2}\left(q^1 + q^0\right) = -\omega^3{}_2. \tag{11.12}$$

In each equation the index of the base manifold [19] is implied, as usual in differential geometry. Thus Eq.(11.7) for example means:

$$\omega^0{}_{\mu 1} = -\frac{\kappa}{2}\left(q^2{}_\mu + q^3{}_\mu\right) = -\omega^1{}_{\mu 0}. \tag{11.13}$$

For

$$\mu = 0 \tag{11.14}$$

Eqs.(11.7) to (11.12) simplify to:

$$\omega^1{}_{02} = -\omega^2{}_{01} = \frac{\kappa}{2} q^0{}_0, \tag{11.15}$$

$$\omega^1{}_{03} = -\omega^3{}_{01} = \frac{\kappa}{2} q^0{}_0, \tag{11.16}$$

$$\omega^2{}_{03} = -\omega^3{}_{02} = \frac{\kappa}{2} q^0{}_0. \tag{11.17}$$

So for electro-statics there is only one independent spin connection element.

The scalar potential ϕ_0^a must be the same scalar in all frames of reference. It follows that:

$$\phi_0^a = \phi^{(0)} \begin{bmatrix} 1 & 0 & 0 & 0 \\ 0 & 1 & 0 & 0 \\ 0 & 0 & 1 & 0 \\ 0 & 0 & 0 & 1 \end{bmatrix}, \; q_0^a = \begin{bmatrix} 1 & 0 & 0 & 0 \\ 0 & 1 & 0 & 0 \\ 0 & 0 & 1 & 0 \\ 0 & 0 & 0 & 1 \end{bmatrix} \tag{11.18}$$

The product of $\boldsymbol{\omega}^a{}_b$ and ϕ^b must be:

$$\begin{aligned} \omega^a{}_{ib}\phi^b &= \omega^a{}_{i1}\phi^1 + \omega^a{}_{i2}\phi^2 + \omega^a{}_{i3}\phi^3 \\ &= \phi^{(0)}\left(\omega^a{}_{i1} + \omega^a{}_{i2} + \omega^a{}_{i3}\right) \end{aligned} \tag{11.19}$$

From Eq.(11.15)–(11.17) it is seen that:

$$a = 1, 2, 3 \tag{11.20}$$

Thus:

$$\boldsymbol{E}^1 = -\boldsymbol{\nabla}\phi - \boldsymbol{\kappa}\phi, \tag{11.21}$$

$$\boldsymbol{E}_1 := -\boldsymbol{E}^1 \tag{11.22}$$

$$\boldsymbol{E}^2 = -\boldsymbol{\nabla}\phi \tag{11.23}$$

$$\boldsymbol{E}_0 := -\boldsymbol{E}^2 \tag{11.24}$$

$$\boldsymbol{E}^3 = -\boldsymbol{\nabla}\phi + \boldsymbol{\kappa}\phi \tag{11.25}$$

$$\boldsymbol{E}_{-1} := \boldsymbol{E}^3 \tag{11.26}$$

The electric field has the three indices, $-1, 0, +1$ of a boson and develops the character of a vector boson.

In the Coulomb law it is almost always observed with great precision [20] that:

$$\phi = -\frac{e}{4\pi\epsilon_0 r} \tag{11.27}$$

In ECE theory this well known potential is used to identify the vector spin connection in the off resonant condition:

$$\boldsymbol{\nabla}\phi = \pm\boldsymbol{\kappa}\phi \tag{11.28}$$

i.e.

$$\boldsymbol{\kappa} = \pm\frac{1}{r}\boldsymbol{e}_r \tag{11.29}$$

where $\boldsymbol{e}_r$ is the radial unit vector in spherical polar coordinates. At resonance it has been shown [1]- [18] that the spin connection is still given by Eq.(11.29) but that the scalar potential with the units of volts may become infinite. This gives a new source of power if the experimental method were found of tuning to SCR.

In magneto-statics there are four basic equations:

$$A^a{}_\mu = A^{(0)}q^a{}_\mu \tag{11.30}$$

$$\boldsymbol{\nabla}\cdot\boldsymbol{B}^a = \mu_0 j^{0a} \tag{11.31}$$

$$\boldsymbol{\nabla}\times\boldsymbol{B}^a = \mu_0\boldsymbol{J}^a{}_m \tag{11.32}$$

$$\boldsymbol{B}^a = \boldsymbol{\nabla}\times\boldsymbol{A}^a - \boldsymbol{\omega}^a{}_b\times\boldsymbol{A}^b \tag{11.33}$$

where $\boldsymbol{A}^a$ is the vector potential, μ_0 is the vacuum permeability, j^{oa} is the homogeneous charge density of ECE theory and $\boldsymbol{J}^a{}_m$ is the current density including magnetization [1]- [18]. There is no scalar potential in magneto-statics so the space-like elements of $A^a{}_\mu$ are used:

$$A^a{}_\mu = (0, -\boldsymbol{A}^a) \tag{11.34}$$

Therefore only the space-like part of the tetrad is used in magneto-statics:

$$q^a{}_\mu = (0, -\boldsymbol{q}^a) \tag{11.35}$$

When electromagnetism and gravitation are independent:

$$j^{0a} = 0 \tag{11.36}$$

It has been shown that Eq.(11.36) implies [1]- [18]:

$$\boldsymbol{B}^1 = \boldsymbol{\nabla}\times\boldsymbol{A}^1 - g\boldsymbol{A}^2\times\boldsymbol{A}^3 \tag{11.37}$$

$$\boldsymbol{B}^2 = \boldsymbol{\nabla}\times\boldsymbol{A}^2 - g\boldsymbol{A}^3\times\boldsymbol{A}^1 \tag{11.38}$$

$$\boldsymbol{B}^3 = \boldsymbol{\nabla}\times\boldsymbol{A}^3 - g\boldsymbol{A}^1\times\boldsymbol{A}^2 \tag{11.39}$$

$$\boldsymbol{\nabla}\times\boldsymbol{A}^0 = \boldsymbol{\nabla}\times\boldsymbol{A}^3 \tag{11.40}$$

The basic equations of ECE magneto-statics are therefore:

$$\boldsymbol{\nabla}\cdot\boldsymbol{B}^a = 0 \tag{11.41}$$

$$\boldsymbol{\nabla} \times \boldsymbol{B}^a = \mu_0 \boldsymbol{J^a}_m \tag{11.42}$$

and

$$\boldsymbol{B}^a = \boldsymbol{\nabla} \times \boldsymbol{A}^a - g\boldsymbol{A}^b \times \boldsymbol{A}^c \tag{11.43}$$

Here:

$$g = \frac{\kappa}{A^{(0)}} \tag{11.44}$$

SCR in magneto-statics is therefore defined by the simultaneous equations:

$$\boldsymbol{\nabla} \times \left(\boldsymbol{\nabla} \times \boldsymbol{A}^a - g\boldsymbol{A}^b \times \boldsymbol{A}^c\right) = \mu_0 \boldsymbol{J^a}_m \tag{11.45}$$

and

$$\boldsymbol{\nabla} \cdot \boldsymbol{A}^b \times \boldsymbol{A}^c = 0 \tag{11.46}$$

Eq.(11.45) can be developed with the vector identities [21]:

$$\boldsymbol{\nabla} \times (\boldsymbol{\nabla} \times \boldsymbol{A}^a) = -\nabla^2 \boldsymbol{A}^a + \boldsymbol{\nabla}(\boldsymbol{\nabla} \cdot \boldsymbol{A}^a) \tag{11.47}$$

and

$$\begin{aligned} \boldsymbol{\nabla} \times \left(\boldsymbol{A}^b \times \boldsymbol{A}^c\right) &= \\ \boldsymbol{A}^b \boldsymbol{\nabla} \cdot \boldsymbol{A}^c &- \boldsymbol{A}^c \boldsymbol{\nabla} \cdot \boldsymbol{A}^b + \\ (\boldsymbol{A}^c \cdot \boldsymbol{\nabla})\boldsymbol{A}^b &- \left(\boldsymbol{A}^b \cdot \boldsymbol{\nabla}\right)\boldsymbol{A}^c \end{aligned} \tag{11.48}$$

To simplify the problem for the sake of illustration only, assume that the vector potential has no divergence:

$$\boldsymbol{\nabla} \cdot \boldsymbol{A}^a = \boldsymbol{\nabla} \cdot \boldsymbol{A}^b = \boldsymbol{\nabla} \cdot \boldsymbol{A}^c = 0 \tag{11.49}$$

and assume that $\boldsymbol{A}^c$ is space independent so that:

$$\left(\boldsymbol{A}^b \cdot \boldsymbol{\nabla}\right)\boldsymbol{A}^c = \boldsymbol{0} \tag{11.50}$$

Eq.(11.45) then becomes:

$$\nabla^2 \boldsymbol{A}^a + g(\boldsymbol{A}^c \cdot \boldsymbol{\nabla})\boldsymbol{A}^b = -\mu_0 \boldsymbol{J^a}_m \tag{11.51}$$

Considering the $\boldsymbol{k}$ component:

$$\begin{aligned} &\left(\frac{\partial^2}{\partial X^2} + \frac{\partial^2}{\partial Y^2} + \frac{\partial^2}{\partial Z^2}\right) {A^a}_Z + \\ g&\left({A^c}_X \frac{\partial}{\partial X} + {A^c}_Y \frac{\partial}{\partial Y} + {A^c}_Z \frac{\partial}{\partial Z}\right) {A^b}_Z \\ &= -\mu_0 {J^a}_{mZ} \end{aligned} \tag{11.52}$$

If:

$${A^a}_Z = {A^b}_Z = {A^c}_Z \tag{11.53}$$

and:

$$\frac{\partial {A^a}_Z}{\partial Z} = \frac{\partial {A^a}_Z}{\partial Y} = 0 \tag{11.54}$$

then:

$$\frac{\partial^2 A^a_Z}{\partial X^2} + g\left(\frac{\partial A^b_Z}{\partial X}\right) A^c_Z = -\mu_0 J^a_{mZ} \tag{11.55}$$

Finally if it is assumed that:

$$g\left(\frac{\partial A^{b}{}_{Z}}{\partial X}\right)=\kappa_0^2 \tag{11.56}$$

Eq.(11.55) becomes the undamped driven oscillator [22]:

$$\frac{\partial^2 A_Z^a}{\partial X^2}+\kappa_0^2 A_Z^a=\mu_0 J_{mZ}^a(0)\cos(\kappa X) \tag{11.57}$$

if there is an initial oscillatory driving current density:

$$J^{a}{}_{mZ}=-J^{a}{}_{mZ}(0)\cos(\kappa X) \tag{11.58}$$

The solution of Eq.(11.57) is:

$$A^{a}{}_{Z}=\frac{\mu_0 J^{a}{}_{mZ}(0)}{\kappa^{2}{}_{0}-\kappa^2}\cos(\kappa X) \tag{11.59}$$

Spin connection resonance occurs at:

$$\kappa_0=\kappa \tag{11.60}$$

at which point $A^{a}{}_{Z}$ goes to infinity and there is a surge or spike in the magnetic field. Eq.(11.53) is compatible with:

$$\boldsymbol{\nabla}\cdot\boldsymbol{A}^b\times\boldsymbol{A}^a=0 \tag{11.61}$$

and

$$\boldsymbol{A}^a=\boldsymbol{A}^b \tag{11.62}$$

so this is a self-consistent development.

11.3 Spin connection resonance in the Faraday Law of Induction

The Faraday Law of induction in ECE theory is [1]- [18]:

$$\boldsymbol{\nabla}\times\boldsymbol{E}^a+\frac{\partial\boldsymbol{B}^a}{\partial t}=\mu_0\boldsymbol{j}^a \tag{11.63}$$

where the electric field is defined now by both the scalar and vector potentials:

$$\boldsymbol{E}^a=-\frac{\partial\boldsymbol{A}^a}{\partial t}-c\boldsymbol{\nabla}A^{0a}-c\omega^{0a}{}_{b}\boldsymbol{A}^b+c\boldsymbol{\omega}^{a}{}_{b}A^{0b} \tag{11.64}$$

and where the magnetic field is defined by Eq.(11.33). The homogeneous current density [1]- [18] is non-zero if and only if there is interaction between the electromagnetic and gravitational fields. Therefore SCR in the Faraday Law of induction can exist only when the two fields interact, as in light bending by gravity. In the standard model (special relativity) the Faraday law of induction in S.I. units is well known to be:

$$\boldsymbol{\nabla}\times\boldsymbol{E}+\frac{\partial\boldsymbol{B}}{\partial t}=\boldsymbol{0} \tag{11.65}$$

where:

$$\boldsymbol{E} = -\frac{\partial \boldsymbol{A}}{\partial t} - \boldsymbol{\nabla}\phi \tag{11.66}$$

and

$$\boldsymbol{B} = \boldsymbol{\nabla} \times \boldsymbol{A} \tag{11.67}$$

Using the vector identity:

$$\boldsymbol{\nabla} \times \boldsymbol{\nabla}\phi := \boldsymbol{0} \tag{11.68}$$

then:

$$-\boldsymbol{\nabla} \times \frac{\partial \boldsymbol{A}}{\partial t} + \frac{\partial}{\partial t}\boldsymbol{\nabla} \times \boldsymbol{A} = \boldsymbol{0} \tag{11.69}$$

and there is no SCR in the standard model Faraday law of induction and no homogeneous current density. In consequence the effect of gravitation on electromagnetism cannot be described in special relativity, and the well known bending of light by gravity cannot be described on the classical level. Its well known prediction by Einstein rests on the assumption of photon mass, which gravitates with the mass of an object such as the sun. This is therefore a semi-classical description. The photon does not exist on the classical level by definition. ECE explains the bending of light by gravity [1]- [18] using the fact that it is a generally covariant unified field theory able to describe the interaction of electromagnetism and gravitation through the first Bianchi identity of Cartan geometry. This becomes the homogeneous electromagnetic field equation, whose Hodge transform gives the inhomogeneous field equation. The Faraday law of induction is part of the homogeneous field equation and thus part of the first Bianchi identity.

For the sake of illustration only, a simple resonance equation can be developed if it is assumed that there is no scalar potential and also assume that the time-like part of the spin connection is zero:

$$\phi^a = 0, \quad \omega^{0a}{}_b = 0 \tag{11.70}$$

In this approximation the electric field reduces to:

$$\boldsymbol{E}^a = -\frac{\partial \boldsymbol{A}^a}{\partial t} \tag{11.71}$$

and Eq.(11.63) becomes:

$$\frac{\partial \boldsymbol{A}^b}{\partial t} \times \boldsymbol{\omega}^a{}_b + \boldsymbol{A}^b \times \frac{\partial \boldsymbol{\omega}^a{}_b}{\partial t} = \mu_0 \boldsymbol{j}^a \tag{11.72}$$

Now assume that the homogeneous current density is oscillatory in time:

$$\boldsymbol{j}^a = \boldsymbol{j}^a\,(0)\sin\omega t \tag{11.73}$$

and differentiate Eq.(11.72) to obtain:

$$\frac{\partial^2 \boldsymbol{A}^b}{\partial t^2} \times \boldsymbol{\omega}^a{}_b + 2\frac{\partial \boldsymbol{A}^b}{\partial t} \times \frac{\partial \boldsymbol{\omega}^a{}_b}{\partial t} + \boldsymbol{A}^b \times \frac{\partial^2 \boldsymbol{\omega}^a{}_b}{\partial t^2} = \mu_0 \omega \boldsymbol{j}^a\,(0)\cos\omega t \tag{11.74}$$

This is a second order differential equation which under certain circumstances [1]- [18] can show resonance. In this case there must be interaction between gravitation and electromagnetism for resonance to occur. For example light emitted by a pulsar could be explained as pulses of electromagnetic radiation at resonance. The pulsar has a very strong gravitational field and in this situation the homogeneous current density is non-zero. A very small initial driving force from this current is greatly amplified at resonance.

Acknowledgements The Head of State (Queen Elizabeth II), Parliament, Prime Minister, and Royal Society are thanked for the award of a Civil List pension in recognition of distinguished contributions to Britain and the Commonwealth in science. The staff and environment of AIAS are thanked for many interesting discussions.

Bibliography

[1] M. W. Evans, Generally Covariant Unified Field Theory: The Geometrization of Physics (Abramis Academic, 2005, 2006) vols. 1 and 2.

[2] ibid. vols 3 and 4 (Abramis Academic, 2006 and 2007, in press, preprints on www.aias.us and www.atomicprecision.com).

[3] L. Felker, The ECE Equations of Unified Field Theory (preprints on www.aias.us and www.atomicprecision.com).

[4] L. Felker and H. Eckardt, papers on www.aias.us and www.atomicprecision.com.

[5] M. W. Evans, Generally Covariant Dynamics (paper 55 (volume 4 chapter 1) of the ECE series, preprints on www.aias.us and www.atomicprecision.com).

[6] M. W. Evans, Geodesics and the Aharonov Bohm Effects (paper 56).

[7] M. W. Evans, Canonical and Second Quantization in Generally Covariant Quantum field Theory (paper 57).

[8] M. W. Evans, The Effect of Torsion on the Schwarzschild Metric and Light Deflection due to Gravitation (paper 58).

[9] M. W. Evans, The Resonance Coulomb Law form ECE Theory: Application to the Hydrogen Atom (paper 59).

[10] M. W. Evans, Application of Einstein Cartan Evans (ECE) Theory to Atoms and Molecules: Free Electrons at Resonance (paper 60).

[11] M. W. Evans and H. Eckardt, Space-time Resonances in the Coulomb Law (paper 61).

[12] M. W. Evans, Application of the ECE Lemma to the Fermion and Electromagnetic Fields (paper 62).

[13] M. W. Evans and H. Eckardt, The Resonant Coulomb Law of Einstein Cartan Evans Field Theory (paper 63).

[14] M. W. Evans, Spin Connection Resonance in Gravitational General Relativity(paper 64).

[15] M. W. Evans, papers and letters in Foundations of Physics Letters and Foundations of Physics from 1994 to present on O(3) electrodynamics and ECE theory.

[16] M. W. Evans (ed.), Modern Nonlinear Optics, a special topical issue in three parts of I. Prigogine and S. A. Rice (series eds.), Advances in Chemical Physics (Wiley-Interscience, New York, 2001, 2nd ed.) vols. 119(1) to 119(3), endorsed by the Royal Swedish Academy; ibid., first edition (Wiley Interscience, 1992, 1993 and 1997), vols. 85(1) to 85(3), award for excellence form Polish Government.

[17] M. W. Evans and L. B. Crowell, Classical and Quantum Electrodynamics and the $\boldsymbol{B}^{(3)}$ Field (World Scientific, 2001).

[18] M. W. Evans and J.-P. Vigier, The Enigmatic Photon (Kluwer, Dordrecht, 1994 to 2002, hardback and softback), in five volumes.

[19] S. P. Carroll, Space-time and Geometry, and Introduction to General Relativity (Addison-Wesley, New York, 2004).

[20] J. D. Jackson, Classical Electrodynamics (Wiley Interscience, 1999, 3rd ed.).

[21] E. G. Milewski, Chief Editor), The Vector Analysis Problem Solver (Research and Education Association, New York, 1987).

[22] J. B. Marion and S. T. Thornton, Classical Dynamics of Particles and Systems (Harcourt Brace College Publishers, 3rd. Ed., 1988), chapter 3.

Chapter 12

Vector boson character of the static electric field

(Paper 66)
by
Myron W. Evans
Alpha Institute for Advanced Study (AIAS).
(emyrone@aol.com, www.aias.us, www.atomicprecision.com)

Abstract

The field equations of Einstein Cartan Evans (ECE) are used to develop the concept of the static electric field as a vector boson with spin indices -1, 0, $+1$, which occur in addition to the vector character of the electric field. The existence of the electric vector boson in physics is inferred directly from Cartan geometry, using the concept of a spinning space-time that defines the electromagnetic field. When the electromagnetic field is independent of the gravitational field the spin connection is dual to the tetrad, producing a set of equations with which to define the electric vector boson. Angular momentum theory is used to develop the basic concept.

Keywords: Einstein Cartan Evans (ECE) field theory, generally covariant unified field theory, electric vector boson, spinning space-time, spin connection, tetrad, angular momentum theory.

12.1 Introduction

Recently, the concept of electromagnetic field as spinning space-time in general relativity has been developed using standard Cartan geometry [1]– [20]. This produces a generally covariant unified field theory in which the electromagnetic field is the well known Cartan torsion within a factor $A^{(0)}$, where $cA^{(0)}$ has the units of volts. The theory is known as Einstein Cartan Evans (ECE) field theory because it is based on the work of Einstein and Cartan. The ECE theory has been tested extensively for technical correctness and also against a range of experimental data. It has therefore been accepted [21] as a valid theory of physics.

In Section 12.2, the duality equations between spin connection and tetrad are developed in a set of six equations, which are valid when the electromagnetic field is independent of the gravitational field. The basic ECE Ansatz [1]- [19] is used to define the scalar potential and from this the electric field is defined in terms of the scalar potential and the spin connection. It is shown that in ECE theory, the static electric field develops a vector boson character. In Section 12.3, angular momentum theory is used to develop the concept of electric vector boson and the complex circular basis [1]- [19] introduced for the internal indices of the electric vector boson.

12.2 Scalar potential and electric field

In the absence of gravitation the spin connection and tetrad forms [1]- [20] are dual in the tangent space-time, so:

$$\omega^a{}_{\mu b} = -\frac{\kappa}{2}\epsilon^a{}_{bc}q^c_\mu \tag{12.1}$$

Here $\omega^a{}_{\mu b}$ is the spin connection form of Cartan geometry, κ has the units of wave-number, $\eta^a{}_{bc}$ is the Levi-Civita tensor in the tangent space-time, and q^c_μ is the tetrad form of Cartan geometry. Eq.(12.1) may be expanded in six equations as follows:

$$\omega^0{}_{\mu 1} = -\frac{\kappa}{2}(q^2_\mu + q^3_\mu) \tag{12.2}$$

$$\omega^0{}_{\mu 2} = -\frac{\kappa}{2}(q^3_\mu + q^1_\mu) \tag{12.3}$$

$$\omega^0{}_{\mu 3} = -\frac{\kappa}{2}(-q^1_\mu - q^2_\mu) \tag{12.4}$$

$$\omega^1{}_{\mu 2} = \frac{\kappa}{2}(q^0_\mu + q^3_\mu) \tag{12.5}$$

$$\omega^1{}_{\mu 3} = -\frac{\kappa}{2}(-q^2_\mu + q^0_\mu) \tag{12.6}$$

$$\omega^2{}_{\mu 3} = -\frac{\kappa}{2}(q^1_\mu + q^0_\mu) \tag{12.7}$$

The basic ECE Ansatz is:

$$A^a_\mu = A^{(0)}q^a_\mu \tag{12.8}$$

where A^a_μ is the electromagnetic potential form. The scalar potential may be defined in two ways:

$$A^a_\mu \rightarrow (A^a_0, \mathbf{0})\text{or}(A^0_\mu, \mathbf{0}) \tag{12.9}$$

However, these two definitions are equivalent as shown as follows.

Use the fundamental definitions:

$$U^a = q^a_0 U^0 \tag{12.10}$$

$$U^0 = q^0_\mu U^\mu \tag{12.11}$$

and assume that:

$$U^0 U_0 = 1 \tag{12.12}$$

for simplicity of argument. Multiply both sides of Eqs.(12.10) and (12.11) by U_0 to obtain:

$$U^a U_0 = q_0^a \tag{12.13}$$

and

$$q_\mu^0 U^\mu U_0 = 1 \tag{12.14}$$

The fundamental normalization property of the tetrad [20] is:

$$q_\mu^a q_a^\mu = 1 \tag{12.15}$$

so

$$q_\mu^0 q_0^\mu = 1 \tag{12.16}$$

and:

$$q_0^a q_a^0 = 1 \tag{12.17}$$

Using Eqs.(12.13), (12.14) and (12.17):

$$q_a^0 U^a = q_\mu^0 U^\mu \tag{12.18}$$

The a and μ indices are therefore equivalent and interchangeable if one index of the tetrad is fixed at zero. Therefore the scalar potential is either:

$$A_0^a = A^{(0)} q_0^a \tag{12.19}$$

or:

$$A_\mu^0 = A^{(0)} q_\mu^0 \tag{12.20}$$

where

$$\phi^{(0)} = cA^{(0)} \tag{12.21}$$

The scalar component is $\phi^{(0)}$, but q_0^a and q_μ^0 are mixed index rank two tensors. The equivalence of (12.19) and (12.20) means that one frame spinning with respect to a second is indistinguishable from a second spinning with respect to the first.

In special relativity (Maxwell Heaviside field theory) the index a is missing and the scalar potential is the time-like part of A_μ:

$$A_\mu = \left(\frac{\phi}{c}, \mathbf{0}\right) \tag{12.22}$$

In general relativity (ECE theory):

$$A_\mu^a = \left(\frac{\phi_0^a}{c}, \mathbf{0}\right) \tag{12.23}$$

where A_0^a has the four components A_0^0, A_0^1, A_0^2, and A_0^3. Each of these four components is a generally covariant quantity. In special relativity there is only one quantity (12.21). That quantity is not generally covariant, it is Lorentz covariant. From the basic ECE Ansatz (12.8):

$$A_0^0 = A^{(0)} q_0^0, \cdots, A_0^3 = A^{(0)} q_0^3 \tag{12.24}$$

where $A^{(0)}$ is frame invariant and a scalar. The four tetrad elements in Eq.(12.24) are generally covariant, i.e. the tetrad q_0^a retains its tensorial character under any type of coordinate transformation. This property represents objectivity in science. All four elements A_0^0, A_0^1, A_0^2, A_0^3 exist in general and are not the same in general. Thus ECE theory contains more information than Maxwell Heaviside (MH) theory and is ECE theory is rigorously objective. In general, all four A_0^a $(a = 0, ..., 3)$ may be observed experimentally. At present, nothing is known about them experimentally, the only thing that is known is the Coulomb Law's potential:

$$\phi = -\frac{e}{4\pi\epsilon_0 r}\epsilon_0 \tag{12.25}$$

in spherical polar coordinates. Here η_0 is the vacuum permittivity, $-e$ is the charge on the electron, and r is the radial coordinate. In special relativity, ϕ is the time-like part of A_μ in electro-statics. As such ϕ is Lorentz covariant, part of the Lorentz covariant A_μ. In MH theory the electric field is well known to be represented by:

$$\boldsymbol{E} = -\boldsymbol{\nabla}\phi - \frac{\partial A}{\partial t} \tag{12.26}$$

where

$$A_\mu = \left(\frac{\phi}{c}, -\boldsymbol{A}\right) \tag{12.27}$$

In MH the electric field is therefore a 3-D vector and part of the 4-D electromagnetic tensor $F_{\mu\nu}$. In electro-statics [22]:

$$\frac{\partial A}{\partial t} = \boldsymbol{0} \tag{12.28}$$

so:

$$\boldsymbol{E} = -\boldsymbol{\nabla}\phi = \frac{e}{4\pi\epsilon_0 r^2}\boldsymbol{e}_r \tag{12.29}$$

where $\boldsymbol{e}_r$ is the radial unit vector of the spherical polar coordinate system. In ECE theory the electric field is obtained from the first Cartan structure equation:

$$T^a = d \wedge q^a + \omega^a{}_b \wedge q^b \tag{12.30}$$

This is transformed into:

$$F^a = d \wedge A^a + \omega^a{}_b \wedge A^b \tag{12.31}$$

by the ECE Ansatz:

$$A^a = A^{(0)} q^a \tag{12.32}$$

$$F^a = A^{(0)} T^a \tag{12.33}$$

Here F^a is the electromagnetic field form and T^a is the Cartan torsion form. These are vector valued two-forms carrying an index a. It is this index that gives the electric field its vector boson character. From Eq.(12.31) [1]– [19]:

$$\boldsymbol{E}^a = -\boldsymbol{\nabla}\phi^a - \frac{\partial \boldsymbol{A}^a}{\partial t} - c\omega^{0a}{}_b \boldsymbol{A}^b + \boldsymbol{\omega}^a{}_b \phi^b \tag{12.34}$$

and it is seen that the electric field is a vector with $a = 0, \ldots, 3$ and can be considered in general as four vectors, $\boldsymbol{E}^0$, $\boldsymbol{E}^1$, $\boldsymbol{E}^2$ and $\boldsymbol{E}^3$. This is the correctly objective description of the electric field, but again, almost nothing is known experimentally about. In special relativity the entity known as the static electric field is the 3-D vector $\boldsymbol{E}$, which is Lorentz covariant. This has no internal label a.

In special relativity the work done on a charge against the action of a static electric field is:

$$W = e(\phi_A - \phi_B) \tag{12.35}$$

in transporting a charge from A to B. In general relativity the amount of work done depends on the index a:

$$W_0^a = e(\phi_{0A}^a - \phi_{0B}^a) \tag{12.36}$$

and this reflects the structure of ECE space-time. The Minkowski space-time of MH theory has no structure. However, the work may still be expressed in ECE theory as the difference of two potentials. Electro-statics may therefore be developed in general relativity using the potential. It may be assumed that:

$$\boldsymbol{A}^a = \boldsymbol{0} \tag{12.37}$$

and therefore the electric field is defined as:

$$\boldsymbol{E}^a = -\boldsymbol{\nabla}\phi^a + \boldsymbol{\omega}^a{}_b\phi^b \tag{12.38}$$

From this point onwards it is necessary to rely on the only information available experimentally about ϕ, and this is the Coulomb Law, Eq.(12.25). Although this is known to be very accurate [22], it is objectively only a special case of general relativity. In Eq.(12.25), ϕ is the time-like component of A_μ, which is Lorentz covariant. Objectivity demands that there must be $\phi_0^a(a = 0, ..., 3)^\mu$ in physics. In the absence of any experimental data on ϕ_0^a it is assumed for simplicity of argument only that:

$$\phi := \phi_0^0 = \phi_0^1 = \phi_0^2 = \phi_0^3 \tag{12.39}$$

Thus:

$$\phi = \phi^{(0)} q_0^0 \tag{12.40}$$

where

$$q_0^0 = \phi_0^1 = \phi_0^2 = \phi_0^3 \tag{12.41}$$

has been assumed for simplicity. From units analysis:

$$q_0^0 = 1 \tag{12.42}$$

Note carefully however that q_0^0 takes this value only in a given frame. In any other frame it is still unit-less but may become different from unity in general. This is the result of general covariance in physics, the basic ansatz of general relativity. The spin connection elements responsible for the scalar potential in ECE may now be defined in terms of q_0^0 using Eqs.(12.2)–(12.7), as follows:

$$\omega_{02}^1 = -\omega_{01}^2 = \omega_{03}^1 = -\omega_{01}^3 = \omega_{03}^2 = -\omega_{02}^3 = \frac{\kappa}{2} q_0^0 \tag{12.43}$$

There is only one independent element, and its value is $\kappa/2$. This greatly simplifies the treatment of the scalar potential in terms of the spin connection.

This treatment relies on the Coulomb potential as the only experimentally determined scalar potential to date for the static electric field. In general, the latter is defined from Eq.(12.38) as:

$$\boldsymbol{E}^a = -\boldsymbol{\nabla}\phi^a + \boldsymbol{\omega}^a{}_0\phi^0 + \cdots + \boldsymbol{\omega}^a{}_3\phi^3 \tag{12.44}$$

$$\boldsymbol{E}^a_1 = -\frac{\partial\phi^a}{\partial x_1} + \boldsymbol{\omega}^a{}_{10}\phi^0 + \cdots + \boldsymbol{\omega}^a{}_{13}\phi^3 \tag{12.45}$$

and so on. From Eq.(12.1), it is seen that the spin connection elements that contribute to Eq.(12.45) are of the type:

$$\mu = 1, 2, 3 \tag{12.46}$$

Furthermore, only tetrad elements of the type q^0_μ may contribute in Eqs.(12.2)–(12.7). These are defined, as we have seen at the start of this Section, by:

$$q^\mu_0 q^0_\mu = 1 \tag{12.47}$$

and

$$U^0 = q^0_\mu U^\mu \tag{12.48}$$

The tetrad elements of type q^0_μ in Eqs.(12.2)–(12.7), with $\mu = 1, 2, 3$, define:

$$\omega^1_{\mu 2} = \omega^1_{\mu 3} = \omega^2_{\mu 3} = \frac{\kappa}{2} q^0_\mu \tag{12.49}$$

with:

$$\omega^2_{\mu 3} = -\omega^1_{\mu 2} \quad \text{etc.} \tag{12.50}$$

This means that in Eq.(12.45):

$$a = 1, 2, 3 \tag{12.51}$$

and the electric field takes on a vector boson character in general relativity. Thus:

$$\boldsymbol{E}^1 = -\boldsymbol{\nabla}\phi + \boldsymbol{\omega}^1\phi \tag{12.52}$$

$$\boldsymbol{E}^2 = -\boldsymbol{\nabla}\phi + \boldsymbol{\omega}^2\phi \tag{12.53}$$

$$\boldsymbol{E}^3 = -\boldsymbol{\nabla}\phi + \boldsymbol{\omega}^3\phi \tag{12.54}$$

where the three vector boson spin connections are defined by:

$$\omega^1{}_1 = \omega^1{}_{10} + \omega^1{}_{11} + \omega^1{}_{12} + \omega^1{}_{13} \tag{12.55}$$

$$\omega^1{}_2 = \omega^1{}_{20} + \omega^1{}_{21} + \omega^1{}_{22} + \omega^1{}_{23} \tag{12.56}$$

$$\omega^1{}_3 = \omega^1{}_{30} + \omega^1{}_{31} + \omega^1{}_{32} + \omega^1{}_{33} \tag{12.57}$$

$$\boldsymbol{\omega}^1 = \omega^1{}_1\boldsymbol{i} + \omega^1{}_2\boldsymbol{j} + \omega^1{}_3\boldsymbol{k} \tag{12.58}$$

and so on for $\boldsymbol{\omega}^2$ and $\boldsymbol{\omega}^2$. From Eq.(12.49):

$$\omega^1{}_1 = \omega^1{}_{12} + \omega^1{}_{13} = \kappa q^0_1 = \kappa \tag{12.59}$$

$$\omega^1{}_2 = \omega^1{}_{21} + \omega^1{}_{23} = \kappa q^0_2 = \kappa \tag{12.60}$$

$$\omega^1{}_3 = \omega^1{}_{32} + \omega^1{}_{33} = \kappa \tag{12.61}$$

$$\omega^2{}_1 = \omega^2{}_2 = \omega^2{}_3 = 0 \tag{12.62}$$

and:

$$E_1^1 = -\frac{\partial \phi}{\partial x_1} + (\omega^1{}_{10} + \omega^1{}_{11} + \omega^1{}_{12} + \omega^1{}_{13})\phi = -\frac{\partial \phi}{\partial x_1} + \kappa\phi \tag{12.63}$$

etc

Thus:

$$\boldsymbol{E}^1 = -\boldsymbol{\nabla}\phi + \boldsymbol{\kappa}\phi \tag{12.64}$$

Similarly:

$$E_1^2 = -\frac{\partial \phi}{\partial x_2} + (\omega_{10}^2 + \omega_{11}^2 + \omega_{12}^2 + \omega_{13}^2)\phi = -\frac{\partial \phi}{\partial x_2} \tag{12.65}$$

$$\boldsymbol{E}^2 = -\boldsymbol{\nabla}\phi \tag{12.66}$$

and:

$$E_1^3 = -\frac{\partial \phi}{\partial x_3} + (\omega_{10}^3 + \omega_{11}^3 + \omega_{12}^3 + \omega_{13}^3)\phi = -\frac{\partial \phi}{\partial x_3} - \kappa\phi \tag{12.67}$$

$$\boldsymbol{E}^3 = -\boldsymbol{\nabla}\phi - \boldsymbol{\kappa}\phi \tag{12.68}$$

It is seen that the spin connections and tetrads are always space-time properties but that ϕ is defined by the scalar amplitude $\phi^{(0)}$. The spin connection plays the role of an "additional $\bar{\boldsymbol{V}}$". Using experimental measurements to date the three vector boson components $\boldsymbol{E}^1$,$\boldsymbol{E}^2$,$\boldsymbol{E}^3$ have been observed only as the static field $\boldsymbol{E}$ without realizing that it has three labels. However, using the concept of spin connection resonance [1]– [19] they become distinguishable, because $\boldsymbol{E}^1$ and $\boldsymbol{E}^3$ cause resonance but $\boldsymbol{E}^3$ does not. For convenience of development in angular momentum theory we re-label the three components as:

$$\boldsymbol{E}^1 = \boldsymbol{E}_1 \tag{12.69}$$

$$\boldsymbol{E}^2 = \boldsymbol{E}_0 \tag{12.70}$$

$$\boldsymbol{E}^3 = \boldsymbol{E}_{-1} \tag{12.71}$$

12.3 Development of the electric vector boson with angular momentum theory

Angular momentum theory is developed in ref. [18], volume one, chapter five. In ECE theory, the electric field is a vector boson defined by:

$$\boldsymbol{E}_1 = -\boldsymbol{\nabla}\phi + \boldsymbol{\omega}\phi \tag{12.72}$$

$$\boldsymbol{E}_0 = -\boldsymbol{\nabla}\phi \tag{12.73}$$

$$\boldsymbol{E}_{-1} = -\boldsymbol{\nabla}\phi - \boldsymbol{\omega}\phi \tag{12.74}$$

The components are written as $(-1, 0, +1)$ to emphasize the angular momentum character of the electric vector boson. In the simplest instance, angular momentum theory can be developed by considering $O(3)$ vector relations. In the Cartesian basis:

$$\boldsymbol{i} \times \boldsymbol{j} = \boldsymbol{k} \tag{12.75}$$

$$\boldsymbol{k} \times \boldsymbol{i} = \boldsymbol{j} \tag{12.76}$$

$$\boldsymbol{j} \times \boldsymbol{k} = \boldsymbol{i} \tag{12.77}$$

where $\boldsymbol{i}$, $\boldsymbol{j}$ and $\boldsymbol{k}$ are unit vectors. Defining the operator:

$$\hat{J}_Z := \boldsymbol{k}\times \tag{12.78}$$

then:

$$\hat{J}_Z \boldsymbol{i} = 1\boldsymbol{j} \tag{12.79}$$

$$\hat{J}_Z \boldsymbol{j} = -1\boldsymbol{i} \tag{12.80}$$

$$\hat{J}_Z \boldsymbol{k} = 0\boldsymbol{k} \tag{12.81}$$

so the eigenvalues of $\hat{J}_Z$ are -1, 0 and $+1$. In the complex circular basis [1]- [19]:

$$\boldsymbol{e}^{(1)} = \boldsymbol{e}^{(2)*} = \frac{1}{\sqrt{2}}(\boldsymbol{i} - i\boldsymbol{j}) \tag{12.82}$$

$$\boldsymbol{e}^{(2)} = \boldsymbol{e}^{(1)*} = \frac{1}{\sqrt{2}}(\boldsymbol{i} + i\boldsymbol{j}) \tag{12.83}$$

$$\boldsymbol{e}^{(3)} = \boldsymbol{e}^{(3)*} = \boldsymbol{k} \tag{12.84}$$

Thus, $O(3)$ symmetry is represented by:

$$\boldsymbol{e}^{(1)} \times \boldsymbol{e}^{(2)} = i\boldsymbol{e}^{(3)*} \tag{12.85}$$

$$\boldsymbol{e}^{(2)} \times \boldsymbol{e}^{(3)} = i\boldsymbol{e}^{(1)*} \tag{12.86}$$

$$\boldsymbol{e}^{(3)} \times \boldsymbol{e}^{(1)} = i\boldsymbol{e}^{(2)*} \tag{12.87}$$

If we now define the operator:

$$\hat{J}^{(3)} := i\boldsymbol{e}^{(3)}\times \tag{12.88}$$

then:

$$\hat{J}^{(3)}\boldsymbol{e}^{(1)} = -1\boldsymbol{e}^{(1)} \tag{12.89}$$

$$\hat{J}^{(3)}\boldsymbol{e}^{(2)} = 1\boldsymbol{e}^{(2)} \tag{12.90}$$

$$\hat{J}^{(3)}\boldsymbol{e}^{(3)} = 0\boldsymbol{e}^{(3)} \tag{12.91}$$

Thus $\hat{J}^{(3)}$ is a type of angular momentum operator with eigenvalues -1, 0 and 1. These eigenvalues signify the existence of $O(3)$ symmetry. These eigenvalues, -1, 0 and 1, appear again in the matrix equivalents of $\boldsymbol{i}$, $\boldsymbol{j}$ and $\boldsymbol{k}$:

$$\boldsymbol{i} = \begin{bmatrix} 0 & 0 & 0 \\ 0 & 0 & 1 \\ 0 & -1 & 0 \end{bmatrix}, \boldsymbol{j} = \begin{bmatrix} 0 & 0 & -1 \\ 0 & 0 & 1 \\ 1 & 0 & 0 \end{bmatrix}, \boldsymbol{k} = \begin{bmatrix} 0 & 1 & 0 \\ -1 & 0 & 1 \\ 0 & 0 & 0 \end{bmatrix} \tag{12.92}$$

In the theory of irreducible tensorial sets, these rotation operators in Euclidean space are first rank $\hat{T}$ operators, which are irreducible tensor operators and under rotations transform into linear combinations of each other. The $\hat{T}$ operators are directly proportional to the scalar spherical harmonic operators. The rotation operators of the full rotation group are related to the $\hat{T}$ operators as follows:

$$\hat{T}^1_{-1} = i\hat{J}^{(1)}\,,\ \hat{T}^1_1 = i\hat{J}^{(2)}\,,\ \hat{T}^1_0 = i\hat{J}^{(3)} \tag{12.93}$$

and are also related to the spherical harmonics and vector spherical harmonics [18].

The complex circular basis (1), (2) and (3) is also a convenient basis with which to represent the electric vector boson as follows:

$$\boldsymbol{E}^{(1)} = -\boldsymbol{\nabla}\phi - \boldsymbol{\omega}^{(1)}\phi \tag{12.94}$$

$$\boldsymbol{E}^{(2)} = -\boldsymbol{\nabla}\phi + \boldsymbol{\omega}^{(1)}\phi \tag{12.95}$$

$$\boldsymbol{E}^{(3)} = -\boldsymbol{\nabla}\phi \tag{12.96}$$

and the vector boson character of the static electric field may be represented as:

$$\hat{J}^{(3)}\boldsymbol{E}^{(1)} = -\phi\boldsymbol{\omega}^{(1)} \tag{12.97}$$

$$\hat{J}^{(3)}\boldsymbol{E}^{(2)} = \phi\boldsymbol{\omega}^{(2)} \tag{12.98}$$

$$\hat{J}^{(3)}\boldsymbol{E}^{(3)} = \phi\boldsymbol{0} \tag{12.99}$$

For plane waves the circularly polarized electric field is:

$$\boldsymbol{E}^{(1)} = \frac{E^{(0)}}{\sqrt{2}}(\boldsymbol{i} - i\boldsymbol{j})e^{i(\omega t - \kappa Z)} = \boldsymbol{E}^{(2)*} \tag{12.100}$$

and its static limit may be defined as the limit:

$$\omega t - \kappa Z \to 0 \tag{12.101}$$

in which case:

$$\text{Real}\,(\boldsymbol{E}^{(1)}) = \text{Real}\,(\boldsymbol{E}^{(2)}) = \frac{E^{(0)}}{\sqrt{2}}\boldsymbol{i} \tag{12.102}$$

$$\text{Im}\,(\boldsymbol{E}^{(1)}) = -\text{Im}\,(\boldsymbol{E}^{(2)}) = -\frac{E^{(0)}}{\sqrt{2}}\boldsymbol{j} \tag{12.103}$$

Therefore the spin connection for plane waves is:

$$\boldsymbol{\omega}^{(1)} = \boldsymbol{\omega}^{(2)*} = \frac{\omega^{(0)}}{\sqrt{2}}(\boldsymbol{i} - i\boldsymbol{j})e^{i(\omega t - \kappa Z)} \tag{12.104}$$

This means that the frame itself is both spinning and translating. This is an example of the more general spin connection of Cartan geometry.

Acknowledgements The British Government is thanked for the award of a Civil List pension for distinguished services to science, and the staff of AIAS are thanked for many interesting discussions.

Bibliography

[1] M. W. Evans, Generally Covariant Unified Field Theory: The Geometrization of Physics (Abramis Academic, 2005 and 2006), vols. 1 and 2.

[2] ibid., in press, volumes 3 and 4 (preprints on www.ais.us and www.atomicprecision.com).

[3] L. Felker, The Evans Equations of Unified Field Theory (preprints on www.aias.us and www.atomicprecision.com)

[4] H. Eckardt and L. Felker, articles on www.aias.us and www.atomicprecision.com, with translations into several languages.

[5] M. W. Evans, Generally Covariant Dynamics (paper 55 (volume 4 chapter 1) of the ECE series, preprints on www.aias.us and www.atomicprecision.com).

[6] M. W. Evans, Geodesics and the Aharonov Bohm Effects, (paper 56).

[7] M. W. Evans, Canonical and Second Quantization in Generally Covariant Quantum Field Theory (paper 57).

[8] M. W. Evans, The Effect of Torsion on the Schwarzschild Metric and Light Deflection due to Gravitation (paper 58).

[9] M. W. Evans, The Resonance Coulomb Law from ECE Theory: Application to the Hydrogen Atom (paper 59).

[10] M. W. Evans, Application of Einstein Cartan Evans (ECE) Theory to Atoms and Molecules: Free Electrons at Resonance (paper 60).

[11] M. W. Evans and H. Eckardt, Space-time Resonances in the Coulomb Law (paper 61).

[12] M. W. Evans, Application of the ECE Lemma to the Fermion and Electromagnetic Fields (paper 62).

[13] M. W. Evans and H. Eckardt, The Resonance Coulomb Law of Einstein Cartan Evans Field Theory (paper 63).

[14] M. W. Evans, Spin Connection Resonance in Gravitational General Relativity (paper 64).

[15] M. W. Evans, Spin Connection Resonance in Magneto-statics (paper 65).

[16] M. W. Evans, letters and papers in Found. Phys. Lett., and Found. Phys., 1994 to present.

[17] M. W. Evans, (ed.), Modern Non-linear Optics, a special topical issue in three parts of I. Prigogine and S. A. Rice (series eds.), Advances in Chemical Physics (Wiley-Interscience, New York, 2001, 2nd edition), vols 119(1) to 119(3), endorsed by the Royal Swedish Academy; ibid first edition edited by M. W. Evans and S. Kielich (Wiley Interscience, 1992, reprinted 1993 and 1997), vols 85(1) to 85(3), Prize for Excellence, Polish Government.

[18] M. W. Evans and J.-P. Vigier, The Enigmatic Photon (Kluwer, Dordrecht, 1994 to 2002, hardback and softback), in five volumes.

[19] M. W. Evans and L. B.Crowell, Classical and Quantum Electrodynamics and the $\boldsymbol{B}^{(3)}$ Field (World Scientific, Singapore, 2001).

[20] S. P. Carroll, Space-time and Geometry, an Introduction to General Relativity (Addison-Wesley, new York, 2004).

[21] Feedback sites to www.aias.us showing essentially unanimous acceptance of ECE theory internationally for three years.

[22] J. D. Jackson, Classical Electrodynamics (Wiley, New York, 3rd edition, 1999).

Chapter 13

ECE Theory of gravity induced polarization changes

(Paper 67)
by
Myron W. Evans
Alpha Institute for Advanced Study (AIAS).
(emyrone@aol.com, www.aias.us, www.atomicprecision.com)

Abstract

Gravity induced polarization effects recently observed form a white dwarf are explained straightforwardly in terms of the Einstein Cartan Evans (ECE) unified field theory. The homogeneous current of ECE theory is shown to change circular to elliptical polarization when electromagnetism is affected by gravitation on the classical level. In the Einstein Hilbert (EH) field theory of gravitation, such an effect does not exist. Therefore ECE is preferred to EH using these experimental data, and is preferred to other contemporary theories by Ockham's Razor.

Keywords: Einstein Cartan Evans (ECE) field theory, gravity induced polarization changes in light, interaction of gravitation and electromagnetism.

13.1 Introduction

Einstein Cartan Evans (ECE) field theory is essentially a straightforward extension of Riemann to Cartan geometry [1]– [21] using the well known Cartan torsion. The latter is the electromagnetic field within a scalar factor $A^{(0)}$, where $cA^{(0)}$ has the units of volts and where c is the vacuum speed of light. ECE theory is a standard theory of general relativity and unifies the gravitational and electromagnetic fields using the well known Cartan structure equations and Bianchi identities. It therefore uses a universally constant speed of light c, and conforms with the Noether Theorem and the Einstein Equivalence Principle.

In order to unify gravitation and electromagnetism on the classical level, there is no need for any fundamental hypothesis of general relativity not already proposed by Einstein. ECE transforms the Cartan torsion directly into the electromagnetic field, resulting in field equations capable of describing both the kinematic and electrodynamic aspects of gravitational lensing. The original Einstein Hilbert (EH) theory of this effect is a semi-classical theory based on the gravitation between a quantized photon and an object such as the sun. As is well known, the EH theory produces a light deflection angle which is twice that given by the Newtonian theory. The photon mass does not enter into the final expression for the light deflection, because it cancels out in the calculation. Therefore the deflection of light by gravitation in the EH theory is independent of frequency and polarization, and does not produce changes in refractive index, and related effects such as reflection, refraction, diffraction and so on.

In ECE theory it is shown in Section 13.2 that the polarization of the deflected light is changed in general from circular to elliptical. This effect is due to the homogeneous current of ECE theory [1]- [20], the current that measures the extent to which gravitation affects electromagnetism on the classical level. The concepts and results of EH theory are obtained in a well defined limit of ECE theory, so the latter describes both the kinematic and electrodynamic aspects of gravitational lensing, on classical, semi-classical and fully quantized levels. The EH theory describes only the semi-classical angle of deflection as already argued. As might be expected by the term "lensing", ECE theory produces polarization, refractive index, reflection, refraction and diffraction effects in general.

In Section 13.2, attention is confined to the change of polarization from circular to elliptical, a change caused by the homogeneous current. In Section 13.3, a discussion is given of a recent paper [22] by Preuss et al. in which gravity induced polarization changes are measured from a white dwarf. A comparison is made of ECE theory with the Poincaré gauge theory used by Preuss et al. [22], showing that ECE is a far simpler description of the data, and is therefore preferred by Ockham's Razor.

13.2 Polarization change due to the homogeneous current

This change originates in the homogeneous ECE field equations [1]- [20]:

$$\nabla \cdot \boldsymbol{B}^a = \mu_0 \tilde{j}^a \tag{13.1}$$

$$\nabla \cdot \boldsymbol{E}^a + \frac{\partial \boldsymbol{B}^a}{\partial t} = \mu_0 \tilde{\boldsymbol{j}}^a \tag{13.2}$$

Here $\boldsymbol{B}^a$ is the magnetic flux density in tesla, $\boldsymbol{E}^a$ is the electric field strength in volt m^{-1}, μ_0 is the vacuum permeability in S.I. units, and j is the homogeneous current. The latter is defined in the notation of Cartan geometry [1]- [21] by:

$$j^a = \frac{A^{(0)}}{\mu_0}(R^a{}_b \wedge q^b - \omega^a{}_b \wedge T^b) \tag{13.3}$$

where $R^a{}_b$ is the Riemann form, q^b is the tetrad form, $\omega^a{}_b$ is the spin connection form, and T^b is the torsion form. Eq. (13.1) is the generally covariant Gauss law of magnetism, and Eq. (13.2) is the generally covariant Faraday law of induction. Interaction of gravitation and electromagnetism occurs when j^a and $\boldsymbol{j}^a$ are non-zero. The interaction is therefore described by the following condition of Cartan geometry:

$$R^a{}_b \wedge q^b \neq \omega^a{}_b \wedge T^b \tag{13.4}$$

This condition has been developed in great detail [1]– [20] elsewhere. For the purposes of this section it is sufficient to note that the current $\widetilde{\boldsymbol{j}}^a$ is non zero when electromagnetism is affected by gravitation on the classical level.

It has been shown [1]– [20] that Eq. (13.2) may be written as:

$$\boldsymbol{\nabla} \times (n\boldsymbol{E}^a) + \frac{\partial}{\partial t}\left(\frac{\boldsymbol{B}^a}{n}\right) = \boldsymbol{0} \tag{13.5}$$

where the refractive index n is defined by:

$$n^2 = \frac{c}{v} \tag{13.6}$$

and where v is the phase velocity. The plane wave solution of Eq. (13.5) is:

$$\boldsymbol{E}_1 = \frac{E^{(0)}}{\sqrt{2}}(\boldsymbol{i} - i\boldsymbol{j})e^{i\phi_1} \tag{13.7}$$

$$\boldsymbol{B}_1 = \frac{B^{(0)}}{\sqrt{2}}(i\boldsymbol{i} + \boldsymbol{j})e^{i\phi_1} \tag{13.8}$$

where:

$$\phi_1 = \frac{\omega}{n}t - n\kappa Z \tag{13.9}$$

The effect of gravity is to change the original $\boldsymbol{E}$ and $\boldsymbol{B}$ of circular polarization as follows:

$$\boldsymbol{E}_1 = n\boldsymbol{E} \tag{13.10}$$

$$\boldsymbol{B}_1 = \frac{1}{n}\boldsymbol{B} \tag{13.11}$$

It has been shown [1]– [20] that this mechanism produces the well known red-shift.

The real and physical part of Eq. (13.7) is:

$$\boldsymbol{E}_1 = \frac{E^{(0)}}{\sqrt{2}}(\boldsymbol{i}\cos\phi_1 + \boldsymbol{j}\sin\phi_1) \tag{13.12}$$

and the real and physical part of the circularly polarized $\boldsymbol{E}$ is:

$$\boldsymbol{E} = \frac{E^{(0)}}{\sqrt{2}}(\boldsymbol{i}\cos\phi + \boldsymbol{j}\sin\phi) \tag{13.13}$$

It is seen that if:

$$\cos\phi_1 = a\cos\phi \tag{13.14}$$

$$\sin\phi_1 = b\sin\phi \tag{13.15}$$

then:

$$\boldsymbol{E}_1 = \frac{E^{(0)}}{\sqrt{2}}(a\boldsymbol{i}\cos\phi + b\boldsymbol{j}\sin\phi) \tag{13.16}$$

for example, if $\phi = 45°$,and $\phi_1 = 60°$, then $a = 1.414$, $b = 0.816$. This means that gravity changes the original circular polarization of $\boldsymbol{E}$ to elliptical polarization, in which a is not the same as b. This effect is not present in EH theory, but has been observed by Preuss et al. [22], whose work is discussed in the next section and compared with ECE theory.

13.3 Gravity induced birefringence from a white dwarf

Polarization changes from a white dwarf have been reported recently by Preuss et al. [22]. The experimental data consist of flux of net circular polarization at a given wavelength divided by the total emitted stellar flux. This is denoted [22] V / I. A gravity induced depolarization is observed experimentally, and measured by a fall in V / I from 26.5% to 20.0%. There is a difference between the linear polarization perpendicular and parallel to the local stellar limb. The theory of Section 13.2 describes the change of polarization qualitatively and straightforwardly and a quantitative description from Section 13.2 could be developed for given $R^a{}_b$, q^b, $\omega^a{}_b$ and T^b.

The theory used by Preuss et al. [22] is far more abstract and convoluted than ECE theory and is called [22] Poincaré gauge theory (PGT). Being a gauge theory, it is at a disadvantage to ECE from the outset because gauge theory introduces concepts such as gauge invariance and an abstract internal gauge space. These concepts are not needed in ECE theory, which relies on general covariance and geometry as in the original theory of general relativity by Einstein. This level of abstraction in PGT is compounded by the use of string theory. The latter has been heavily criticized, as is well known, and can be applied only to cosmology and particle physics. Even in those areas it is rarely predictive, and even when something new is predicted by string theory, many adjustable parameters must be used, rendering the prediction essentially meaningless.

ECE is a foundational theory of general relativity which can be applied more straightforwardly and far more simply [1]- [20] to all areas of physics and chemistry; and ECE theory uses only the four physical dimensions of relativity. The theory used by Preuss et al. [22] is typical of the hugely abstracted levels of string theory and gauge theory in contemporary physics. However PGT and ECE have one thing in common, they use both the Cartan curvature and torsion, but the latter is in PGT a massless, anti-symmetric, Kalb Rammond field of super-gravity and superstring theory.

The torsion in PGT is thus a gravitational torsion and is introduced as a second order Lagrangian term in Eq. (1) of [22]. This results in the authors [22] Lorentz field (their Eq. (11)) and translation field (their Eq. (12)). These are both electromagnetic fields. Their Lorentz field is a Riemann tensor which contains terms to second order in the electromagnetic potential. These terms may be used to define the ECE spin field, $\boldsymbol{B}^{(3)}$ [1]- [20], but are derived in a much more complicated manner. Nevertheless, the spin field does emerge form

PGT [22]. This gets us away from the Maxwell Heaviside field of the standard model. Their translation field [22] is a torsion tensor of Riemann geometry, equivalent to a torsion form of Cartan geometry [1]- [21].

In ECE theory the electromagnetic field is the Cartan torsion itself within the factor $A^{(0)}$, as suggested independently in 2003 by Evans [1]- [20] and Cartan himself in well known correspondence with Einstein. In PGT the torsion is a gravitational torsion which is introduced through gauge invariance. PGT does not recognise that the Cartan torsion is itself the electromagnetic field. Therefore PGT cannot produce the ECE spin field observed in non-linear optics in the inverse Faraday effect [1]- [20].

String theory cannot be applied to non-linear optics, but ECE theory has been extensively applied to non-linear optics [1]- [20]. There is another conceptual difference between PGT and ECE. The former assumes a violation of the Einstein Equivalence Principle in gravity induced polarization changes, but as seen in Section 13.2, ECE produces the phenomenon without violation of this fundamental principle. Finally, PGT also assumes a non-constant c, while ECE uses the constant c advocated originally by Einstein in 1905.

It is concluded that ECE is preferred to EH on the grounds of experimental data, and is preferred to PGT and all string theories by Ockham's Razor.

Acknowledgements The British Government is thanked for the award of a Civil List pension for distinguished services to science, and the staff of AIAS are thanked for many interesting discussions.

Bibliography

[1] M. W. Evans,Generally Covariant Unified Field Theory (Abramis Academic, 2005), vol. 1.

[2] M. W. Evans, ibid., vol. 2, (2006).

[3] M. W. Evans, ibid., vol. 3, (2006).

[4] M. W. Evans, ibid., vol. 4 (2006 / 2007, in prep., preprint on www.aias.us and www.atomicprecision.com).

[5] L. Felker,The Evans Equations of Unified Field Theory (preprint on www.aias.us and www.atomicprecision.com).

[6] H. Eckardt and L. Felker , papers on www.aias.us and www.atomicprecision.com.

[7] M. W. Evans,Generally Covariant Dynamics (55th paper of the ECE Series, paper 1 of volume 4, preprint on www.aias.us and www.atomicprecision.com).

[8] M. W. Evans,Geodesics and the Aharonov Bohm Effects in ECE Theory (ibid., paper 56).

[9] M. W. Evans,Canonical and Second Quantization in Generally Covariant Quantum Field Theory (ibid., paper 57).

[10] M. W. Evans,The Effect of Torsion on the Schwarzschild Metric and Light Deflection due to Gravitation (ibid., paper 58).

[11] M. W. Evans, The Resonant Coulomb Law from ECE Theory: Application to the Hydrogen Atom (ibid., paper 59).

[12] M. W. Evans, Application of Einstein Cartan Evans (ECE) Theory to Atoms and Molecules: Free Electrons at Resonance (ibid., paper 60).

[13] M. W. Evans and H. Eckardt,Space-time Resonances in the Coulomb Law (ibid., paper 61).

[14] M. W. Evans,Application of the ECE Lemma to the Fermion and Electromagnetic Fields. (ibid., paper 62).

[15] M. W. Evans and H. Eckardt,The Resonant Coulomb Law of Einstein Cartan Evans Field Theory (ibid., paper 63).

[16] M. W. Evans,Spin Connection Resonance in Gravitational General Relativity (ibid., paper 64).

[17] M. W. Evans,Spin Connection Resonance in Magneto-statics (ibid., paper 65).

[18] M. W. Evans,Vector Boson Character of the Electric Field (ibid., paper 66).

[19] M. W. Evans (ed.),Modern Non-Linear Optics, a special topical issue in three parts of I. Prigogine and S. A. Rice (series eds.),Advances in Chemical Physics (Wiley Inter-science, new York, 2001, 2nd. Ed.), vols. 119(1) to 199(3), endorsed by the Royal Swedish Academy; ibid., M. W. Evans and S. Kielich (eds.), first edition, (Wiley Inter-science, New York, 1992, reprinted 1993 and 1917 (softback), vols. 85(1) to 85(3), prize for excellence from the Polish Government.

[20] M. W. Evans and L. B. Crowell,Classical and Quantum Electrodynamics and the $\boldsymbol{B}^{(3)}$ Field (World Scientific, 2001); M. W. Evans and J.-P. Vigier,The Enigmatic Photon (Kluwer, Dordrecht, 1994 to 2002, hardback and softback), in five volumes.

[21] S. P. Carroll,Space-time and Geometry, an Introduction to General Relativity (Addison-Wesley, New York, 2004).

[22] O. Preuss, S. K. Solanki, M. P. Haugan and S. Jordan,Gravity-induced Birefringence within the framework of Poincaré gauge Theory (arXiv: gr-qc / 0507071 v2 28 Jul 2006).

Chapter 14

Spin connection resonance in counter gravitation

(Paper 68)
by
Horst Eckardt
Alpha Institute for Advanced Study (AIAS).
(horsteck@aol.com, www.aias.us, www.atomicprecision.com)
and
Myron W. Evans
Alpha Institute for Advance Study (AIAS).
(emyrone@aol.com, www.aias.us, www.atomicprecision.com)

Abstract

The equations for counter gravitation are developed in objective, or generally covariant, physics. The latter is represented by Einstein Cartan Evans (ECE) field theory and based on Cartan geometry (Riemann geometry extended with torsion). It is shown that the effect of counter gravitation can be greatly enhanced at spin connection resonance. The equation for resonance is solved numerically, and numerical results also obtained for counter gravitational resonance in the Newtonian force. The latter is greatly enhanced at resonance in a direction opposite to the force due to gravity, resulting in counter gravitation.

Keywords: Counter gravitation, spin connection resonance, Einstein Cartan Evans (ECE) unified field theory, objective physics, generally covariant unified field theory.

14.1 Introduction

It is well known that both the Newton and Coulomb inverse square laws hold to high precision in the laboratory, so the interaction between the two laws must be insignificant under normal laboratory conditions. If two interacting charged masses are considered, the interaction between them is the sum of the electrostatic interaction between charges (the dominant term by many orders of magnitude in the laboratory) and the gravitational interaction between the two masses. The interaction term is very small, and has never been measured in the laboratory. Nevertheless, a generally covariant unified field theory such as the Einstein Cartan Evans (ECE) theory [1]- [21] allows for the existence of interaction terms because in general the gravitational and electromagnetic fields may interact in ECE theory. Their interaction is controlled by the homogeneous current of ECE theory, which is defined in Section 14.2. In order to develop counter gravitational technology, the very small interaction term must be amplified by resonance. The equation controlling resonance is developed in Section 14.2, and solved numerically in Section 14.3 for the scalar potential of the electrostatic field and for the Newtonian force. At resonance, termed "spin connection resonance" (SCR) the electric field induced Newtonian force may be amplified by the necessary many orders of magnitude in a direction opposite to the force due to gravity. This must be the basis of practical counter gravitation. Without resonance, the electric field induced Newtonian force is far too small to be measured and far too small to be of any practical utility. This much is obvious from the precision of the Coulomb and Newton inverse square laws under laboratory conditions, but artifactual claims about counter gravitation continue to proliferate. In this paper we make a fresh start by developing resonant counter gravitational technology based on circuit designs governed by resonance equations. The gravitational field between two charged masses in the laboratory is very well known to be many orders of magnitude smaller than the electric field, so claims about large cross effects are obvious experimental artifacts.

14.2 Homogeneous current and resonance equation

The interaction of electromagnetism and gravitation in ECE theory [1]- [21] is described by the homogeneous current:

$$j^a = \frac{A^{(0)}}{\mu_0}(R^a{}_b \wedge q^b - \omega^a{}_b \wedge T^b) \tag{14.1}$$

and the geometrical condition for interaction of the two fields is therefore:

$$R^a{}_b \wedge q^b \neq \omega^a{}_b \wedge T^b \tag{14.2}$$

Here $R^a{}_b$ is the curvature form, q^b is the tetrad form, $\omega^a{}_b$ is the spin connection form and T^b is the torsion form. If condition (14.2) is fulfilled the electromagnetic field can affect the gravitational field on the classical level, and vice versa. An example is polarization effects due to light deflected by gravity [1]- [21]. The homogeneous current is governed by the homogeneous field equation of ECE theory:

$$d \wedge F^a = \mu_0 j^a \tag{14.3}$$

where

$$F^a = d \wedge A^a + \omega^a{}_b \wedge A^b \tag{14.4}$$

Here A^a is the electromagnetic potential form, defined by:

$$A^a = A^{(0)} q^a \tag{14.5}$$

and F^a is the electromagnetic field form, defined by:

$$F^a = A^{(0)} T^a \tag{14.6}$$

The factor $cA^{(0)}$ has the units of volts, where c is the speed of light, and $cA^{(0)}$ is the primordial voltage present in the universe. In Eq.(14.3) μ_0 is the vacuum permeability in S.I. units. The Hodge dual [1]– [21] of Eq.(14.3) is:

$$d \wedge \widetilde{F}^a = \mu_0 J^a = \mu_0 \widetilde{j}^a \tag{14.7}$$

and the objective or generally covariant Coulomb Law is part of Eq.(14.7), in which the Hodge dual current, the inhomogeneous current, is:

$$\widetilde{j}^a = \frac{A^{(0)}}{\mu_0} (\widetilde{R}^a{}_b \wedge q^b - \omega^a{}_b \wedge \widetilde{T}^b) \tag{14.8}$$

For a given initial driving voltage $cA^{(0)}$, the inhomogeneous current, and the quantity $\widetilde{R}^a{}_b \wedge q^b - \omega^a{}_b \wedge \widetilde{T}^b$ are greatly amplified at SCR. This means that the effect of the electromagnetic field on the gravitational field is greatly amplified, and for the Coulomb law, the effect of the electric field on the Newtonian force is greatly amplified. Such an effect does not exist in the standard model, because in the latter, the classical electromagnetic field is Lorentz covariant only, and not generally covariant as needed for objectivity in physics.

From Eqs.(14.3) and (14.4) the structure of the resonance equation is:

$$d \wedge (d \wedge A^a + \omega^a{}_b \wedge A^b) = \mu_0 j^a \tag{14.9}$$

and its Hodge dual gives another resonance equation. It has been shown [1]– [21] that the Newtonian force between two masses m_1 and m_2 is:

$$f^a = -m_1 m_2 G (R^a{}_b \wedge q^b - \omega^a{}_b \wedge T^b) \tag{14.10}$$

where G is the Newton constant. The Newtonian force is therefore:

$$f^a = -m_1 m_2 \frac{\mu_0 G}{A^{(0)}} j^a \tag{14.11}$$

When the electromagnetic and gravitational fields are independent:

$$R^a{}_b \wedge q^b = \omega^a{}_b \wedge T^b \tag{14.12}$$

the only contribution to $\widetilde{j}^a$ [1]– [21] is from the source mass, so:

$$\widetilde{j}^a = \frac{A^{(0)}}{\mu_0} (\widetilde{R}^a{}_b \wedge q^b)_{source} \tag{14.13}$$

Under this condition it has been shown [1]– [21] that the Coulomb law is:

$$\begin{gathered} \boldsymbol{\nabla} \cdot \boldsymbol{E}^a = -\phi^{(0)} R^a{}_i{}^{i0} \\ i = 1, 2, 3 \end{gathered} \tag{14.14}$$

where:

$$\phi^{(0)} = cA^{(0)} \tag{14.15}$$

The curvature elements appearing in Eq.(14.14) are those due to the source mass, the mass carrying the source charge. It has also been shown [1]- [21] that the electric field in ECE theory is in general a vector boson defined by:

$$a = 1, 2, 3 \tag{14.16}$$

so there are three equations of type (14.14):

$$\boldsymbol{\nabla} \cdot \boldsymbol{E}^1 = -\phi^{(0)} R^1{}_i{}^{i0} \tag{14.17}$$

$$\boldsymbol{\nabla} \cdot \boldsymbol{E}^2 = -\phi^{(0)} R^2{}_i{}^{i0} \tag{14.18}$$

$$\boldsymbol{\nabla} \cdot \boldsymbol{E}^3 = -\phi^{(0)} R^3{}_i{}^{i0} \tag{14.19}$$

Here, summation over repeated indices is implied, so:

$$R^1{}_i{}^{i0} = R^1{}_1{}^{10} + R^1{}_2{}^{20} + R^1{}_3{}^{30} \tag{14.20}$$

and so on. In Eqs.(14.17) to (14.20) the Riemann form elements are generated purely by the source mass, so:

$$\boldsymbol{\nabla} \cdot \boldsymbol{E}^a = \mu_0 \widetilde{j}^a = -\phi^{(0)} R^a{}_i{}^{i0} \tag{14.21}$$

If the electric field induces a Newtonian force, the Coulomb law is changed to:

$$\boldsymbol{\nabla} \cdot \boldsymbol{E}^a = \mu_0 \widetilde{j}^a = -\phi^{(0)} (R^a{}_i{}^{i0} + \omega^a{}_{ia} T^{bi0}) \tag{14.22}$$

The effect of the electric field on the elements $R^a{}_i{}^{i0}$ is given by $\omega^a{}_{ib} T^{bi0}$. Therefore the complete Coulomb Law becomes:

$$\boldsymbol{\nabla} \cdot \boldsymbol{E}^a = \mu_0 (\widetilde{j}^a_{source} + \widetilde{j}^a_{int}) \tag{14.23}$$

where the interaction current is defined by:

$$\widetilde{j}^a_{int} = -\frac{\phi^{(0)}}{\mu_0} (R^a{}_i{}^{i0} + \omega^a{}_{ib} T^{bi0})_{int} \tag{14.24}$$

The interaction current in the absence of SCR is very tiny, and has never been measured experimentally. Claims to the contrary are clearly artifactual, because otherwise the Coulomb Law would not hold to very high precision in the laboratory, contrary to well known [22] and accurately reproducible and repeatable experimental data on the Coulomb Law.

The interaction Coulomb Law can therefore be written as:

$$\boldsymbol{\nabla} \cdot \boldsymbol{E}^a = -\boldsymbol{\omega}^a{}_{bint} \cdot \boldsymbol{E}^b \tag{14.25}$$

where:

$$\boldsymbol{\omega}^a{}_{bint} \cdot \boldsymbol{E}^b \sim 0 \tag{14.26}$$

Here $\omega^a{}_{bint}$ is the interaction spin connection. This is non-zero if and only if the electric field induces changes in the Newtonian force. Such changes do not exist in the standard model. For simplicity of argument only it is assumed that the indices a and b are the same, so Eq.(14.25) is simplified to:

$$\nabla \cdot \boldsymbol{E} = -\boldsymbol{\omega}_{int} \cdot \boldsymbol{E} \tag{14.27}$$

The electric field can be defined [1]- [21] as:

$$\boldsymbol{E} = -(\nabla + \boldsymbol{\omega})\phi \tag{14.28}$$

where $\boldsymbol{\omega}$ is the spin connection in the absence of interaction between the electric field and Newtonian gravitation. A positive sign has been adopted for illustration only in Eq.(14.28) for $\boldsymbol{\omega}$, but in general [1]- [21]:

$$\omega_r = 0 \quad , \quad \pm\frac{1}{r} \tag{14.29}$$

in spherical polar coordinates. In the absence of interaction between the electric field and the Newtonian force [1]- [21]:

$$\frac{d^2\phi}{dr^2} + \frac{1}{r}\frac{d\phi}{dr} - \frac{1}{r^2}\phi = -\frac{\rho}{\epsilon_0} \tag{14.30}$$

but in the presence of interaction:

$$\frac{d^2\phi}{dr^2} + \frac{1}{r}\frac{d\phi}{dr} - \frac{1}{r^2}\phi - \omega_{int} \cdot (\nabla + \boldsymbol{\omega})\phi = -\frac{\rho}{\epsilon_0} \tag{14.31}$$

i.e. there is an extra term due to the interaction spin connection $\boldsymbol{\omega}_{int}$. Eq.(14.31) is the resonance equation:

$$\frac{d^2\phi}{dr^2} + \left(\frac{1}{r} - \omega_{int}\right)\frac{d\phi}{dr} - \left(\frac{1}{r^2} + \frac{\omega_{int}}{r}\right)\phi = -\frac{\rho}{\epsilon_0} \tag{14.32}$$

and is solved numerically in Section 14.3 for various models of the interaction spin connection. Here ρ is the charge density and ϵ_0 is the vacuum permeability.

Resonant counter gravitation works by amplifying ϕ at resonance from Eq.(14.32). It is seen from Eq.(14.25) that the interaction term is amplified, meaning that the effect of the electric field on the Newtonian force is maximized in a direction opposite to the gravitational field of the Earth.

The basic structure of the interaction Coulomb Law is:

$$\nabla \cdot \boldsymbol{E}^a = \frac{\rho^a}{\epsilon_0} - \boldsymbol{\omega}^a{}_{bint} \cdot \boldsymbol{E}^b \tag{14.33}$$

If it is assumed for the sake of simplicity of development that only the diagonal elements of the interaction spin connection exist:

$$(\nabla + \boldsymbol{\omega}^1{}_1 int) \cdot \boldsymbol{E}^1 = \frac{\rho^1}{\epsilon_0} \qquad \text{etc.} \tag{14.34}$$

where the vector electric boson [1]- [21] is defined by:

$$\left.\begin{aligned} \boldsymbol{E}^1 &= -\nabla\phi + \boldsymbol{\omega}\phi \\ \boldsymbol{E}^2 &= -\nabla\phi \\ \boldsymbol{E}^3 &= -\nabla\phi - \boldsymbol{\omega}\phi \end{aligned}\right\} \tag{14.35}$$

Therefore there are three resonance equations in general:

$$(\boldsymbol{\nabla} + \boldsymbol{\omega}^1{}_{1int}) \cdot (-\boldsymbol{\nabla}\phi + \boldsymbol{\omega}\phi) = \frac{\rho^1}{\epsilon_0} \tag{14.36}$$

$$(\boldsymbol{\nabla} + \boldsymbol{\omega}^2{}_{2int}) \cdot (-\boldsymbol{\nabla}\phi) = \frac{\rho^2}{\epsilon_0} \tag{14.37}$$

$$(\boldsymbol{\nabla} + \boldsymbol{\omega}^3{}_{3int}) \cdot (-\boldsymbol{\nabla}\phi - \boldsymbol{\omega}\phi) = \frac{\rho^3}{\epsilon_0} \tag{14.38}$$

The labels on the charge densities indicate that there are three different charge densities present in general. If it is assumed that:

$$\rho = \rho^1 = \rho^2 = \rho^3 \tag{14.39}$$

and that:

$$\boldsymbol{\omega}_{int} = \boldsymbol{\omega}^1{}_{1int} = \boldsymbol{\omega}^2{}_{2int} = \boldsymbol{\omega}^3{}_{3int} \tag{14.40}$$

the three resonance equations simplify to:

$$\nabla^2\phi + \boldsymbol{\omega}_{int} \cdot \boldsymbol{\nabla}\phi + \boldsymbol{\nabla} \cdot (\boldsymbol{\omega}\phi) - \boldsymbol{\omega}_{int} \cdot \boldsymbol{\omega}\phi = -\frac{\rho}{\epsilon_0} \tag{14.41}$$

$$\nabla^2\phi + \boldsymbol{\omega}_{int} \cdot \boldsymbol{\nabla}\phi = -\frac{\rho}{\epsilon_0} \tag{14.42}$$

$$\nabla^2\phi + \boldsymbol{\omega}_{int} \cdot \boldsymbol{\nabla}\phi + \boldsymbol{\nabla} \cdot (\boldsymbol{\omega}\phi) + \boldsymbol{\omega}_{int} \cdot \boldsymbol{\omega}\phi = -\frac{\rho}{\epsilon_0} \tag{14.43}$$

The resonance patterns from these three equations are obtained numerically and discussed in Section 14.3. The ability of the electric field to affect the Newtonian force is represented by the interaction spin connection $\boldsymbol{\omega}_{int}$. Off resonance this effect is very tiny, as argued, but at resonance its effect may be amplified enough to make counter gravitation feasible.

The effect of the electric field on the Newtonian force is conveniently demonstrated through the Hodge dual of Eq.(14.10):

$$\widetilde{f}^a = -m_1 m_2 G(\widetilde{R}^a{}_b \wedge q^b - \omega^a{}_b \wedge \widetilde{T}^{bi0}) \tag{14.44}$$

In the absence of any effect of the electric field on the Newtonian force:

$$\widetilde{f}^a = -m_1 m_2 G(R^0{}_1{}^{10} + R^0{}_2{}^{20} + R^0{}_3{}^{30}) \tag{14.45}$$

which is the ordinary Newtonian force between m_1 and m_2, the well known inverse square distance dependence being given by:

$$\frac{1}{r^2} = R^0{}_1{}^{10} + R^0{}_2{}^{20} + R^0{}_3{}^{30} \tag{14.46}$$

In the presence of interaction, the Newtonian force due to the electric field is:

$$\widetilde{f}^a = -m_1 m_2 G \omega^a{}_{ib} T^{bi0} \tag{14.47}$$

and simplifies to:

$$\widetilde{f} = -m_1 m_2 G \omega_{int} \cdot \boldsymbol{T} \tag{14.48}$$

where $\boldsymbol{T}$ is a torsion vector defined by:

$$\boldsymbol{E} = \phi \boldsymbol{T} = \nabla\phi \quad , \quad (\nabla \pm \boldsymbol{\omega})\phi \tag{14.49}$$

Thus:

$$\left.\begin{aligned}(\nabla + \omega)\phi = \boldsymbol{T}\phi,\\ \nabla\phi = \boldsymbol{T}\phi\end{aligned}\right\} \tag{14.50}$$

and:

$$\left.\begin{aligned}\widetilde{f} = &- m_1 m_2 G\omega_{int} \cdot \frac{1}{\phi}(\nabla + \omega)\phi\,,\\ &- m_1 m_2 G\omega_{int} \cdot \frac{1}{\phi}\nabla\phi\end{aligned}\right\} \tag{14.51}$$

where:

$$\frac{d^2\phi}{dr^2} + \left(\frac{1}{r} - \omega_{int}\right)\frac{d\phi}{dr} - \left(\frac{1}{r^2} - \frac{\omega_{int}}{r}\right)\phi = -\frac{\rho}{\epsilon_0} \tag{14.52}$$

Therefore to maximize the effect of the electric field on the Newtonian force at SCR, the following term must be maximized:

$$\widetilde{f} = -m_1 m_2 G\omega_{int}\frac{\phi'}{\phi} \tag{14.53}$$

where

$$\phi' = \frac{d\phi}{dr} \tag{14.54}$$

and where:

$$\frac{d\phi'}{dr} + \left(\frac{1}{r} - \omega_{int}\right)\phi' - \left(\frac{1}{r^2} - \frac{\omega_{int}}{r}\right)\phi = -\frac{\rho}{\epsilon_0} \tag{14.55}$$

This is illustrated numerically and discussed in detail in Section 14.3.

14.3 Numerical results

The resonance equation

$$\frac{d^2\phi}{dr^2} + \left(\frac{1}{r} - \omega_{r,int}\right)\frac{d\phi}{dr} - \left(\frac{1}{r^2} + \frac{\omega_{r,int}}{r}\right)\phi = -\frac{\rho}{\epsilon_0} \tag{14.56}$$

has been solved numerically. This is the radial equation of the resonant Coulomb law of Chapter 9, Eq.(9.34) therein, enhanced by an additional spin connection term which describes the interaction of gravitation and electromagnetism (see Eq. (14.52) of Section 14.2). In order to study the solutions of this equation, we first consider the interaction-free case, $\omega_{r,int} = 0$, with no driving charge density ρ. Eq.(14.56) contains a singularity for $r = 0$. Therefore we integrate numerically from right to left. There are three types of solutions, depending on the initial or boundary conditions at the right-hand side. Starting at $r = 10$, we have chosen the three combinations

$$\phi(r = 10) = 0.11, \qquad \frac{d\phi(r = 10)}{dr} = 0.01 \tag{14.57}$$

$$\phi(r = 10) = 0.10, \qquad \frac{d\phi(r = 10)}{dr} = 0.01 \tag{14.58}$$

$$\phi(r=10)=0.09, \quad \frac{d\phi(r=10)}{dr}=0.01 \tag{14.59}$$

In Fig. 14.1 the three solutions are graphed. In the second case the initial conditions define a straight line which hits the coordinate origin. If the direction of this line does not point to the center, the solution $\phi(r)$ turns to plus or minus inifinity for $r \rightarrow 0$. In the following we choose the initial conditions in such a way that the ordinary Coulomb solution for a charge at $r=0$ is obtained. Then the solution (setting $4\pi\epsilon_0$ to unity) is

$$\phi(r)=-\frac{1}{r} \tag{14.60}$$

$$\frac{d\phi}{dr}=\frac{1}{r^2} \tag{14.61}$$

For $r=10$ we obtain $\phi(10)=-0.1$ and $d\phi/dr=0.01$. Now we switch on the interacting spin connection. Its physical form is unknown, we only know that it is a function of r and possibly a fuctional of ϕ. In the limit $r \rightarrow \infty$ it should vanish. In the following we assume it to have the same form as for the resonant Coulomb law: $\omega_{r,int}=\pm 1/r$. Then we get the following result for the three possible values in vector boson notation (see Chapter 12):

$$\omega_{r,int[-1]}=-\frac{1}{r} \tag{14.62}$$

$$\omega_{r,int[0]}=0 \tag{14.63}$$

$$\omega_{r,int[+1]}=\frac{1}{r} \tag{14.64}$$

Inserting this form into Eq.(14.56) with no stimulation of resonance ($\rho=0$) leads in all cases (14.62–14.64) to nearly the same solutions. Differences are not visible in the graph (Fig. 14.2). This is in accordance with our finding in Chapter 7, Table 1, where the Coulomb spin connection did not lead to any remarkable deviations from the ordinary Coulomb law.

The situation changes as soon as we apply a driving term ρ which is dependent on a predefined wave number κ. In the simple case

$$\rho=A cos(\kappa r) \tag{14.65}$$

(depicted in Fig. 14.6) we get a clear dependence of ϕ from the wave number. Even the characteristic of the solution changes as can be seen from Fig. 14.3. In this diagram ϕ was plotted for four κ values $(0.25, 0.5, 1., 2.)$ and an interacting spin connection (Eq.(14.62)). Another interesting question is how the three spin connections (14.62–14.64) lead to different solutions ϕ if the wave number is the same. This result is presented in Figs. 14.4 and 14.5. Obviously the characteristic remains the same, but the $\omega_{r,int[-1]}$ leads to larger values of $|\phi|$ while $\omega_{r,int[+1]}$ effects a reduction.

Another interesting question is how the driving force in combination with the interacting spin connection creates resonances of ϕ. In contrast to the results described in Chapter 7 and Chapter 9, an oscillatory ρ does not lead to oscillatory resonances in ϕ. The main effect is the enhancement of the rate of increase or decrease for $r \rightarrow 0$. This behaviour is concentrated on the center

Type	Equation
1	$f_1(r) = A\cos(\kappa r)$
2	$f_2(r) = A\cos(\kappa r)e^{-0.25r}$
3	$f_3(r) = A(\sin(\kappa r) + \sin(\kappa 2r))^2$
4	$f_4(r) = A\cos(\kappa r)\cos(\cos(\kappa r))\cosh(\sin(\kappa r)) +$ $\sin(2\kappa r)\sin(\cos(\kappa r))\sinh(\sin(\kappa r))$
5	$f_5(r) = \begin{cases} 0.5A & \text{if} \quad f_4(r) > -0.2A \\ -1.5A & \text{elsewhere} \end{cases}$

Table 14.1: Models for the driving force

of the charge, which is plausible since gravity comes from the central mass and interaction with electromagnetism is highest where both have their strongest values.

Before looking at the results in detail we list the models for the driving force used (Table 14.1). The first type is a pure cosine term which is folded by an exponentially decreasing function in the second case. Type 3 is a combination of two frequencies while type 4 is the driving force obtained for the equivalent circuit in Chapter 9(essentially a combination of three frequencies). Finally we have changed this model to a rectangular signal in type 5. The signal forms are shown in Figs. 14.6–14.10.

The resonance curves show some maximal amplitude of ϕ in dependence of κ. Since we do not have an oscillatory maximum difference as in Chapters 7 and 9, we have chosen the value of ϕ at a grid point near to the center ($r = 1/30$) as an indication of resonance. This value is plotted against κ in Figs. 14.11-14.15. The five diagrams correspond to the driving forces of Fig. 14.6–14.10. Resonance is not sharply structured but more oscillatory in nature. Figs. 14.11 and 14.12 show a harmonic form with maximum at $\kappa = 0$. This means that a constant ρ produces the highest resonance. Other wave forms (Figs. 14.13, 14.14) lead to anharmonic resonance curves. It is remarkable that the driving force of the Coulomb resonant circuit (Chapter 9) produces also a very high effect (compare the ordinate values of the diagrams). This may be a hint that both the Coulomb and gravito-electromagnetic interactional resonance are connected.

The last example (Fig. 14.15) of this group shows the result of a rectangular signal (Fig. 14.10). The signal amplitudes have been adjusted to Fig. 14.9. The rectangular form obviously enhances the first minimum at $\kappa = 0.2$. According to the examples considered here, this is a very effective form of the driving force to evoke resonance effects. For an exact comparison the driving forces would have to be normalized precisely, which was not the case in this calculation.

We have changed the spin connection form of Eq.(14.62–14.64) from $1/r$ to $1/r^3$ type. This gives extremely high values in the resonance diagram Fig. 14.16. For $r \to 0$ the numerical solution becomes unstable so we have left out some grid points near to 0. The effects of the vector bosons [0] and [1] are nearly identical on this scale, the boson [−1] produces a giant resonance. This last example shows that the form of the spin connection is most important for the size of the resonance effect, while the form of the driving force determines its shape. According to Eq.(14.53) the Hodge dual of the Newtonian gravitational force is proportional to

$$\widetilde{f} \sim -\omega_{r,int} \frac{\phi'}{\phi} \tag{14.66}$$

For comparison with measurements we would have to take the Hodge dual once more to obtain the force itself, but this would require knowledge of the metric. Therefore we restrict to the form (14.66) to give a qualitative impression of the effects. The expression ϕ'/ϕ is shown in Fig. 14.17 for type 5, $\kappa = 0.25$. Due to zeros in ϕ there are poles in ϕ'/ϕ at certain radii. These poles cannot occur in an off-resonant Coulomb potential since this remains always positive or negative, depending of the sign of the central charge. So the poles are an effect of resonance. Their position can mainly be influenced by the form of the driving force and the boundary conditions of the potential. For $r \to 0$ we have $\phi \to \pm\infty$ as was shown earlier. For both asymptotes the sign of ϕ' is different from that of ϕ so that the ratio of both is always negative as is the case in Fig. 14.17.

The last graph (Fig. 14.18) shows the full force term (14.66) where the ratio ϕ'/ϕ has been multiplied by the interacting spin connection. Since $\omega_{r,int[0]}$ is zero, the force for this vector boson type always disappears. The other two differ in sign and result to an attractive and a repulsive force near to the center when gravitation is impacted by an electrical potential. The zeros of ϕ at certain radii lead to sharp peaks. For practical applicability of counter gravitation this means that one would get a gravitational instability at these radii. It would be advisable to avoid these effects by design and utilize the smooth range near to the center. How the different modes of the vector boson can be evoked is not clear yet. The numerical results show that the driving term and the boundary conditions of the potential are most important for designing counter-gravitation devices. It should be possible to design an equivalent circuit for Eq.(14.56). The form of the interacting spin connection is not known but seems not to change the results qualitatively.

Acknowledgements The British Government is thanked for the award of a Civil List pension for distinguished services to science, and the staff of AIAS are thanked for many interesting discussions.

Bibliography

[1] M. W. Evans, Generally Covariant Unified Field Theory (Abramis Academic, 2005), vol. 1.

[2] M. W. Evans, ibid., vol. 2 (2006).

[3] M. W. Evans, ibid., vol. 3 (2006).

[4] M. W. Evans, ibid., vol. (2006 / 2007, in prep., preprint on www.aias.us).

[5] L. Felker, The Evans Equations of Unified Field Theory (preprint on www.aias.us).

[6] H. Eckardt and L. Felker, papers on www.aias.us.

[7] M. W. Evans, Generally Covariant Dynamics (55th paper on www.aias.us, chapter 1 of volume four).

[8] M. W. Evans, Geodesics and the Aharonov Bohm Effects in ECE Theory (ibid., paper 56).

[9] M. W. Evans, Canonical and Second Quantization in Generally Covariant Unified Field Theory (ibid., paper 57).

[10] M. W. Evans, The Effect of Torsion on the Schwarzschild Metric and Light Deflection due to Gravitation (paper 58).

[11] M. W. Evans, The Resonant Coulomb Law from ECE Theory: Application to the Hydrogen Atom (paper 59).

[12] M. W. Evans, Application of Einstein Cartan Evans (ECE) Theory to Atoms and Molecules: Free Electrons at Resonance (ibid., paper 60).

[13] M. W. Evans and H. Eckardt, Space-time Resonances in the Coulomb Law (ibid., paper 61).

[14] M. W. Evans, Application of the ECE Lemma to the Fermion and Electromagnetic Fields (ibid., paper 62).

[15] M. W. Evans and H. Eckardt, The Resonant Coulomb Law of Einstein Cartan Evans Field Theory (ibid., paper 63).

[16] M. W. Evans, Spin Connection Resonance in Gravitational General Relativity (ibid., paper 64).

[17] M. W. Evans, Spin Connection Resonance in Magneto-statics (ibid., vol. 65).

[18] M. W. Evans, Vector Boson Character of the Electric Field (ibid., paper 66).

[19] M. W. Evans, ECE Theory of Gravity Induced Polarization Changes (ibid., paper 67).

[20] M. W. Evans (ed.), Modern Non-Linear Optics, a special topical issue in three parts of I. Prigogine and S. A. Rice (series editors), Advances in Chemical Physics (Wiley-Inter-science, New York, 2001, 2nd. Ed.,), vols. 119(1) - 119(3); ibid., M. W. Evans and S. Kielich (eds.), first edition (Wiley-Interscience, New York, 1991, reprinted 1992 and 1997), vols. 85(1) - 85(3).

[21] M. W. Evans and L. B. Crowell, Classical and Quantum Electrodynamics and the $\boldsymbol{B}^{(3)}$ Field (World Scientific, 2001); M. W. Evans and J.-P. Vigier, The Enigmatic Photon (Kluwer, Dordrecht, 1994 - 2002, hardback and softback), in five volumes.

[22] J. D. Jackson, Classical Electrodynamics (Wiley, New York, 1999, 3rd. Ed.).

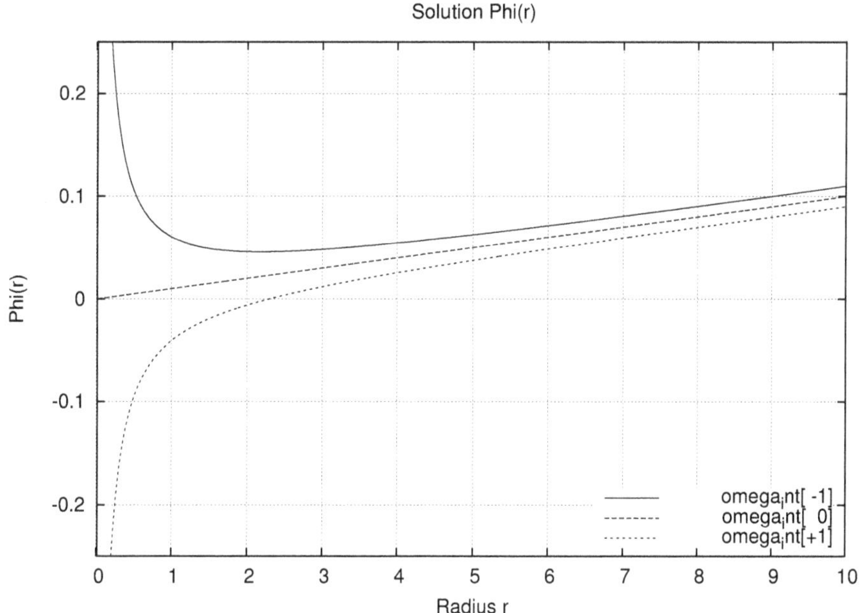

Figure 14.1: Solution types for Eq.(14.56 for $\omega_{r,int} = 0$, no driving force, dependent on initial conditions at the right ($d\phi/dr(10) = 0.01$)

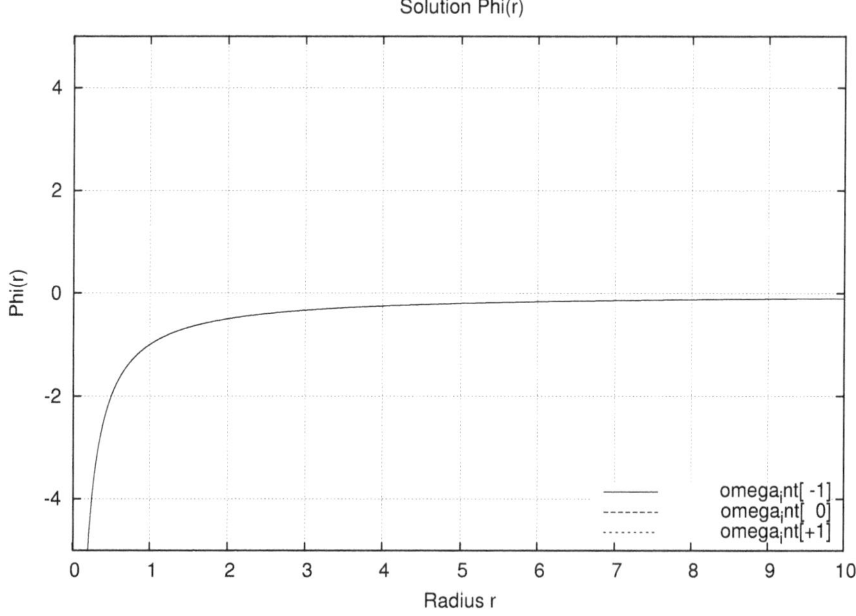

Figure 14.2: Solution for $\omega_{r,int} = \omega_{r,Coul}$, no driving force

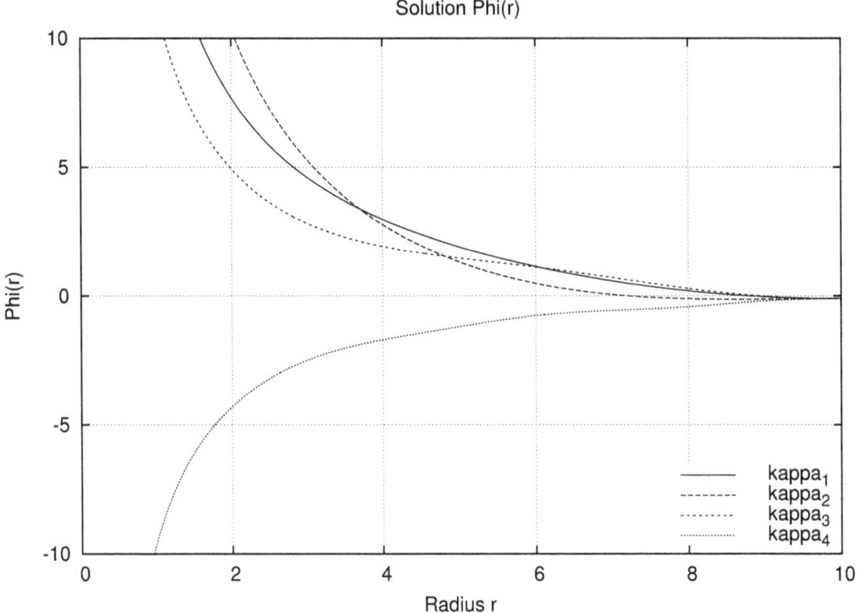

Figure 14.3: κ-dependence for ϕ for type=1,$\omega_{r,int[-1]}$, $\kappa = 0.25, 0.5, 1.0, 2.0$

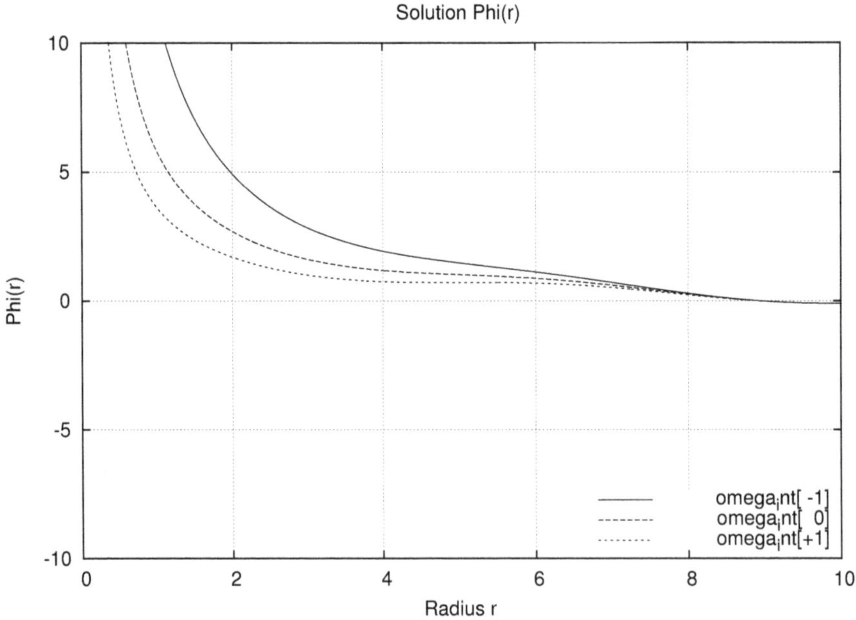

Figure 14.4: κ-dependence for ϕ for type=1,$\omega_{r,int[-1]}$, $\kappa = 0.25, 0.5, 1.0, 2.0$

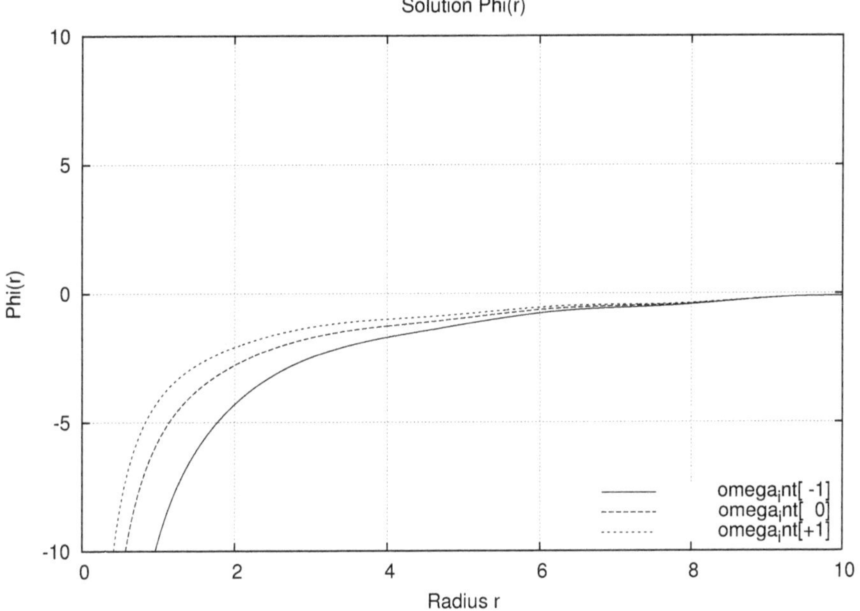

Figure 14.5: $\omega_{r,int}$-dependence of ϕ for type=1, $\kappa = 2$.

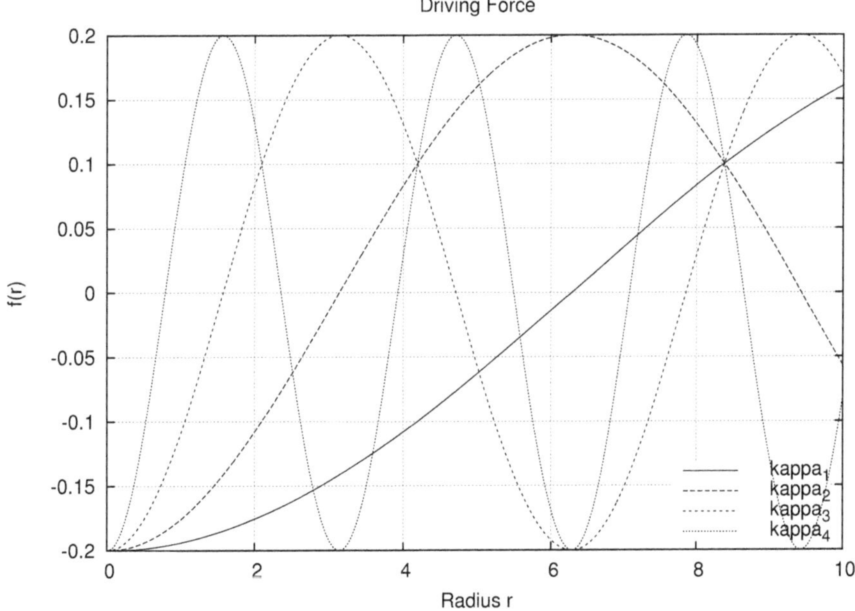

Figure 14.6: Driving force, type 1, for four κ values: $\kappa = 0.25, 0.5, 1.0, 2.0$

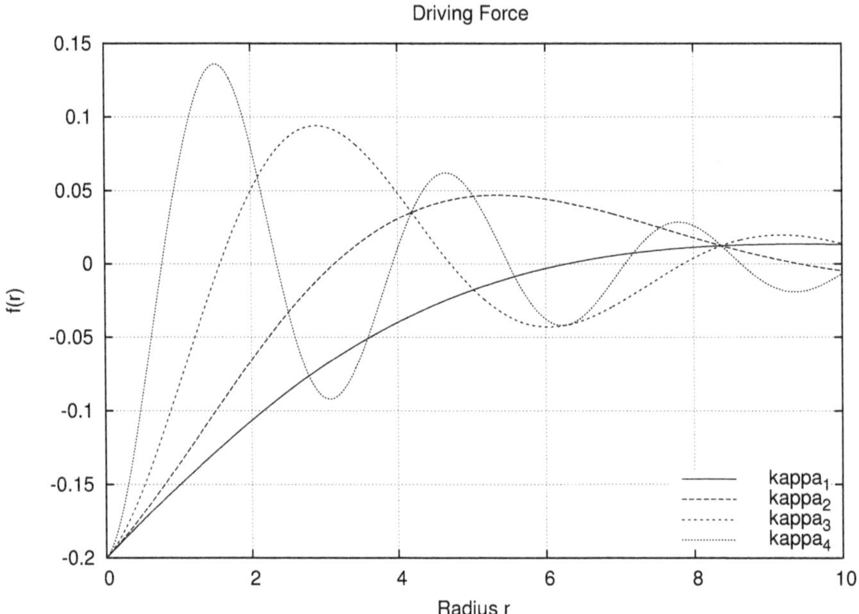

Figure 14.7: Driving force, type 2, for four κ values: $\kappa = 0.25, 0.5, 1.0, 2.0$

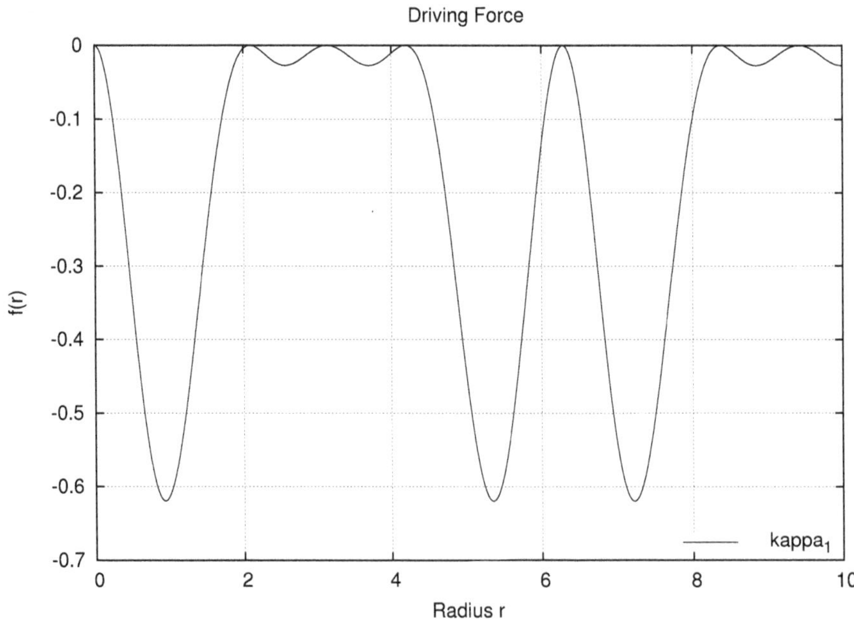

Figure 14.8: Driving force, type 3, $\kappa = 1.0$

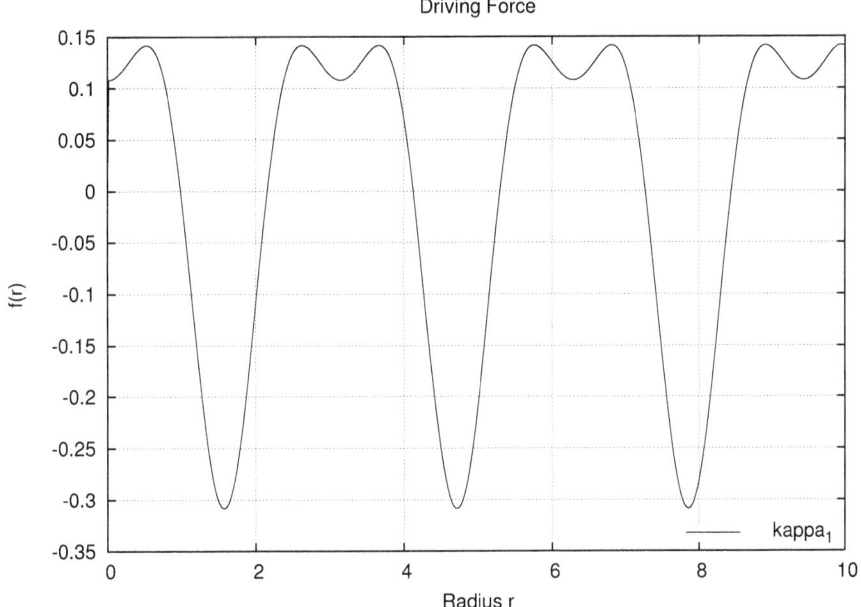

Figure 14.9: Driving force, type 4, $\kappa = 1.0$

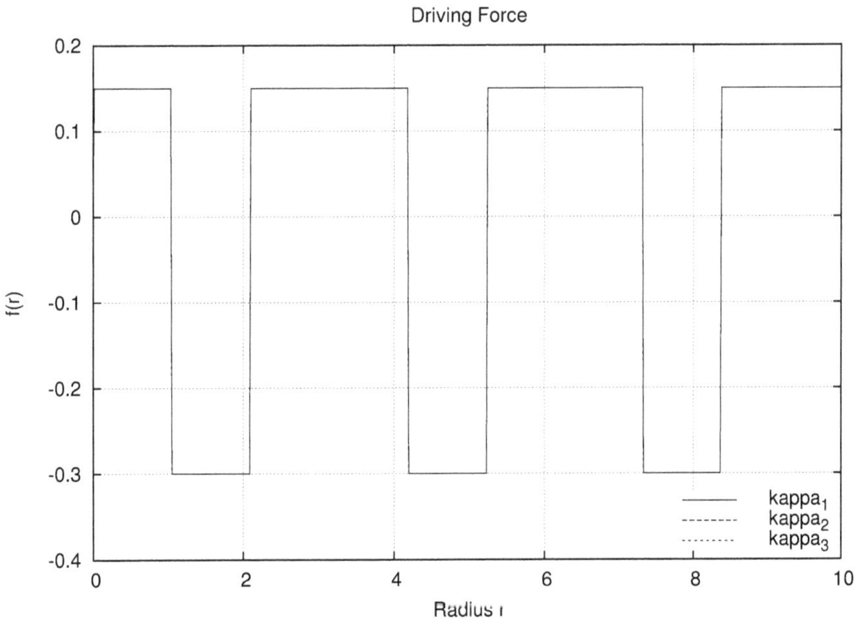

Figure 14.10: Driving force, type 5, $\kappa = 1.0$

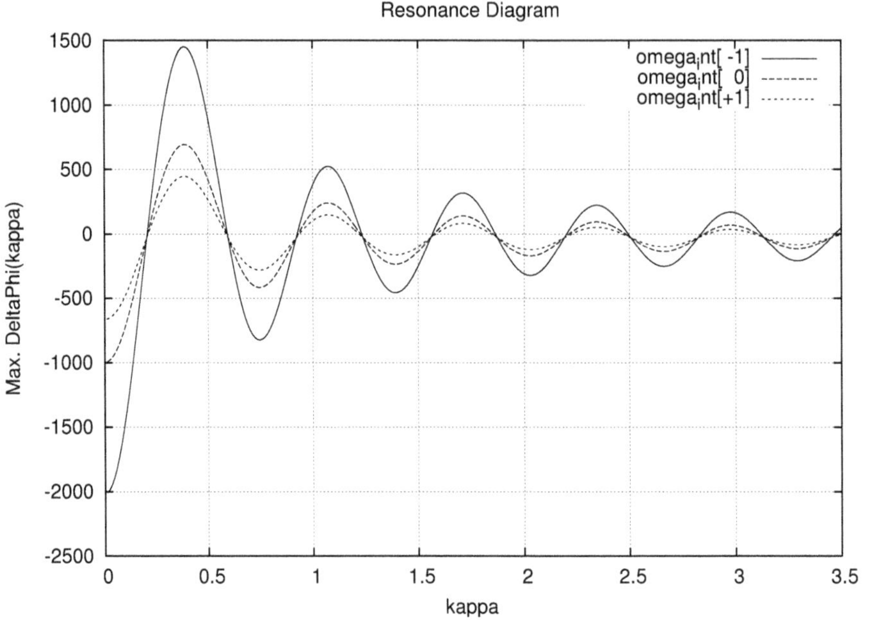

Figure 14.11: Resonance diagram, type 1

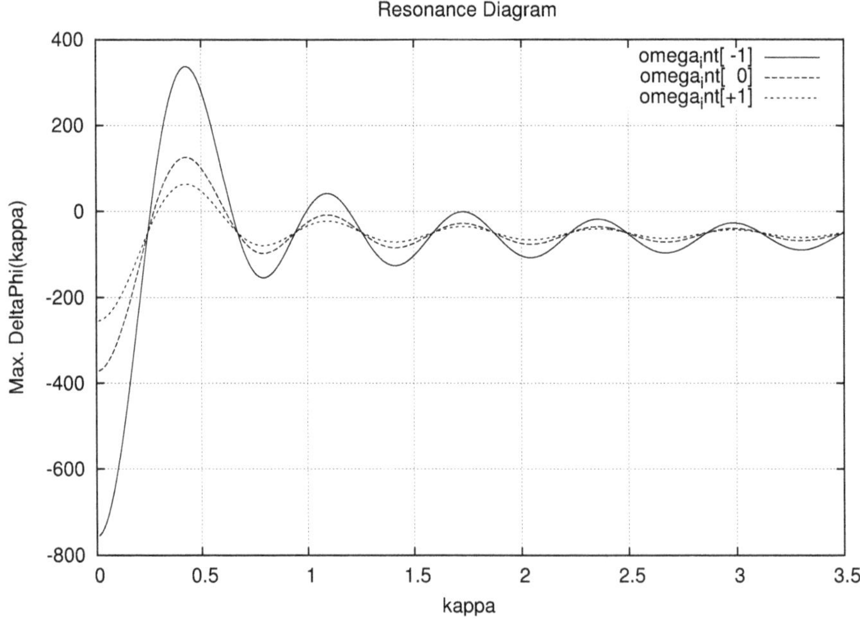

Figure 14.12: Resonance diagram, type 2

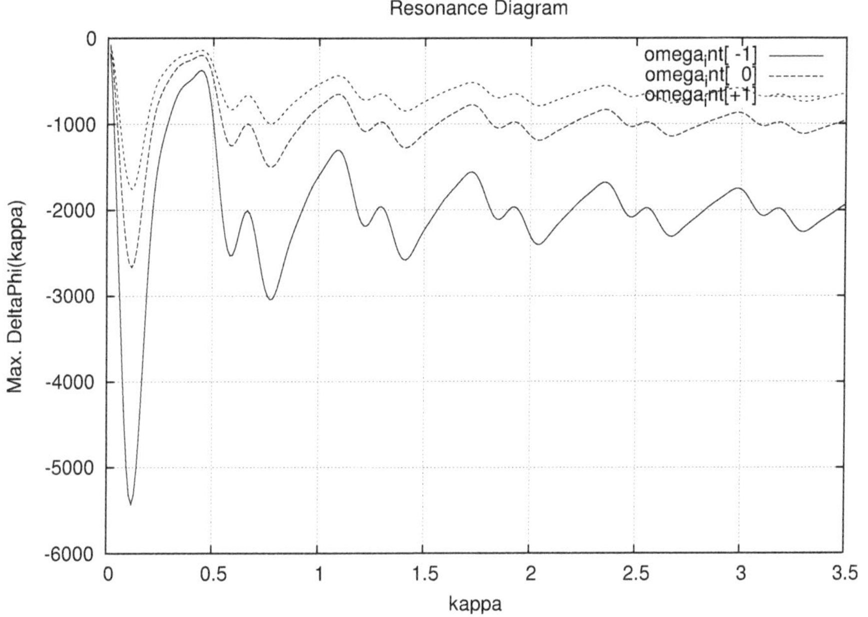

Figure 14.13: Resonance diagram, type 3

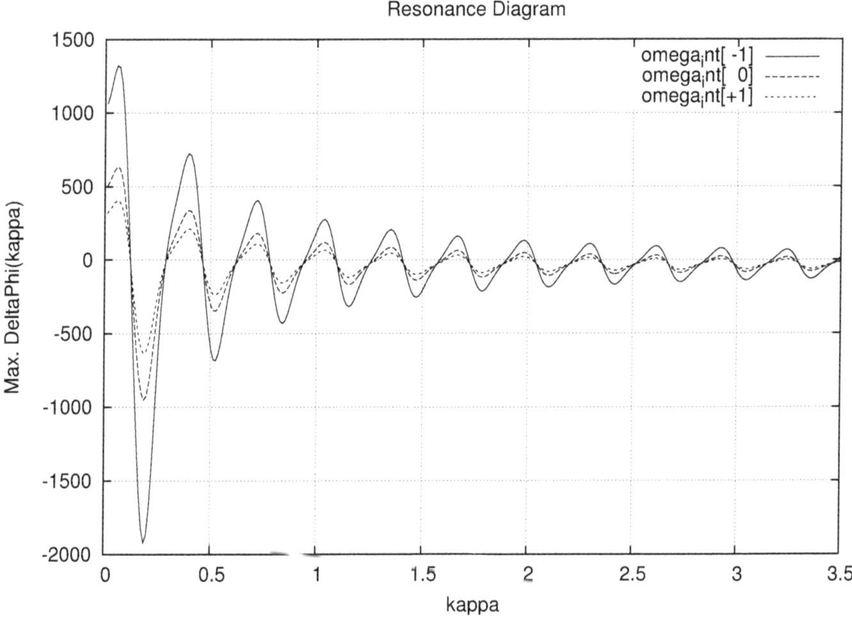

Figure 14.14: Resonance diagram, type 4

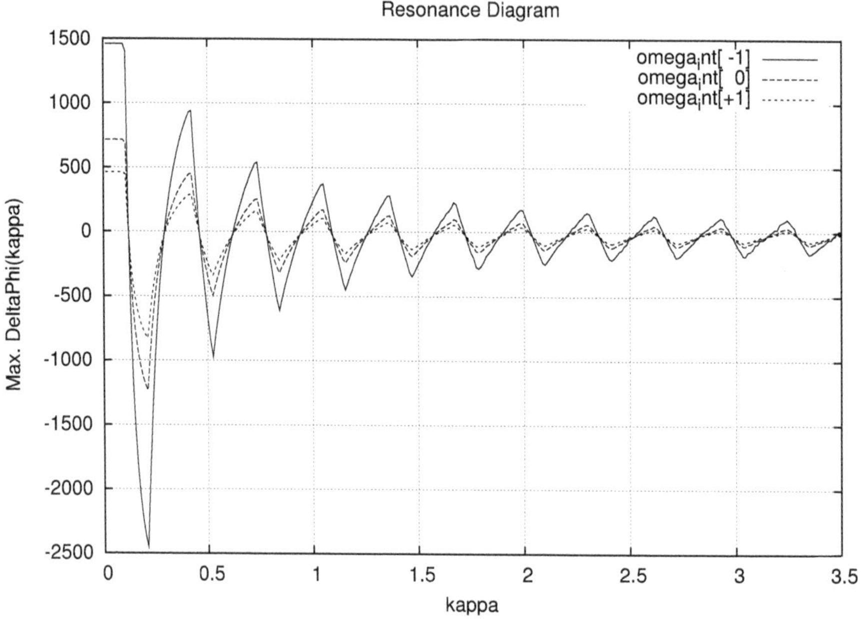

Figure 14.15: Resonance diagram, type 5

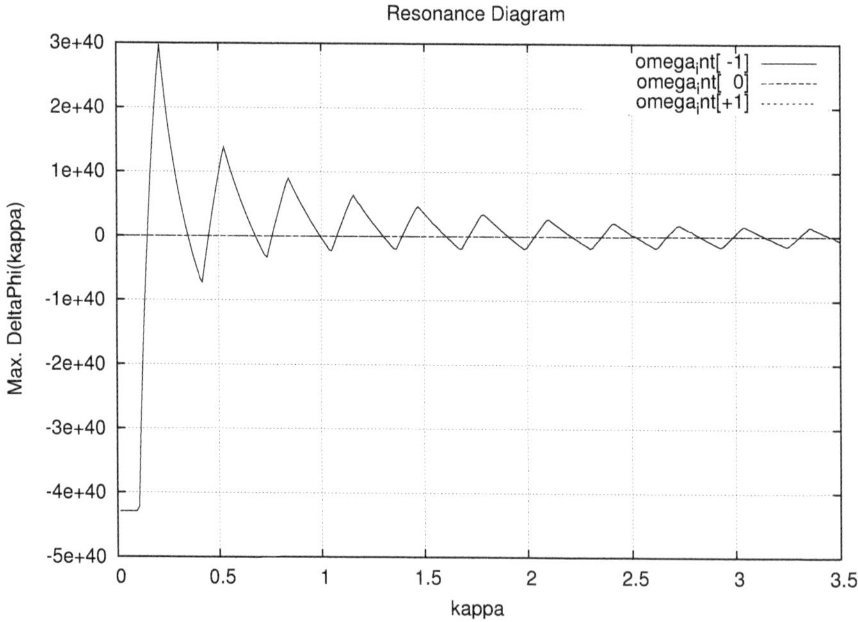

Figure 14.16: Resonance diagram for interacting spin connection $1/r^3$, type 5

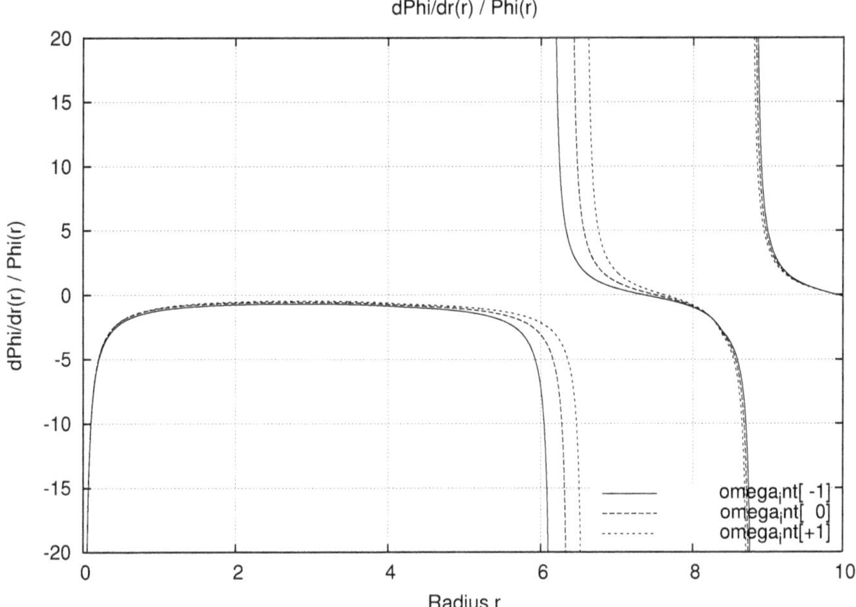

Figure 14.17: $(d\phi/dr)/\phi$ for type=5,$\kappa = 0.25$

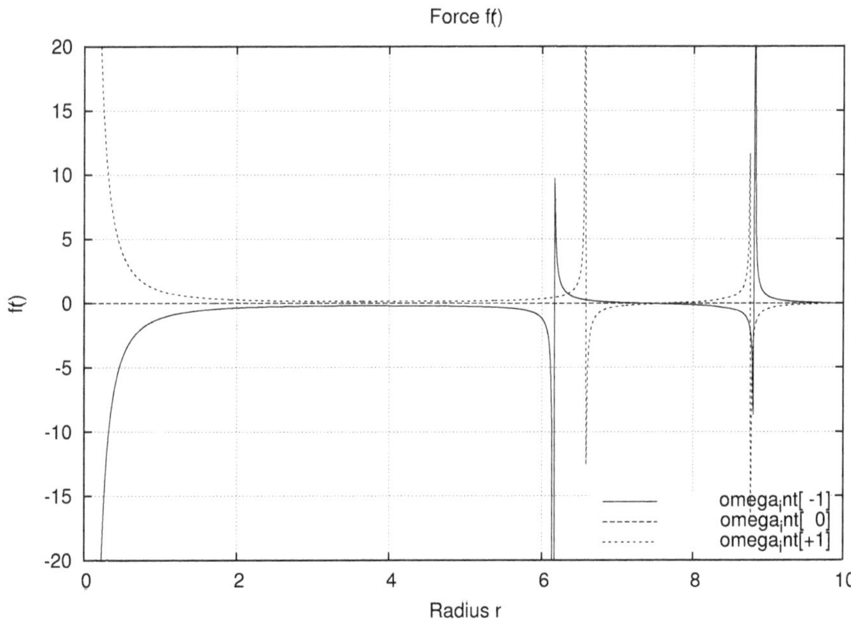

Figure 14.18: Force term $f\ (r)$ for type=5, $\kappa = 0.25$

Chapter 15

Effect of gravitation on radiatively induced fermion resonance

(Paper 69)
by
Myron W. Evans
Alpha Institute for Advanced Study (AIAS).
(emyrone@aol.com, www.aias.us, www.atomicprecision.com)

Abstract

Radiatively induced fermion resonance (RFR) is the resonance equivalent of the inverse Faraday effect (IFE), which is the magnetization of matter by circularly polarized radiation. The effect of gravitation on RFR is developed in this paper by considering the spin connection of Einstein Cartan Evans (ECE) field theory to be approximately dual to the tetrad. At spin connection resonance (SCR) the effect of gravitation is amplified, so the resulting gravitational shift in resonance frequency of RFR may become measurable. Similar considerations apply to all other forms of resonant spectroscopy.

Keywords: Radiatively induced fermion resonance (RFR), inverse Faraday effect (IFE), Einstein Cartan Evans (ECE) field theory, spin connection resonance (SCR), gravitationally induced shifts in atomic and molecular spectroscopy.

15.1 Introduction

The inverse Faraday effect (IFE) is well known to be the bulk magnetization of matter by circularly or elliptically polarized radiation. It is observable at all frequencies in all materials, and on the simplest level, one electron. The resonant equivalent of IFE is radiatively induced fermion resonance (RFR), which is fermion resonance induced by a circularly polarized electromagnetic field. Analogously, the resonant equivalent of bulk magnetization by a magnet (static magnetic field) is ESR or NMR. The resolution of RFR is much higher than ESR or NMR, and the RFR technique has a characteristic chemical shift pattern different from that of ESR or NMR. In this paper, Einstein Cartan Evans (ECE) field theory [1]- [10] is used to investigate the effect of gravitation on RFR and on atomic and molecular resonance spectra in general. In Section 15.2, the general theory is reviewed, particularly in respect to the ECE spin field. In Section 15.3, levels of approximation are described in the solution of the basic RFR equations, and in Section 15.4, spin connection resonance (SCR) is described as a possible means of amplifying gravitational induced spectral shifts so that they become observable in the laboratory.

15.2 General theory

Using indexless notation [1]- [10] for clarity of concepts, the effect of gravitation may be measured through the equations:

$$F = d \wedge A + \omega \wedge A \tag{15.1}$$

and

$$d \wedge F = \mu_0 j \tag{15.2}$$

where

$$j = \frac{A^{(0)}}{\mu_0}(R \wedge q - \omega \wedge T) \tag{15.3}$$

and

$$A = A^{(0)} q \tag{15.4}$$

$$F = A^{(0)} T \tag{15.5}$$

Here, the various differential forms are expressed without their indices [1]- [10] so that the basic structure of the equations is revealed the most clearly. They are as follows: F is the electromagnetic field, A is the electromagnetic potential, ω is the spin connection, j is the homogeneous current, R is the curvature, q is the tetrad, and T is the torsion. In these equations $cA^{(0)}$ is the primordial voltage, and μ_0 is the vacuum permeability in the SI. System of units. [11].

The spin connection resonance (SCR) equation is, from Eqs.(15.1) and (15.2):

$$d \wedge (d \wedge A + \omega \wedge A) = \mu_0 j = A^{(0)}(R \wedge q - \omega \wedge T) \tag{15.6}$$

The effect of gravitation on the electromagnetic field is governed by j, the homogeneous current of ECE theory. If there is no effect:

$$j = 0 \tag{15.7}$$

and

$$d \wedge (d \wedge A + \omega \wedge A) = 0 \tag{15.8}$$

In this case, translational and rotational motions are independent. The translational motion governs the gravitational field and the rotational motion governs the electromagnetic field. Einstein Hilbert (EH) field theory is governed by translational motion defined through the following Cartan geometry:

$$R \wedge q = 0 \tag{15.9}$$

$$T = 0 \tag{15.10}$$

$$D \wedge R = 0 \tag{15.11}$$

Eq.(15.9) is the Ricci cyclic equation, and Eq.(15.11) is the second Bianchi identity. The well known EH field equation is obtained from the second Bianchi identity ([11]) and Noethers Theorem. As can be seen from Eq.(15.10) there is no Cartan torsion T in the EH theory. From Eqs. (15.9) and (15.10):

$$j = 0 \tag{15.12}$$

self-consistently, because EH is a theory of gravitation upon which there is no electromagnetic influence. Note that if $\widetilde{R}$ be the Hodge dual [1]– [10] of R, then:

$$\widetilde{R} \wedge q \neq 0 \tag{15.13}$$

The rotational motion defines the electromagnetic field by:

$$d \wedge F = 0 \tag{15.14}$$

and its Hodge dual:

$$d \wedge \widetilde{F} = \mu_0 J \tag{15.15}$$

where:

$$J = \frac{A^{(0)}}{\mu_0}(\widetilde{R} \wedge q - \omega \wedge \widetilde{T}) \tag{15.16}$$

is the inhomogeneous current of ECE theory. For rotational motion:

$$R \wedge q = \omega \wedge T \tag{15.17}$$

$$\widetilde{R} \wedge q = \omega \wedge \widetilde{T} \tag{15.18}$$

and

$$j_{rotation} = \widetilde{j}_{rotation} = 0 \tag{15.19}$$

but from Eq.(15.13)

$$J = \frac{A^{(0)}}{\mu_0}(\widetilde{R} \wedge q)_{translation} \neq 0 \tag{15.20}$$

Eqs.(15.14) and (15.15) give the ECE laws [1]– [10] of electrodynamics unaffected by gravitation. Eqs.(15.17) and (15.18) indicate that for pure rotational motion the rotational curvature R is the dual of the Cartan torsion T, and the tetrad is the dual of the spin connection [1]– [10].

From Eqs.(15.6) and (15.8) it is seen that the influence of gravitation on electromagnetism is to change A and ω through the presence of j. This change also introduces the possibility of SCR through Eq.(15.6) [1]– [10] and its Hodge dual. If the effect of gravitation is very weak, (as in the laboratory), then, for rotational motion, ω is dual to A in Eq.(15.6) to an excellent approximation. We may thus consider the effect of gravitation to be a change in A produced by j. The ECE spin field is defined by [1]– [10]:

$$\boldsymbol{B}^{(3)*} = -igA \wedge A^* \tag{15.21}$$

where A^* is the complex conjugate of A, and where:

$$g = \frac{\kappa}{A^{(0)}} \tag{15.22}$$

where κ is, in free space, a wave-number. Thus $\boldsymbol{B}^{(3)}$ (switching to vector notation) is changed by gravitation, and the RFR resonance frequency is shifted by gravitation. At resonance from Eq.(15.6) it is seen that this RFR shift is greatly amplified, so may become measurable in the laboratory. Similar considerations apply for all types of atomic and molecular spectroscopy. We may also bring into consideration quantum electrodynamics (QED) through the ECE Lemma applied to A:

$$\Box A = RA \tag{15.23}$$

where:

$$R = -kT \tag{15.24}$$

Here $\Box$ is the d'Alembertian, R is the scalar curvature, k is Einsteins constant and T is the index contracted canonical energy-momentum density. The latter in ECE contains in general contributions from all fields and interaction terms.

15.3 Levels of approximation in IFE and RFR

In general relativity both effects originate in the $\omega \wedge A$ term of:

$$F = d \wedge A + \omega \wedge A \tag{15.25}$$

and there are various levels of approximation that can be used to evaluate $\omega \wedge A$: classical, semi-classical, special relativistic QED, and general relativistic ECE. For rigorously objective physics [1]– [10] general relativity must be applied to all equations and concepts without exception. This is the basic ECE philosophy needed to produce a generally covariant unified field theory. For electromagnetism free of gravitational influence the spin connection is dual to the potential, defining the ECE spin field. The IFE and RFR follow directly from the spin field, which in vector notation in the complex circular basis [1]– [10] is defined as follows:

$$\boldsymbol{B}^{(3)*} = -ig\boldsymbol{A}^{(1)} \times \boldsymbol{A}^{(2)} \tag{15.26}$$

The IFE is magnetization due to the radiated spin field $\boldsymbol{B}^{(3)}$, and RFR is resonance due to $\boldsymbol{B}^{(3)}$. Analogously, bulk magnetization is due to a static magnetic field, and ESR and NMR are resonance phenomena due to a static magnetic field. In free space (no interaction with matter such an electron), the spin field is defined [1]– [10] by:

$$\boldsymbol{B}^{(3)} = \boldsymbol{B}^{(0)}\boldsymbol{k} = \kappa A^{(0)}\boldsymbol{k} \tag{15.27}$$

where:

$$g = \frac{\kappa}{A^{(0)}} \tag{15.28}$$

The conjugate product of non-linear optics [1]– [10] is defined by:

$$\boldsymbol{A}^{(1)} \times \boldsymbol{A}^{(2)} = \boldsymbol{A} \times \boldsymbol{A}^* \tag{15.29}$$

where $\boldsymbol{A}^*$ is the complex conjugate of $\boldsymbol{A}$. When $\boldsymbol{B}^{(3)}$ interacts with matter, on the simplest level an electron, then:

$$g \to g' \tag{15.30}$$

where g' is to be determined as follows from dynamics. The IFE is the magnetization:

$$\boldsymbol{M}^{(3)} = \frac{1}{\mu_0}\boldsymbol{B}^{(3)} = -ig\boldsymbol{A}^{(1)} \times \boldsymbol{A}^{(2)} = g'\boldsymbol{A}^{(0)2}\boldsymbol{k} \tag{15.31}$$

The factor $\boldsymbol{A}^{(0)2}$ can be related to the power density I (watts per square meter) of the electromagnetic field using the standard optical equation [12]:

$$\boldsymbol{A}^{(0)2} = \frac{\mu_0}{c}\left(\frac{I}{\omega^2}\right) \tag{15.32}$$

Therefore:

$$\boldsymbol{M}^{(3)} = \frac{g'}{c}\left(\frac{I}{\omega^2}\right)\boldsymbol{k} \tag{15.33}$$

The factor g' must also be calculated from general relativity in a fully self-consistent development. It must be calculated from the ECE wave equation in the presence of interaction between the electromagnetic field and one electron [1]- [10]:

$$(\gamma^a(i\hbar\partial_a - eA_a) - mc)q^c = 0 \tag{15.34}$$

where γ^a is the Dirac matrix, $\hbar$ is the reduced Planck constant, e is the charge on the electron, m is the mass of the electron, c is the vacuum speed of light and q^c is the tetrad. In the absence of interaction between the fermion and the gravitational field this equation reduces to the well known Dirac equation [1]-[10]:

$$(\gamma^\mu(i\hbar\partial_\mu - eA_\mu) - mc)\psi = 0 \tag{15.35}$$

with the minimal prescription used to describe the interaction between the free fermion and the electromagnetic potential. This equation is well known to be the basis of QED and to successfully describe the Zeeman effect through the half integral spin of the fermion. It is the basis of ESR and NMR. In the non-relativistic quantum limit Eq.(15.35) reduces to the Schrödinger-Pauli equation:

$$H\psi = E\psi \tag{15.36}$$

in which the hamiltonian is described using the Pauli matrices:

$$H = \frac{1}{2m}\boldsymbol{\sigma}\cdot(\boldsymbol{p} + e\boldsymbol{A})\boldsymbol{\sigma}\cdot(\boldsymbol{p} + e\boldsymbol{A}^*)\boldsymbol{\sigma} + V \tag{15.37}$$

Here $\boldsymbol{p}$ is the classical momentum, $\boldsymbol{A}$ is the classical vector potential, $\boldsymbol{\sigma}$ denotes a Pauli matrix and V denotes the potential energy. For a static magnetic field [1]-[10]:

$$H\psi = \frac{e\hbar}{2m}\boldsymbol{\sigma}\cdot\boldsymbol{B}\psi \tag{15.38}$$

and for an electromagnetic field:

$$H\psi = \frac{\mu_0 ce^2}{2m}\left(\frac{I}{\omega^2}\right)\sigma_Z\psi \tag{15.39}$$

Electron spin resonance (ESR) is described by:

$$\hbar\omega_{res} = \frac{e\hbar}{2m}(1-(-1))B \tag{15.40}$$

with resonance angular frequency:

$$\omega_{res} = \frac{e}{m}B \tag{15.41}$$

RFR is described by [1]- [10]:

$$\hbar\omega_{res} = \frac{\mu_0 ce^2}{2m}\left(\frac{I}{\omega^2}\right)(1-(-1)) \tag{15.42}$$

with resonance angular frequency:

$$\omega_{res} = \left(\frac{\mu_0 ce^2}{\hbar m}\right)\frac{I}{\omega^2} \tag{15.43}$$

The IFE can be approximated by classical special relativistic limit of Eq.(15.35), which is the special relativistic Hamilton Jacobi equation [1]- [10]:

$$(p^\mu - eA^\mu)(p_\mu - eA^*_\mu) = m^2c^2 \tag{15.44}$$

The solution of Eq.(15.44) for N electrons in a sample volume V is:

$$\boldsymbol{B}^{(3)}_{sample} = \frac{N}{V}\frac{\mu_0 e^3c^2}{2m\omega^2}\left(\frac{B^{(0)}}{\sqrt{m^2\omega^2+e^2B^{(0)2}}}\right)\boldsymbol{B}^{(3)} \tag{15.45}$$

In the limit:

$$m\omega \ll eB^{(0)} \tag{15.46}$$

we obtain:

$$\boldsymbol{B}^{(3)}_{sample} = \mu_0\boldsymbol{M}^{(3)}_{sample} \to \frac{N}{V}\left(\frac{e^3c}{2m^2}\right)\frac{I}{\omega^2}\boldsymbol{k} \tag{15.47}$$

Comparing equations (15.33) and (15.47):

$$g' = \frac{N}{V}\frac{e^3c^2}{2m^2} \tag{15.48}$$

in this approximation.

15.4 Spin connection resonance

In the presence of gravitation the key resonance equation is:

$$d\wedge(d\wedge A + \omega\wedge A) = \mu_0 j \tag{15.49}$$

In the off resonant condition the effect of gravitation in the laboratory is very small, so to an excellent approximation, and for rotational motion [1]- [10]:

$$\omega\wedge = -gA \tag{15.50}$$

Thus Eq.(15.49) becomes:

$$d \wedge (d \wedge A - igA \wedge A^*) = \mu_0 j \tag{15.51}$$

For circularly polarized radiation in free space [1]– [10];

$$d \wedge A = -igA \wedge A^* \tag{15.52}$$

so:

$$d \wedge (d \wedge A) = \frac{\mu_0}{2} j \tag{15.53}$$

Therefore there are equations such as:

$$d \wedge \boldsymbol{B}^{(3)*} = \frac{\mu_0}{2} j \tag{15.54}$$

The effect of gravitation is to make $B^{(3)}$ space and time dependent, for example to make it precess in a cone as follows. Considering the space part of Eq.(15.54):

$$\boldsymbol{\nabla} \times \boldsymbol{B}^{(3)} = \frac{\mu_0}{2} \boldsymbol{j} \tag{15.55}$$

produces the precessional equations:

$$\frac{\partial B_Z^{(3)}}{\partial Y} - \frac{\partial B_Y^{(3)}}{\partial Z} = \frac{\mu_0}{2} j_X \tag{15.56}$$

$$\frac{\partial B_X^{(3)}}{\partial Z} - \frac{\partial B_Z^{(3)}}{\partial X} = \frac{\mu_0}{2} j_Y \tag{15.57}$$

$$\frac{\partial B_Y^{(3)}}{\partial X} - \frac{\partial B_X^{(3)}}{\partial Y} = \frac{\mu_0}{2} j_Z \tag{15.58}$$

Resonance equations can be derived from Eqs.(15.56) to (15.58) as follows. At resonance, the $\boldsymbol{B}^{(3)}$ field would be greatly amplified, meaning that the RFR line would be shifted enough by gravitation for the shift to become measurable in the laboratory. This might lead to a practical way of measuring the effect of gravitation on a spectrum in the laboratory. At present this is only possible by astronomy (red shifts for example).

Differentiating Eq.(15.55):

$$\frac{\partial^2 B_Z^{(3)}}{\partial Y^2} - \frac{\partial}{\partial Y}\left(\frac{\partial B_Y^{(3)}}{\partial Z}\right) = \frac{\mu_0}{2} \frac{\partial j_X}{\partial Y} \tag{15.59}$$

This becomes a resonance equation under the mathematical conditions:

$$\frac{\partial}{\partial Y}\left(\frac{\partial B_Y^{(3)}}{\partial Z}\right) = -\kappa_0^2 B_Z^{(3)} \tag{15.60}$$

i.e.:

$$B_Y^{(3)} = -\kappa_0^2 \iint B_Z^{(3)} dZ dY \tag{15.61}$$

and

$$\frac{\partial j_X}{\partial Y} = j^{(0)} \cos(\kappa Y) \tag{15.62}$$

under which Eq.(15.59) becomes the undamped oscillator equation:

$$\frac{\partial^2 B_Z^{(3)}}{\partial Y^2} + \kappa_0^2 B_Z^{(3)} = \frac{\mu_0}{2} j^{(0)} \cos(\kappa Y) \tag{15.63}$$

At SCR from this equation, $B_z^{(3)}$ is greatly amplified.

Acknowledgements The British Government is thanked for the award of a Civil List pension for distinguished services to science, and the staff of AIAS are thanked for many interesting discussions.

Bibliography

[1] M. W. Evans, Generally Covariant Unified Field Theory (Abramis Academics, 2005), vol. One.

[2] M. W. Evans, Generally Covariant Unified Field Theory (Abramis Academic, 2006), vol. Two.

[3] M. W. Evans, Generally Covariant Unified Field Theory (Abramis Academic, 2006), vol. Three.

[4] M. W. Evans, Generally Covariant Unified Field Theory (Abramis Academic, 2007), vol. Four, preprint on www.aias.us and www.atomicprecision.com .

[5] L. Felker, The Evans Equations of Unified Field Theory (preprint on www.aias.us and www.atomicprecision.com).

[6] H. Eckardt and L. Felker, paper on homepage of www.aias.us.

[7] M. W. Evans, (Ed.), Modern Non-Linear Optics, a special topical issue in three parts of I. Prigogine and S. A. Rice (Series Eds.), Advances in Chemical Physics, (Wiley-Inter-science, New York, 2001, 2nd. Ed.), vols. 119(1) to 119(3), endorsed by the Royal Swedish Academy;. M. W. Evans and S. Kielich (Eds.), ibid., first edition (Wiley-Inter-science, New York, 1992, reprinted 1993 and 1997), vols. 85(1) to 85(3), awarded a prize for excellence by the Polish Government.

[8] M. W. Evans and L. B. Crowell, Classical and Quantum Electrodynamics and the $\boldsymbol{B}^{(3)}$ Field (World Scientific, 2001).

[9] M. W. Evans and J.-P. Vigier, The Enigmatic Photon (Kluwer, Dordrecht, 1994 to 2002, hardback and softback), in five volumes.

[10] M. W. Evans and A. A. Hasanein, The Photomagenton in Quantum Field Theory (World Scientific, 1994).

[11] J. D. Jackson, Classical Electrodynamics (Wiley, New York, 1999, 3rd. ed.).

[12] P. W. Atkins, Molecular Quantum Mechanics (Oxford Univ. Press, 1983, 2nd. ed.).

Chapter 16

Chirality and spin vectors in ECE Theory

(Paper 70)
by
Myron W. Evans
Alpha Institute for Advanced Study (AIAS).
(emyrone@aol.com, www.aias.us, www.atomicprecision.com)

Abstract

The fundamental chirality and spin vectors of ECE theory are identified using the basic definition of the tetrad field as the rank two mixed index tensor that links two column vectors. The chirality vector defines the direction of spin (left or right handedness or chirality) and the spin vector indicates the existence of spin through a phase factor. The tetrad tensor is therefore made up of both handedness and spin, for example the components of the electromagnetic potential field are components of the tetrad tensor within a scalar factor $A^{(0)}$ where $cA^{(0)}$ is the primordial voltage of ECE theory. Similarly the fermion field is defined by a chirality two-spinor and a spin two-spinor. For the fermion, the tetrad field is a 2×2 mixed index tensor. The weak and strong fields can be developed similarly in terms of fundamental chirality and spin column vectors. Each field can be represented using inter-convertible representation spaces, for example the space part of the electromagnetic field can be represented by the $O(3)$ or $SU(n)$ groups, were $n = 2, .., n$. The $SU(2)$ representation of the electromagnetic field is the Majorana representation. This allows for field unification in any representation space.

Keywords: Einstein Cartan Evans (ECE) field theory, handedness, chirality, spin.

16.1 Introduction

In Einstein Cartan Evans (ECE) field theory [1]– [8] the fundamental field is the tetrad, which is a rank two mixed index tensor [9] that transforms as such under the general coordinate transformation, and is thus generally covariant. Therefore the fundamental fields of physics are tetrads of various kinds: the gravitational, electromagnetic, weak , strong and matter fields. The tetrad is defined as follows:

$$V^a = q^a_\mu V^\mu \tag{16.1}$$

where V^a and V^μ are column vectors which are also generally covariant. The tetrad field is therefore defined by the way in which V^a and V^μ are related geometrically, and the tetrad in turn defines the torsion tensor $T^a{}_{\mu\nu}$. In ECE theory the electromagnetic field for example is defined by the ansatz:

$$A^a_\mu = A^{(0)} q^a_\mu \tag{16.2}$$

$$F^a{}_{\mu\nu} = A^{(0)} T^a{}_{\mu\nu} \tag{16.3}$$

where $cA^{(0)}$ is the primordial voltage, c being the speed of light in vacuo and $A^{(0)}$ the potential magnitude of the electromagnetic field. The gravitational field is also defined by the tetrad, the symmetric metric being:

$$g_{\mu\nu} = q^a_\mu q^b_\nu \eta_{ab} \tag{16.4}$$

where η_{ab} is the metric in the tangent spacetime of Cartan geometry [1]– [9] at point P in the base manifold. In Section 16.2, V^a is defined as the chirality vector, and V^μ as the spin vector for the electromagnetic and fermion fields. The electromagnetic tetrad A^a_μ is therefore made up both of chirality (handedness) and spin - it can be left or right circularly polarized for example. Components of the tetrad tensor A^a_μ are denoted [1]– [9] $A^{(1)}_X$, and so on, and are components of the electromagnetic potential field. The electromagnetic field tensor is then defined by the first Cartan structure equation:

$$F^a{}_{\mu\nu} = (d \wedge A^a)_{\mu\nu} + (\omega^a{}_b \wedge A^b)_{\mu\nu} \tag{16.5}$$

where $\omega^a{}_b$ is the spin connection. For free rotation [1]– [8], the spin connection is dual to the tetrad:

$$\omega^a{}_{\mu b} = -\frac{\kappa}{2} \epsilon^a{}_{bc} q^c_\mu \tag{16.6}$$

and the spin connection can therefore be identified as being itself a potential field component. In this special case of pure rotation the spin connection becomes a generally covariant mixed index tensor (a tetrad tensor). In general however the spin connection is not a tensor [9]. Similarly the Christoffel connection of Riemann geometry is not a tensor in general because it does not transform covariantly under the general coordinate transformation. The tetrad in contrast always transforms covariantly because it is a rank two mixed-index tensor [9]. In general relativity any quantity with this property of general covariance may be a physical quantity (for example the Riemann tensor and the metric). In the standard model in contrast the electromagnetic potential field is a vector (i.e rank one tensor) A_μ and is developed with gauge theory in which it is not gauge invariant. A great deal of confusion results in the standard model for this reason, because it is held that a quantity that is not gauge invariant is not a physical quantity, a view that predates quantum theory and relativity and goes back to Heaviside. Faraday and Maxwell in contrast regarded A_μ as physical. At the same time in the standard model, A_μ is considered physical in the minimal prescription. The standard model is therefore self-inconsistent, in that A_μ is at once non-physical and physical, and is also incomplete, because it is special relativity, i.e. Lorentz covariant but not generally covariant as required by Einsteinian general relativity. ECE theory clears up this confusion by regarding A_μ as a generally covariant tetrad field, which is always a physical field. In ECE theory the tetrad is also the gravitational field, and in the latter, a is the index that defines the Minkowski or flat tangent spacetime of Cartan geometry [1]–[9] and μ is the index of a curving base manifold. As seen in Eq.(16.4), the symmetric metric of gravitational theory is made up of two tetrads multiplied together. The tetrad is therefore the fundamental gravitational field, and not the symmetric metric. The gravitational tetrad is therefore defined as the rank two tensor that links the flat spacetime column four-vector V^a with the curved spacetime column four-vector V^μ. These are four-vectors because there are four dimensions, time and three space dimensions. The gravitational tetrad therefore has 16 components. In the well known Einstein Hilbert (EH) theory there is no consideration given to torsion, only to curvature. For this reason EH is not a unified field theory as is well known. ECE is a unified field theory in the well defined sense that it is governed not by Riemann geometry without torsion, but by Cartan geometry with inclusion of both the Riemann or curvature form $R^a{}_b$ and the Cartan torsion form T^a. These are governed by the two well known Cartan structure equations:

$$T^a = D \wedge q^a := d \wedge q^a + \omega^a{}_b \wedge q^b \tag{16.7}$$

$$R^a{}_b = D \wedge \omega^a{}_b := d \wedge \omega^a{}_b + \omega^a{}_c \wedge \omega^c{}_b \tag{16.8}$$

and the two Bianchi identities of Cartan (i.e. differential) geometry:

$$D \wedge T^a := R^a{}_b \wedge q^b \tag{16.9}$$

$$D \wedge R^a{}_b := 0 \tag{16.10}$$

It is seen that the curvature and torsion are inter-related ineluctably by the basic geometry. The EH theory is the limit:

$$T^a = 0 \tag{16.11}$$

In Section 16.2 therefore the electromagnetic and fermion fields are developed as tetrad fields governed by Eqs.(16.7) to (16.10), and thus linked to the gravitational field by these equations of Cartan geometry, thus synthesizing a generally covariant unified field theory as required by the basic philosophy of objectivity (Bacon) and relativity (Einstein and others). In so doing, Section 16.2 defines the index a of the electromagnetic and fermion fields as that of chirality and the index μ of these fields as that of spin. Therefore the same overall method is used for the gravitational, electromagnetic and fermion fields, in that the tetrad definition links one column vector to another. For the fermion field, the $SU(2)$ representation space is used as is well known, and so the column vectors have two entries, i.e. are two-spinors. The spinor field is therefore a 2×2 tetrad with four components. The tetrad is therefore one two-component row vector superimposed on another. If each row vector is transposed to a column two-vector, the result is a column four-vector, the Dirac spinor [1]– [8] made up of two Pauli spinors. It is shown in Section 16.2 that the a index of the tetrad in this case represents handedness (right or left fermion) and the μ index represents spin. The Dirac spinor contains chirality (referred to in this case as helicity). The effect of any other field on the fermion field is then governed by the geometry of Eqs.(16.7) to (16.10) and by the minimal prescription. Finally Section 16.3 is a discussion of how these concepts can be extended to the weak and strong fields using the appropriate representation spaces, and how fields can be inter-related using ECE theory using any representation space such as $O(3)$ or $SU(n)$.

16.2 Definition of the chirality and spin vectors

We first review the development [1] of the Cartesian vector:

$$\mathbf{R} = X\mathbf{i} + Y\mathbf{j} + Z\mathbf{k} \tag{16.12}$$

in the $SU(2)$ basis, giving:

$$\begin{aligned} R = \boldsymbol{\sigma}\cdot\mathbf{R} &= X\sigma_1 + Y\sigma_2 + Z\sigma_3 = \\ &= \begin{bmatrix} Z & X - iY \\ X + iY & -Z \end{bmatrix} \end{aligned} \tag{16.13}$$

Thus:

$$R^2 = X^2 + Y^2 + Z^2 = \begin{bmatrix} 1 & 0 \\ 0 & 1 \end{bmatrix} \tag{16.14}$$

The $SU(2)$ group is that of the unitary, unimodular matrices:

$$UU^{+} = 1, \quad det\, U = 1 \tag{16.15}$$

which have the general form:

$$U = \begin{bmatrix} a & b \\ -b^* & a^* \end{bmatrix} \tag{16.16}$$

with

$$aa^* + bb^* = 1 \tag{16.17}$$

Define the two component spinor with complex valued elements:

$$\zeta = \begin{bmatrix} \zeta_1 \\ \zeta_2 \end{bmatrix} \tag{16.18}$$

with hermitian conjugate:

$$\zeta^{+} = [\zeta_1^*, \quad \zeta_2^*] \tag{16.19}$$

We obtain the invariant:

$$X^2 + Y^2 + Z^2 = \zeta_1\zeta_1^* + \zeta_2\zeta_2^* \tag{16.20}$$

and therefore a relation between the Cartesian $O(3)$ elements and the $SU(2)$ elements. The chirality column vector in $SU(2)$ representation is defined as:

$$V^a = \begin{bmatrix} \zeta_1 \\ \zeta_2 \end{bmatrix} = \begin{bmatrix} e^R \\ e^L \end{bmatrix} \tag{16.21}$$

with elements:

$$\zeta_1 = \zeta_2^* = \frac{1}{\sqrt{2}}(1-i) \tag{16.22}$$

The spin column vector in $SU(2)$ representation is defined as:

$$V^\mu = e^{-i\phi} \begin{bmatrix} 1 \\ 1 \end{bmatrix} \tag{16.23}$$

where ϕ is the phase of the fermionic field. The chirality and spin column vectors are related by the $SU(2)$ tetrad field q^a_μ :

$$V^\mu = q^a_\mu V^\mu \tag{16.24}$$

i.e.

$$q^a_\mu = \frac{e^{i\phi}}{\sqrt{2}} \begin{bmatrix} 1 & -i \\ 1 & i \end{bmatrix} \tag{16.25}$$

The tetrad (16.25) is a simple example of a right or left handed spinning field. It can be seen that:

$$\zeta_1\zeta_2 = 1 = \frac{1}{\sqrt{2}}(1-i)\frac{1}{\sqrt{2}}(1+i) \tag{16.26}$$

so the origin of chirality, or left and right handedness, is the factorization in Eq.(16.26). If the phase is defined for a fermionic field propagating for convenience along the Y axis:

$$\phi = \omega t - \kappa Y \tag{16.27}$$

it is seen that:

$$\square q^a_\mu = 0 \tag{16.28}$$

and this is the equation of the hypothetically massless fermion, the Weyl equation. If the two rows of the tetrad matrix in Eq.(16.25) are transposed into column vectors:

$$\psi = \frac{e^{i\phi}}{\sqrt{2}} \begin{bmatrix} 1 \\ -i \\ 1 \\ i \end{bmatrix} \tag{16.29}$$

The four-spinor ψ consists of two Pauli spinors:

$$\phi^R = \frac{e^{i\phi}}{\sqrt{2}} \begin{bmatrix} 1 \\ -i \end{bmatrix}, \quad \phi^L = \frac{e^{i\phi}}{\sqrt{2}} \begin{bmatrix} 1 \\ i \end{bmatrix} \tag{16.30}$$

and obeys the equation:

$$\square \psi = 0 \tag{16.31}$$

If the Y Pauli matrix is denoted by:

$$\sigma_Y = \begin{bmatrix} 0 & -i \\ i & 0 \end{bmatrix} \tag{16.32}$$

the Weyl equations can be expressed as:

$$\sigma_Y \phi^R = -\phi^R \tag{16.33}$$

$$\sigma_Y \phi^L = -\phi^L \tag{16.34}$$

ie:

$$\begin{bmatrix} 0 & -i \\ i & 0 \end{bmatrix} \begin{bmatrix} 1 \\ -i \end{bmatrix} = \begin{bmatrix} -1 \\ i \end{bmatrix} = -\begin{bmatrix} 1 \\ -i \end{bmatrix} \tag{16.35}$$

$$\begin{bmatrix} 0 & -i \\ i & 0 \end{bmatrix} \begin{bmatrix} 1 \\ i \end{bmatrix} = \begin{bmatrix} 1 \\ i \end{bmatrix} \tag{16.36}$$

The helicity eigenvalues in Eqs.(16.35) and (16.36) are $\pm$ 1. The eigenfunctions are the right and left Pauli spinors ϕ^R and ϕ^L and the eigen-operator is σ_Y. Therefore the Pauli spinors ϕ^R and ϕ^R are those of a massless fermion propagating along Y. This is a simple example of how chirality or helicity can be built up from two types of column vector, one static and representing the sense of handedness (right or left) of the spin, and the other the spin itself. The resulting tetrad is therefore a combination of right and left spin, and for a phase of the type (16.27), propagates along Y. This is an example in special relativity because the Weyl equations are the massless Dirac equations. In general relativity [1]– [8]:

$$\begin{bmatrix} e^R \\ e^L \end{bmatrix} = \begin{bmatrix} q_1^R & q_2^R \\ q_1^L & q_2^L \end{bmatrix} \begin{bmatrix} q^1 \\ q^2 \end{bmatrix} \tag{16.37}$$

and:

$$(\Box + kT)q_\mu^a = 0 \tag{16.38}$$

where:

$$q_\mu^a = \begin{bmatrix} q_1^R & q_2^R \\ q_1^L & q_2^L \end{bmatrix} \tag{16.39}$$

If we define:

$$\psi = \begin{bmatrix} q_1^R \\ q_2^R \\ q_1^L \\ q_2^L \end{bmatrix} \tag{16.40}$$

then the ECE wave equation is:

$$(\Box + kt)\psi = 0 \tag{16.41}$$

It is known experimentally that this equation must reduce to the Dirac equation for the free single fermion uninfluenced by any other type of field:

$$\left(\Box + \left(\frac{mc}{\hbar}\right)^2\right)\psi = 0 \tag{16.42}$$

In this limit:

$$kT = \frac{m^2c^2}{\hbar^2} \tag{16.43}$$

Eq.(16.43) is an example of the equivalence principle. The effect of gravitation, or any other type of field, or combination of fields, on a fermion is given by Eq.(16.41). Thus Eq.(16.41) describes for example the gravitational interaction between two fermions such as electrons. The electric interaction between two electrons is of course described by the Coulomb Law, which can also be expressed in terms of general relativity using ECE theory [1]- [8]. The electromagnetic field is described in ECE theory by the 4×4 tetrad defined by:

$$V^a = q^a_\mu V^\mu \tag{16.44}$$

where V^a is the chirality vector of electromagnetism, and V^μ its spin vector. In general:

$$a = (0), (1), (2), (3) \tag{16.45}$$

and

$$\mu = 0, X, Y, Z \tag{16.46}$$

The space indices of a are those of the complex circular basis and those of μ are in the Cartesian basis. It was discovered experimentally by Arago in 1811 that the electromagnetic field is left (L) or right (R) circularly polarized. Therefore each index (1) has L and R components and similarly for (2). The (3) index is longitudinal and is similarly defined for L and R. The (0) index is time-like. Therefore the left handed circularly polarized transverse potential field component is:

$$\boldsymbol{A}^{(1)}_L = \frac{A^{(0)}}{\sqrt{2}}(\mathbf{i} - i\mathbf{j})e^{i\phi} \tag{16.47}$$

and the right handed circularly polarized transverse component is:

$$\boldsymbol{A}^{(1)}_R = \frac{A^{(0)}}{\sqrt{2}}(\mathbf{i} + i\mathbf{j})e^{i\phi} \tag{16.48}$$

where ϕ is the electromagnetic phase and where $*$ denotes complex conjugate [1]- [8]. The complex conjugates of (16.47) and (16.48) are found by reversing the sign of i:

$$\boldsymbol{A}^{(2)}_L = \frac{A^{(0)}}{\sqrt{2}}(\mathbf{i} + i\mathbf{j})e^{-i\phi} \tag{16.49}$$

$$\boldsymbol{A}^{(2)}_R = \frac{A^{(0)}}{\sqrt{2}}(\mathbf{i} - i\mathbf{j})e^{-i\phi} \tag{16.50}$$

The complex circular unit vector basis has $O(3)$ symmetry and is:

$$\mathbf{e}^{(1)} \times \mathbf{e}^{(2)} = i\mathbf{e}^{(3)*} \tag{16.51}$$

where:

$$\mathbf{e}^{(1)} = \mathbf{e}^{(2)*} = \frac{1}{\sqrt{2}}(\mathbf{i} - i\mathbf{j}) \tag{16.52}$$

$$\mathbf{e}^{(3)} = \mathbf{k} \tag{16.53}$$

For left circular polarization the chirality and spin column vectors appropriate to the transverse potential vector (16.47) to (16.50) can therefore be defined by:

$$V^{(1)}_L = \frac{1}{\sqrt{2}}\begin{bmatrix} 0 \\ 1 \\ -i \\ 0 \end{bmatrix}, \quad V^\mu = \begin{bmatrix} 0 \\ 1 \\ 1 \\ 0 \end{bmatrix} e^{-i\phi} \tag{16.54}$$

and the tetrad is therefore:

$$q_{L\mu}^{(1)} = \frac{1}{\sqrt{2}} \begin{bmatrix} 0 & 0 & 0 & 0 \\ 0 & 1 & 0 & 0 \\ 0 & 0 & -i & 0 \\ 0 & 0 & 0 & 0 \end{bmatrix} e^{i\phi} \tag{16.55}$$

with individual scalar components:

$$\begin{aligned} q_{LX}^{(1)} &= \frac{1}{\sqrt{2}} e^{i\phi}, \\ q_{LY}^{(1)} &= \frac{-i}{\sqrt{2}} e^{i\phi}, \end{aligned} \tag{16.56}$$

The complete transverse tetrad vector is:

$$q_L^{(1)} = \frac{1}{\sqrt{2}} (\mathbf{i} - i\mathbf{j}) e^{i\phi} \tag{16.57}$$

and from the ansatz (16.2) the electromagnetic potential's transverse vector is:

$$\mathbf{A}_L^{(1)} = A^{(0)} \mathbf{q}_L^{(1)} = \frac{A^{(0)}}{\sqrt{2}} (\mathbf{i} - i\mathbf{j}) e^{i\phi} \tag{16.58}$$

In right circular polarization:

$$V_R^{(1)} = \frac{1}{\sqrt{2}} \begin{bmatrix} 0 \\ 1 \\ i \\ 0 \end{bmatrix}, \quad V^{\mu} = \begin{bmatrix} 0 \\ 1 \\ 1 \\ 0 \end{bmatrix} e^{-i\phi} \tag{16.59}$$

giving:

$$q_{R\mu}^{(1)} = \frac{1}{\sqrt{2}} \begin{bmatrix} 0 & 0 & 0 & 0 \\ 0 & 1 & 0 & 0 \\ 0 & 0 & i & 0 \\ 0 & 0 & 0 & 0 \end{bmatrix} e^{i\phi} \tag{16.60}$$

and:

$$q_R^{(1)} = \frac{1}{\sqrt{2}} (\mathbf{i} + i\mathbf{j}) e^{i\phi} \tag{16.61}$$

$$\mathbf{A}_R^{(1)} = A^{(0)} \mathbf{q}_L^{(1)} = \frac{A^{(0)}}{\sqrt{2}} (\mathbf{i} + i\mathbf{j}) e^{i\phi} \tag{16.62}$$

The complex conjugates of index (2) are obtained straightforwardly from these equations of index (1) by reversing the sign of i wherever it occurs. It is seen that:

$$q_0^{(1)} = q_Z^{(1)} = q_0^{(2)} = q_Z^{(2)} = 0 \tag{16.63}$$

The chiral and spin vectors in longitudinal polarization are defined by:

$$V_R^{(3)} = \begin{bmatrix} 0 \\ 0 \\ 0 \\ 1 \end{bmatrix}, V_L^{(3)} = \begin{bmatrix} 0 \\ 0 \\ 0 \\ -1 \end{bmatrix}, V^{\mu} = \begin{bmatrix} 0 \\ 0 \\ 0 \\ 1 \end{bmatrix} \tag{16.64}$$

and

$$q_Z^{(3)} = \pm 1, A^{(3)} = \pm A^{(0)}\mathbf{k} \tag{16.65}$$

The sign change depends on whether the field is left or right circularly polarized.

The time-like polarizations are given by:

$$V^{(0)} = \begin{bmatrix} 1 \\ 0 \\ 0 \\ 0 \end{bmatrix} = V^{\mu} \tag{16.66}$$

and:

$$q_{\mu}^{(0)} = \begin{bmatrix} 1 & 0 & 0 & 0 \\ 0 & 0 & 0 & 0 \\ 0 & 0 & 0 & 0 \\ 0 & 0 & 0 & 0 \end{bmatrix} \tag{16.67}$$

ie.

$$q_0^{(0)} = 1, A_0^{(0)} = A^{(0)} q_0^{(0)} \tag{16.68}$$

and are the same for both senses of polarization.

It can be seen that: $\boldsymbol{A}_R^{(1)}$, $\boldsymbol{A}_L^{(1)}$, $\boldsymbol{A}^{(3)}$ and $\boldsymbol{A}^{(0)}$ are solutions of the ECE wave equation [1]– [8] in the approximation:

$$\left(\Box + \left(\frac{mc}{\hbar}\right)^2\right) A_{\mu}^a = 0 \tag{16.69}$$

where m is the mass of the photon. It is known that m is very tiny so to an excellent approximation:

$$\Box A_{\mu}^a = 0 \tag{16.70}$$

which is a generally covariant form of the d'Alembert equation. For finite photon mass free of any other field such as gravitation we obtain a generally covariant form of the Proca equation:

$$(\Box + kT)A_{\mu}^a = 0, kT = \left(\frac{mc}{\hbar}\right)^2 \tag{16.71}$$

In ECE theory both the d'Alembert and Proca equations of special relativity (the standard model) are made generally covariant as required by the most basic principle of general relativity, the principle of general covariance of any equation of physics. The standard model is not generally covariant and breaks this principle. Another problem occurs in the standard model because there the Proca equation is not gauge invariant [10], and is therefore unphysical, conflicting diametrically with photon mass theory in the standard model. Photon mass theory is at the root of the bending of light by gravity, proven experimentally to an accuracy of 1: 100,000 by NASA Cassini. So the standard model is hopelessly self-inconsistent because in one part of it, Einstein Hilbert theory, the photon mass is proven with this accuracy, and in another part, gauge theory, the photon mass is unphysical because the Proca equation is not gauge invariant. The fault lies with the gauge principle because of its abstract and ad hoc introduction of a fibre bundle by Yang and Mills. In fibre bundle theory there is no geometrical interpretation of the a index, as required by general relativity.

The origin of circular polarization of the electromagnetic field on ECE theory are therefore the chirality (or helicity) vectors $V_L^{(1)}$ and $V_R^{(1)}$. These are four vectors, representing one time-like and three space-like polarization. The electromagnetic potential field is the tetrad field within $A^{(0)}$, and the tetrad links the chirality and spin vectors. For the fermion field the tetrad superimposes the chirality and spin vectors to give the Dirac spinor in the limit of a free fermion. In this case:

$$a = L,\ R \tag{16.72}$$

$$\mu = 1,\ 2 \tag{16.73}$$

but the overall method is the same, indicating that ECE is a unified field theory. Indeed, the electromagnetic field can be described [2] in the same way as the fermion field by defining the chiral and spin vectors of the electromagnetic field by:

$$V^a = \frac{1}{\sqrt{2}} \begin{bmatrix} 1-i \\ 1+i \end{bmatrix},\ V^\mu = \begin{bmatrix} 1 \\ 1 \end{bmatrix} e^{-i\phi} \tag{16.74}$$

giving the space-like part of the electromagnetic tetrad:

$$q^a_\mu = \frac{e^{i\phi}}{\sqrt{2}} \begin{bmatrix} 1 & -i \\ 1 & i \end{bmatrix} \tag{16.75}$$

The upper and lower row vectors of the tetrad are transposed to column vectors giving:

$$\psi_{em} = \frac{e^{i\phi}}{\sqrt{2}} \begin{bmatrix} q^{(1)}_{XR} \\ q^{(1)}_{YR} \\ q^{(1)}_{XL} \\ q^{(1)}_{YL} \end{bmatrix} \tag{16.76}$$

This is a column four-vector analogous to the Dirac spinor, and the electromagnetic potential in this massless approximation obeys:

$$\Box A^a_\mu = 0 \tag{16.77}$$

where:

$$A^a_\mu = A^{(0)} \psi_{em} \tag{16.78}$$

It is seen that equation (16.77) is the same for the 2×2 square matrix (16.75) or the column vector (16.76). This shows that the electromagnetic field's transverse components [11] can be put into $SU(2)$ representation as first shown by Majorana in the nineteen twenties.

16.3 The fundamental chiral elements and field unification

The right and left handed spins in the chirality column vector:

$$V^a = \frac{1}{\sqrt{2}} \begin{bmatrix} 1-i \\ 1+i \end{bmatrix} \tag{16.79}$$

are the most fundamental elements of the fermion field. The well known Pauli spinors are constructed from the tetrad as follows:

$$(\phi^R)^T = [1\,0]q^a_\mu \tag{16.80}$$

$$(\phi^L)^T = [0\,1]q^a_\mu \tag{16.81}$$

where the superscript T denotes "transpose" [12]. Thus:

$$(\phi^R)^T = [1\,0]\begin{bmatrix} q_1^R & q_2^R \\ q_1^L & q_2^L \end{bmatrix} = \begin{bmatrix} q_1^R & q_2^R \end{bmatrix} \tag{16.82}$$

and:

$$\phi^R = \begin{bmatrix} q_1^R \\ q_2^R \end{bmatrix} \tag{16.83}$$

Similarly:

$$(\phi^L)^T = [0\,1]\begin{bmatrix} q_1^R & q_2^R \\ q_1^L & q_2^L \end{bmatrix} = \begin{bmatrix} q_1^L & q_2^L \end{bmatrix} \tag{16.84}$$

and:

$$\phi^L = \begin{bmatrix} q_1^L \\ q_2^L \end{bmatrix} \tag{16.85}$$

Thus:

$$\phi^R = (\begin{bmatrix} 1, 0 \end{bmatrix} q^a_\mu)^T \tag{16.86}$$

$$\phi^L = (\begin{bmatrix} 0, 1 \end{bmatrix} q^a_\mu)^T \tag{16.87}$$

and the Dirac spinor is:

$$\psi = \begin{bmatrix} \phi^R \\ \phi^L \end{bmatrix} \tag{16.88}$$

It follows from the ECE equation:

$$(\Box + kT)q^a_\mu = 0 \tag{16.89}$$

that:

$$(\Box + kT)\psi = 0 \tag{16.90}$$

For a free fermion unaffected by any other type of field:

$$kT = \left(\frac{mc}{\hbar}\right)^2 \tag{16.91}$$

and we recover the Dirac equation [1]– [8]:

$$\left(\Box + \left(\frac{mc}{\hbar}\right)^2\right)\psi = 0 \tag{16.92}$$

or

$$\left(\Box + \left(\frac{mc}{\hbar}\right)^2\right)q^a_\mu = 0 \tag{16.93}$$

It is seen that the most fundamental elements of ψ are V^a and V^μ, and that generally covariant Dirac equation is:

$$(\Box + kT)\psi = 0 \tag{16.94}$$

where:

$$\psi = \begin{bmatrix} q^R \\ q^L \end{bmatrix} = \begin{bmatrix} q_1^R \\ q_2^R \\ q_1^L \\ q_2^L \end{bmatrix} \tag{16.95}$$

The tetrad is therefore:

$$q_\mu^a = \begin{bmatrix} (\phi^R)^T \\ (\phi^L)^T \end{bmatrix} \tag{16.96}$$

and the Dirac spinor is:

$$\psi = \begin{bmatrix} \phi^R \\ \phi^L \end{bmatrix} \tag{16.97}$$

For Eqs.(16.82) and (16.84):

$$q_\mu^a = \begin{bmatrix} 1 \\ 0 \end{bmatrix} (\phi^R)^T + \begin{bmatrix} 0 \\ 1 \end{bmatrix} (\phi^L)^T \tag{16.98}$$

i.e.:

$$\begin{aligned} \begin{bmatrix} q_1^R & q_2^R \\ q_1^L & q_2^L \end{bmatrix} &= \begin{bmatrix} 1 \\ 0 \end{bmatrix} \begin{bmatrix} q_1^R & q_2^R \end{bmatrix} + \begin{bmatrix} 0 \\ 1 \end{bmatrix} \begin{bmatrix} q_1^L & q_2^L \end{bmatrix} \\ &= \begin{bmatrix} q_1^R & q_2^R \\ 0 & 0 \end{bmatrix} + \begin{bmatrix} 0 & 0 \\ q_1^L & q_2^L \end{bmatrix} \end{aligned} \tag{16.99}$$

Eq.(16.98) shows that the tetrad can be analyzed as te sum of two transposed Pauli spinors. Therefore the fundamental Pauli spinors are made up of elements of Cartan geometry, the chirality column vector V^a and the spin column vector V^μ.

Now define the state spinors:

$$\zeta_1 = \begin{bmatrix} 1 \\ 0 \end{bmatrix}, \quad \zeta_2 = \begin{bmatrix} 0 \\ 1 \end{bmatrix} \tag{16.100}$$

which are inter-convertible by the parity inversion operator $\hat{P}$:

$$\hat{P}\zeta_1 = \zeta_2 \tag{16.101}$$

and so may be considered as fundamental states of handedness or chirality. The analysis has been carried out for the fermion field with $SU(2)$ symmetry. The $SU(2)$ group is the unitary unimodular group of matrices such as:

$$S = \frac{1}{\sqrt{2}} \begin{bmatrix} 1 & -i \\ -i & 1 \end{bmatrix} \tag{16.102}$$

whose hermitian transpose is:

$$S^+ = \frac{1}{\sqrt{2}} \begin{bmatrix} 1 & i \\ i & 1 \end{bmatrix} \tag{16.103}$$

so that:

$$SS^+ = \frac{1}{2} \begin{bmatrix} 1 & -i \\ -i & 1 \end{bmatrix} \begin{bmatrix} 1 & i \\ i & 1 \end{bmatrix} = \begin{bmatrix} 1 & 0 \\ 0 & 1 \end{bmatrix} \tag{16.104}$$

and $det\ S = 1$. Eq.(16.79) can be developed into the sum:

$$V^a = \frac{1}{\sqrt{2}}\left(\begin{bmatrix} 1 \\ 1 \end{bmatrix} - i(\zeta_1 - \zeta_2)\right) \tag{16.105}$$

and so the chirality vector is made up of the real part:

$$Re\, V^a = \frac{1}{\sqrt{2}}\mathbf{1} := \frac{1}{\sqrt{2}}\begin{bmatrix} 1 \\ 1 \end{bmatrix} \tag{16.106}$$

and the imaginary part:

$$Im\, V^a = \frac{1}{\sqrt{2}}(\zeta_2 - \zeta_1) \tag{16.107}$$

These can be regarded as the fundamental elements of chirality or handedness.

The unitary unimodular matrix S is made up of two Pauli matrices:

$$S = \frac{1}{\sqrt{2}}(\sigma_0 - i\sigma_1) \tag{16.108}$$

where:

$$\sigma_0 = \begin{bmatrix} 1 & 0 \\ 0 & 1 \end{bmatrix}, \quad \sigma_1 = \begin{bmatrix} 0 & 1 \\ 1 & 0 \end{bmatrix} \tag{16.109}$$

The other two Pauli matrices are:

$$\sigma_2 = \begin{bmatrix} 0 & -i \\ i & 0 \end{bmatrix}, \quad \sigma_3 = \begin{bmatrix} 1 & 0 \\ 0 & -1 \end{bmatrix} \tag{16.110}$$

The related matrix:

$$\zeta = \frac{1}{\sqrt{2}}\begin{bmatrix} 1 & -1 \\ 1 & 1 \end{bmatrix} \tag{16.111}$$

has the orthoganility [12] property:

$$\zeta\zeta^T = \begin{bmatrix} 1 & 0 \\ 0 & 1 \end{bmatrix} \tag{16.112}$$

and so:

$$S = \frac{1}{\sqrt{2}}(\mathbf{1} + i(\zeta_2 - \zeta_1)) \tag{16.113}$$

is unitary and unimodular [12] while:

$$\zeta = \frac{1}{\sqrt{2}}\begin{bmatrix} 1 & \zeta_2 - \zeta_1 \\ 1 & \end{bmatrix} = \frac{1}{\sqrt{2}}\begin{bmatrix} \mathbf{1} & \zeta_2 - \zeta_1 \end{bmatrix} \tag{16.114}$$

is orthogonal. The $SU(2)$ matrix:

$$S = \begin{bmatrix} q_X^{(1)} & q_Y^{(1)} \\ -iq_X^{(1)} & -iq_Y^{(1)} \end{bmatrix} e^{-i\phi} \tag{16.115}$$

where:

$$q_X^{(1)} = \frac{1}{\sqrt{2}}e^{i\phi}, \quad q_Y^{(1)} = \frac{-i}{\sqrt{2}}e^{i\phi} \tag{16.116}$$

and

$$\boldsymbol{A}^{(1)} = A^{(1)}\boldsymbol{q}^{(1)} = A^{(0)}(q_X^{(1)}\boldsymbol{i} + q_Y^{(1)}\boldsymbol{j}) \tag{16.117}$$

give and $SU(2)$ representation of the electromagnetic field. Using the further development:

$$\begin{bmatrix} 1 \\ 1 \end{bmatrix} = \begin{bmatrix} 1 \\ 0 \end{bmatrix} + \begin{bmatrix} 0 \\ 1 \end{bmatrix} \tag{16.118}$$

it is found that the chirality vector is made up of the elements ζ_1 and ζ_2 as follows:

$$V^a = \frac{1}{\sqrt{2}}(\zeta_1 + \zeta_2 - i(\zeta_1 - \zeta_2)) \tag{16.119}$$

and that the spin vector is made up of the same elements as follows:

$$V^\mu = (\zeta_1 + \zeta_2)e^{-i\phi} \tag{16.120}$$

So it is concluded that the fermion field's most fundamental elements are:

$$\zeta_1 = \begin{bmatrix} 1 \\ 0 \end{bmatrix}, \quad \zeta_2 = \begin{bmatrix} 0 \\ 1 \end{bmatrix} \tag{16.121}$$

together with the phase factor $e^{-i\phi}$. We define ζ_1 and ζ_2 as the chiral elements of the field. These elements also define the electromagnetic field's transverse elements as follows:

$$V_L^{(1)} = \frac{1}{\sqrt{2}}\left(\begin{bmatrix} 0 \\ \zeta_1 \\ 0 \end{bmatrix} - i\begin{bmatrix} 0 \\ \zeta_2 \\ 0 \end{bmatrix}\right) \tag{16.122}$$

$$V^\mu = \frac{1}{\sqrt{2}}\left(\begin{bmatrix} 0 \\ \zeta_1 \\ 0 \end{bmatrix} + \begin{bmatrix} 0 \\ \zeta_2 \\ 0 \end{bmatrix}\right)e^{-i\phi} \tag{16.123}$$

and are therefore unifying elements of the fermionic and electro-magnetic fields. The tetrad elements of the electromagnetic field can be put in $SU(2)$ form by using:

$$\boldsymbol{q}^{(1)} \cdot \boldsymbol{q}^{(2)} + \boldsymbol{q}^{(2)} \cdot \boldsymbol{q}^{(1)} + \boldsymbol{q}^{(3)} \cdot \boldsymbol{q}^{(3)} = \boldsymbol{q}_1\boldsymbol{q}_1^* + \boldsymbol{q}_2\boldsymbol{q}_2^* \tag{16.124}$$

For circular polarization:

$$\begin{aligned} \boldsymbol{q}^{(1)} &= \frac{1}{\sqrt{2}}(\boldsymbol{i} - i\boldsymbol{j})e^{i\phi} = \boldsymbol{q}^{(2)*}, \\ \boldsymbol{q}^{(3)} &= \boldsymbol{k} \end{aligned} \tag{16.125}$$

and so:

$$q_1q_1^* + q_2q_2^* = 3 \tag{16.126}$$

A possible solution is:

$$q_1 = \sqrt{\frac{3}{2}}e^{i\phi}, \quad q_2 = -\sqrt{\frac{3}{2}}e^{i\phi} \tag{16.127}$$

Geometrically, the electromagnetic field has the same origin as the fermionic field, as shown by Eq.(16.124).

Finally in this section the minimal prescription as used for example by Dirac to define the well known half integral spin of the fermion in ESR or NMR is defined for use in ECE theory. The ECE Lemma is [1]- [8]

$$\Box q^a_\mu = R q^a_\mu \tag{16.128}$$

where R is a well defined [1]- [8] scalar curvature and where:

$$\Box = \partial_\mu \partial^\mu = \gamma^\mu \gamma^\nu \partial_\mu \partial_\nu \tag{16.129}$$

Here γ^μ is a Dirac matrix. The d'Alembertian operator has been isolated in Eq.(16.128) from an analysis [1]- [8] of the tetrad postulate. The minimal prescription is a momentum addition which describes the effect of the electromagnetic field on the fermion field in Dirac's original analysis [1]- [8]. In gauge theory it is the result of the gauge invariance principle [9], but that is an abstract procedure which as we have argued already, leads to a diametric self inconsistency in the standard model. In ECE theory (general relativity) the minimal prescription is developed as in the original intent - a momentum addition.

In ECE the quantum operator equivalence is used in the same form as special relativity, because the partial four-derivative in ECE is the same as in special relativity [1]- [9]:

$$p^\mu = i\hbar \partial^\mu \tag{16.130}$$

where:

$$p^\mu = \left(\frac{E_n}{c}, \boldsymbol{p} \right), \quad \partial^\mu = \left(\frac{1}{c} \frac{\partial}{\partial t}, -\boldsymbol{\nabla} \right) \tag{16.131}$$

as usual. Now define:

$$A_\mu := A^{(0)}_\mu + A^{(1)}_\mu + A^{(2)}_\mu + A^{(3)}_\mu \tag{16.132}$$

so that all four states of polarization of A^a_μ are accounted for. The semi-classical minimal prescription in ECE theory is then defined as:

$$p_\mu \rightarrow p_\mu + eA_\mu \tag{16.133}$$

$$p^\mu \rightarrow p^\mu + eA^\mu \tag{16.134}$$

This means that:

$$\partial_\mu \rightarrow \partial_\mu - i\frac{e}{\hbar} A_\mu \tag{16.135}$$

$$\partial^\mu \rightarrow \partial^\mu - i\frac{e}{\hbar} A^\mu \tag{16.136}$$

and

$$\Box = \partial_\mu \partial^\mu \rightarrow \left(\partial_\mu - i\frac{e}{\hbar} A_\mu \right) \left(\partial^\mu - i\frac{e}{\hbar} A^\mu \right) \tag{16.137}$$

i.e.

$$\Box' := \Box - i\frac{e}{\hbar} \left(A_\mu \partial^\mu + A^\mu \partial_\mu \right) - i\frac{e^2}{\hbar^2} A_\mu A^\mu \tag{16.138}$$

The wave equation that defines the interaction of the Dirac spinor ψ and A_μ is therefore

$$(\Box' + kT)\psi = 0 \tag{16.139}$$

In the absence of gravitation, this becomes:

$$(\Box' + \frac{m_e c}{\hbar})\psi = 0 \tag{16.140}$$

Eq.(16.140) produces all the familiar half integral spin magnetic effects such as the Stern Gerlach experiment, the Zeeman effect, ESR, NMR and MRI, and for the electromagnetic field, RFR [1]– [8]. The overall effect in these phenomena is described by:

$$\Box \rightarrow \Box' \tag{16.141}$$

Any type of field interaction can be described in this way. If for example we wish to describe the effect of gravitation on interacting electromagnetic and fermionic fields Eq.(16.139) must be used. Both the fermion and electromagnetic fields in this case are affected by gravitation through the Cartan geometry defined in Eqs.(16.7) to (16.10). The gravitational field is represented by the Riemann or curvature form. The fermion field in these equations is the tetrad q^a_μ in $SU(2)$ representation, a 2×2 matrix. If Eqs.(16.7) to (16.10) are then developed in $SU(2)$ representation there is an $SU(2)$ symmetry torsion and curvature which interact with each other. The interaction of the fermion and gravitational fields is then governed in this way - again by Cartan geometry. The minimal prescription used in Eq.(16.139) is semi-classical - the fermion field is quantized but the electromagnetic field is classical. The next level is a fully quantized theory in which the electromagnetic field is governed by the ECE wave equation:

$$(\Box + kT)A^a_\mu = 0 \tag{16.142}$$

From Eqs.(16.133) and (16.134) and conservation of momentum, the photon momentum is changed by the electron momentum and vice versa in such a way that the total momentum (photon plus electron) before and after collision is the same but the individual momenta of photon and electron are changed. If the electron momentum is for example increased by a collision, the photon momentum must be decreased as follows [1]:

$$A_\mu \rightarrow A_\mu - \frac{1}{e}p_\mu \tag{16.143}$$

Thus Eq.(16.142) is changed to:

$$(\Box'' + kT)A^a_\mu = 0 \tag{16.144}$$

where:

$$\Box'' = \Box + \frac{i}{\hbar}(p_\mu \partial^\mu + p^\mu \partial_\mu) - \frac{p_\mu p^\mu}{\hbar^2} \tag{16.145}$$

If there is no gravitation present:

$$kT = \left(\frac{m_p c}{\hbar}\right)^2 \tag{16.146}$$

where m_p is the photon mass. Therefore the problem is to solve simultaneously the following equations:

$$\left(\Box' + \left(\frac{m_p c}{\hbar}\right)^2\right)\psi = 0 \tag{16.147}$$

and

$$\left(\Box'' + \left(\frac{m_p c}{\hbar}\right)^2\right) A^a_\mu = 0 \tag{16.148}$$

The electron mass m_e is many orders of magnitude greater than the photon mass, perhaps as many as forty orders of magnitude greater. Therefore in most textbook treatments of for example the Compton effect, the energy of the electron is represented in the classical special relativistic limit by the Einstein equation:

$$En = (m^2c^4 + p^2c^2)^{\frac{1}{2}} \tag{16.149}$$

but the photon is represented as a pure wave with no mass using the de Broglie equation:

$$En = hv, \quad p = \frac{hv}{c} \tag{16.150}$$

In Feynman's quantum electrodynamics, exchange of a virtual photon is used. However, in general relativity Eqs.(16.147) and (16.148) must be used and solved simultaneously with sufficient numerical precision to give the experimentally known effects of quantum corrections, such as the anomalous magnetic moment of the electron in the Lamb shift, and the Casimir effect.

Acknowledgements The British Government is thanked for the award of a Civil List pension for distinguished services to science, and the staff of AIAS are thanked for many interesting discussions.

Bibliography

[1] M. W. Evans, Generally Covariant Unified Field Theory: the Geometrization of Physics (Abramis Academic, 2005 and 2006), volumes one to three.

[2] M. W. Evans, Generally Covariant Unified Field Theory: the Geometrization of Physics (Abramis Academic, 2007, in prep., preprint on www.aias.us and www.atomicprecision.com, papers 55 to 70).

[3] L. Felker, The Evans Equations of Unified Field Theory (www.aias.us and www.atomicprecision.com). ; H. Eckart and L. Felker, article on home page of these websites.

[4] M. W. Evans, ed., Modern Non-Linear Optics, a special topical issue in three parts of I. Prigogine and S. A. Rice (series eds.), Advances in Chemical Physics (Wiley Interscience, New York, 2001, 2nd ed.), volumes 119(1) to 119(3); ibid., first edition, ed. M. W. Evans and S. Kielich (Wiley-Interscience, New York, 1991, reprinted 1992 and 1997, 1st. ed.,) volumes 85(1) to 85(3).

[5] M. W. Evans and L. B. Crowell, Classical and Quantum Electrodynamics and the $\boldsymbol{B}^{(3)}$ Field (World Scientific, Singapore, 2001).

[6] M. W. Evans and J.-P. Vigier, The Enigmatic Photon (Kluwer, Dordrecht, 1994 to 2002), in five volumes.

[7] M. W. Evans and A. A. Hasanein, The Photomagneton in Quantum Field Theory (World Scientific, Singapore, 1994).

[8] M. W. Evans, The Photon's Magnetic Field (World Scientific, Singapore, 1992).

[9] S. P. Carroll, Space-time and Geometry, an Introduction to General Relativity (Addison-Wesley, New York, 2004).

[10] L. H. Ryder, Quantum Field Theory (Cambridge, 1996, 2nd ed.).

[11] J. D. Jackson, Classical Electrodynamics (Wiley, New York, 1999, 3rd ed.).

[12] G. Stephenson, Mathematical Methods for Science Students (Longmans, London, 1968).

www.ingramcontent.com/pod-product-compliance
Ingram Content Group UK Ltd.
Pitfield, Milton Keynes, MK11 3LW, UK
UKHW060611180726
13836UKWH00011B/2498

9 781845 492489